Modern Petroleum
Refining Processes

Modern Petroleum Refining Processes

Sixth Edition

BK Bhaskara Rao

Department of Chemical Engineering
Indian Institute of Technology, Kharagpur

Oxford & IBH Publishing Co. Pvt. Ltd.

New Delhi

(*A Unit of* CBS Publishers & Distributors Pvt Ltd)

CBSPD

CBS Publishers & Distributors Pvt Ltd

New Delhi • Bengaluru • Chennai • Kochi • Kolkata • Lucknow • Mumbai
Hyderabad • Jharkhand • Nagpur • Patna • Pune • Uttarakhand

Modern Petroleum Refining Processes

Sixth Edition

ISBN-13: 978-81-204-1773-1
ISBN-10: 81-204-1773-9`

© 2014, BK Bhaskara Rao (Sixth Edition)

Previous edition: 1983, 1990, 1997, 2002, 2007

Reprint: 2017, 2018, 2020, 2023

OXFORD & IBH

New Delhi
(A Unit of CBS Publishers & Distributors Pvt Ltd)

Published by **Satish Kumar Jain** and produced by **Varun Jain** for
CBS Publishers & Distributors Pvt Ltd
4819/XI Prahlad Street, 24 Ansari Road, Daryaganj, New Delhi 110 002, India
Ph: 011-23289259, 23266861   Website: www.cbspd.com
                                          e-mail: delhi@cbspd.com

Corporate Office: 204 FIE, Industrial Area, Patparganj, Delhi 110 092, India
Ph: 011-4934 4934   Fax: 011-4934 4935   e-mail: publishing@cbspd.com;
                                                   publicity@cbspd.com

Branches

- **Bengaluru:** Seema House 2975, 17th Cross, KR Road, Banasankari 2nd Stage, Bengaluru 560 070, Karnataka, India
  Ph: +91-80-26771678/79   Fax: +91-80-26771680   e-mail: bangalore@cbspd.com
- **Chennai:** 7, Subbaraya Street, Shenoy Nagar, Chennai 600 030, Tamil Nadu, India
  Ph: +91-44-26680620, 26681266   Fax: +91-44-42032115   e-mail: chennai@cbspd.com
- **Kochi:** 42/1325, 1326, Power House Road, Opp KSEB, Power House, Ernakulum Kochi 682 018, Kerala, India
  Ph: +91-484-4059061-65,67   Fax: +91-484-4059065   e-mail: kochi@cbspd.com
- **Kolkata:** 147, Hind Ceramics Compound, 1st Floor, Nilgunj Road, Belghoria, Kolkata-700056, West Bengal, India
  Ph: +033-25633055, 033-25633056   e-mail: kolkata@cbspd.com
- **Lucknow:** Basement, Khushnuma Complex, 7 Meerabai Marg (Behind Jawahar Bhawan), Lucknow-226001, UP, India
  Ph: +91-522-4000032   e-mail: tiwari.lucknow@cbspd.com
- **Mumbai:** PWD Shed, Gala no 25/26, Ramchandra Bhatt Marg, Next to JJ Hospital Gate no. 2, Opp. Union Bank of India Noorbaug, Mumbai-400009, Maharashtra, India
  Ph: 022-66661880/89   e-mail: mumbai@cbspd.com

Representatives

- Hyderabad   0-9885175004
- Patna   0-9334159340
- Jharkhand   0-9811541605
- Pune   0-9923910676
- Nagpur   0-9421945513
- Uttarakhand   0-9716462459

Printed at Chaman Enterprises, Daryaganj, New Delhi, India

DEDICATED TO
MY PARENTS

Preface to the Sixth Edition

This book was published thirty years ago, since that time, it has been edited five times to accommodate the rapid changes occurring in the field, specially Indian petroleum Industry which registered a fast growth. I am immensely happy to present the Sixth Edition of the same. Two chapters, namely Bio Fuels and Green House Gases the most debated chapters in the World are incorporated in this volume. I acknowledge with a deep sense of gratitude to all the students, engineers in the industry, academicians and technical institutes for their encouragement and suggestions. I hope the efforts of the author will be appreciated by all and fruitful suggestions to further improve the volume are welcome.

20th September, 2013   **B.K. Bhaskara Rao**

Contents

x

1

Origin, Formation and Composition of Petroleum

1.1 ORIGIN AND FORMATION OF PETROLEUM

Today, most of the countries in the world are importers of energy. The fossil fuels, accumulated over aeons of geological activity are irreversibly consumed at a rate more than million times faster than they were formed. This has left us in a precarious position especially for petroleum and its products. The hike in price of petroleum and its products, both in national and international scenes is frequent for two simple reasons; the mounting demands and fast depletion of reserves. The importance of petroleum in present day civilization is ever increasing due to its unmatched contribution for our energy requirements, in lubrication and in petrochemical field. Thus its competence to serve mankind is unquestionable and unique too. Sixty percent of the energy needs of the world are met by petroleum. The advent of I.C. and Jet engines have revolutionalised the techniques of motive power, a fact, without which the rumbling civilisation would have to contend with a snails-pace. Such a premium stock of limited resources is fast depleting, perhaps due to indiscriminate and wanton consumption. The important question today is how long can the reserves meet the demand even with sky high prices? The high degree of conservation and restrictions in consumption may draw out the global reserves to another century at the consumption rate of today. Then, what?

According to Mayer and Hocott "There is no dearth of petroleum and natural gas resources remaining in the earth. As a matter of fact, there is no foreseen shortage of available supplies by present technology until well into the next century".[1]

Of course, every effort is made to locate new prospective fields, and innovation in search of secondary recovery techniques to lift the oils from existing sources, and what are presently uneconomical fields, is in progress. The controversies may be subdued by understanding the formation of petroleum, at least to some extent. Perhaps resources may not be the problem, but availability may be.

1.1.1 Occurrence of Petroleum

Petroleum occurs in the earth's crust, in all possible states and varies in colour from light brown to dark brown or black, exhibiting luminescence in some cases. It is a mixture of various hydrocarbons, of homologous series namely paraffins, naphthenes and aromatics. Thus the main elements are C (84-86%) and H_2 (11-14%) and other elements O_2, N_2, S rarely constitute about 8%, including different metals in traces. Petroleum is more homogeneous than coal and occurs mostly in sedimentary rocks.

1.1.2 Origin and Formation

Scientists till now are entangled with the problem of explaining successfully the origin and formation of huge hydrocarbon deposits and notably a sound theory has yet to be evolved. Mendeleeff and Berthelot, were caught up with the idea of explaining reasons for such deposits. Their explanation was based upon the inorganic reactions, mainly on the activity of acetylene series. Some carbides produce hydrocarbons when reacted with water, such as

$$CaC_2 + 2\,H_2O \rightarrow C_2H_2 + Ca(OH)_2$$
$$Al_4C_3 + 12\,H_2O \rightarrow 3\,CH_4 + 4\,Al(OH)_3$$

Assuming the availability of such carbides in earth's crest, they arrived at this axiom; in fact, the deposits of such magnitudes could never be balanced with these ideas.

The cosmic hypothesis of V.D. Sokolov depicts that the hydrocarbon vapours were already in cosmic clouds. Favourable conditions leading to precipitation of these clouds, rained hydrocarbons, which were either adsorbed or entrapped in earth's crest. A doctrine may possibly be explained if ever man comes across such giant deposits in a distant planet, where there happened to be no life at all.

Truly speaking, hydrocarbon vapours are present in certain planets. Hoyle cited the presence of hydrocarbons in the atmosphere of Venus.[2] These hydrocarbons must be distinguished from the earthly deposits, as purely derived from inorganic materials. Different and conflicting theories appeared in large number to explain the formation of petroleum deposits. The inorganic basis of petroleum formation had to be given up in favour of organic theory due to the following observations and facts as enumerated by J.D. Haun[3]:

(a) The homologous series present in petroleum are found only in organic matters.

(b) Nitrogen compounds in petroleum, especially plant derived porphyrins, comprise a very small amount but their significance is immense in the formation of petroleum from life source.

(c) The range of radio isotopes of carbon C^{12}-C^{13} is within the ranges of natural Carbonaceous materials and not from inorganics, as they do not have any chance to absorb radioactivity.

(d) Optical activity, a pre-requisite of natural organics, is exhibited by petroleum.

(e) Petroleum formation is a result of low temperature activity only.

(f) Petroleum is always associated with sedimentary rocks, (even recent formations too) and not with igneous rocks.

(g) Small quantities of petroleum (hydrocarbons) in recent sediments suggest, that the formation of petroleum is normal, continuous and does not require any severe physico-chemical conditions.

(h) Most organisms, like diatoms are found in petroleum.

1.1.3 Organic Theories

Engler (1886) was of the opinion that the large marine animals, which dwelt on the globe in the pre-historic days were the main fat contributors. After the natural extinction of these mammoths, the body fat was slowly converted to hydrocarbons, as per his version. But the laboratory experiments revealed that the conversion of fat by hydrolysis yielded only acids and not hydrocarbon gases. Further, surprisingly no oil well was ever sighted with any fossil remains of such animals. This is how the long flourished hopes of this natural fat theory had to be given up.

There are other schools of thoughts concerning the origin and formation of petroleum; some of which are listed as:

Thomas-Graham (1843) was of the view that natural naphtha must be a product of action upon vegetable matter of high temperature. Chaptal (1845) was of the opinion, that when plants became entirely decomposed into vegetable moulds these contained certain oils which escaped decomposition. J.W. Draper (1846) suggested the action of natural heat of earth on coal. Popoff (1875) showed that methane could be produced by decomposition of cellulose. Hoppe-Seyler in 1886 showed that bacteria could produce methane, as is ever evident in swampy areas. Treibs (1934) discovered some of the plant pigments and biological matter in crudes; micro-scopic organic remains found in crudes actually prompted API to launch a big project API 43 to investigate thoroughly.[4]

A bright explanation by Mikhailovosky, N. Potering and N.C. Anderson seems to be near the goal. Accordingly, the source of petroleum was not a definite species of flora as initiated by Engler, but the organic matter of sea oozes consisting of remains of plants and animal organisms. The initial decomposition of the vegetable and animal matter was a result of activity of micro-organisms; later the organic matter underwent changes due to pressure and temperature of the crest of the earth. Emphasizing the fact, Arkhagelskey, prophesied that petroleum took birth in argillaceous rocks enriched with organic matter and later migrated and got stored in arenaceous rocks. The extension works of P. Trusk and G. Petrov in USA characterised the composition and content of organic matter in rocks of different ages. Controversies always flare up, one school advocates—temperature—pressure distillation or by tectonic stress. Another school advocates as a biochemical process; yet another sees it purely as a chemical or radioactivity reaction— and so on. However, biochemical[5a] theory has been received with some favour, as it is concerned with low temperature and pressure, and bacteria of versatility. Thus it may be concluded that the formation is via a combination of physical and biological processes rather than individually. Petroleum formation[6a] from organic mass may be expounded by two distinct processes namely physical and biological.

1.1.4 Physical Methods

It conjures all the parameters in an ideal reaction and depends upon the factors like:

1. Heat   2. Pressure   3. Heat and pressure
4. Catalysts   5. Radio activity.

1.1.5 Biological Methods

All the biologic methods are governed by source and environ-ments such as :

1. By preservation of hydrocarbons synthesised by the source sediment organisms.
2. Biological reduction (anerobic or aerobic) sources being:
   (a) Fatty acids,
   (b) Proteins and amino acids,
   (c) Carbohydrates.
3. Biological and physical methods, such as
   (a) By condensation of bacterially produced methane under high pressure and temperature in presence of catalysts.
   (b) By bacterial modification of sediments.

API 43—(A.B.C.) completely relates to the transformation of organic material into petroleum. C.E. Zobell in his report of this project mentioned that conversion is genuinely a micro-biological process.[6b] Bacteria, as described by Zobell can survive in fact in all critical conditions, from 0° to 85°C, under extreme pressures upto 10,000 kg/cm^2 and also at well depths of 10,462 metres.[7]

Laboratory tests confirmed that 'Lipoclastic' anaerobes have the capacity to split long chain hydrocarbons to shorter chains.

Evidence in biogeneicity does not differentiate between vegetable and animal source.[8] Close association of crude oil with marine remains suggests it is from fish molluscs, lamellibranches, diatoms foraminiferae and other sea creatures, known from their fossilised remains to have been present in immense quantities in ancient seas. Preponderance of petroleum known today is derived from ancient biological matter, particularly from lipid rich lower marine plants such as plankton[9]. Terrestrial plants supply terpens, which suggests the phenomenon of atmospheric volatilisation of vegetation. The nature of recent hydrocarbons found, does however support the biogenic origin; the main reason being the presence of odd carbon number. These matters are often referred as kerogens or mother substances for shale.

Kerogen is divided into three classes based on the nature of original organic source material. Hydrogen rich straight chain groups usually are from algae, lipids and planktons. These contribute to Type 1 and Type 11 forms of kerogen while Type 111 kerogen is oxygen rich cyclic carbon structures obtained from higher plants, known as lignins. As organic matter accumulates and is converted to kerogen during diagenesis, it will be mixed with various proportions of inorganics *i.e.* sedimentary soils. The sedimentary deposit is classified both in terms of proportion of kerogen and amounts of Type 1 and Type 11. If the kerogen % is less than 1 the bearing sedimentary rock becomes oil source rock and above 1 and upto 50 the rock is classified as Oil shale, above 50 it is known as Humic kerogen as it is found to contain higher plant materials Kerogen to coal, oil, gas is driven by exposure to temperature, pressure, geological environments. The formation of fuels from kerogen is therefore called as geochemical stage or catagenesis.

According to Kenneth Kobe, petroleum results from a series of biochemical and chemical reactions which start with organic remains of dead micro-organisms.[10] Bacteria are involved in first transformation of the constituents of decaying micro-organisms into hydrocarbons.

Anerobic conditions keep porphyrins to remain in crude; if oxygen had been there these would have been decomposed.[11] Bacteria can also decompose organic matter to CO_2, H_2S, etc. Further, it is discovered that no bacteria can produce more than C_2-compounds, thus C_{11} to C_{14} compounds must have been formed from marine life as is evident from certain Gambian crudes.[12] Most organisms are found to generate petroleum in their metabolic processes too (Meyer and Hocott).

Crude oils found in the younger sedimentary deposits do in fact contain appreciable amounts of oxygen and nitrogen. This shows organic material as the source. McNab *et al.*, divided crudes into three types depending upon the time of deposit[13]

Tertiary crudes	:	$11 \times 10^6 - 74 \times 10^6$ years of age
(Eocene oil)		(asphaltic crudes)
Mesozoicera oil	:	$75 \times 10^6 - 200 \times 10^6$ years of age
Paleozoic crudes	:	$200 \times 10^6 - 500 \times 10^6$ years of age
		(paraffinic crudes)

Brooks and Frost concluded that the organic matter can be decomposed easily under the action of natural catalyst as well as bacteria.[14] The evidence of bacteria in sedimentary rocks is positive and the organics are initially decomposed to acids and gases (CH_4), till the activity of bacteria ceases due to frankensteinism. I.M. Gubkin demonstrated that the formation of petroleum and gas from organics, scattered in argillaceous rocks was a local process and started really with accumulation of organic matter in sea oozes, it may be the fat or carbohydrate. Apart from this, some scientists regard that the hydrogenation of carbon oxides results in -CH_2-chains, which may be congenially taking place under the catalytic activity of earth's crest due to favourable conditions of temperature and pressure. But this is only a derived literature thrusted to explain and can never match with the quantum of oil. Thus every new thought merely resulted in more sterile addendum. The overall explanation hitherto may be summed up in two steps: firstly the action of bacteria in contributing the lighter fractions, secondly continuous catalytic action of earth at depths of 1-2 Km to yield heavier hydrocarbons. It is observed with the depth of mine and time of formation and storage, the API gravity of crude and paraffincicity increase because of severity of reaction, which is in full agreement with the above picture.[9b]

Age of the oil is indicated by CPI (Carbon Preference Index). Hydrocarbons of recent origin show 4 to 5, while ancient bitumen show about 1. The odd carbon number is dominant in recent formations, as the oil ripens the oddity decreases, hence young-deep oil formations CPI is less than young-shallow oils. Age-depth oils do have less CPI than young-shallow deposits.

These are all possible natural reactions occurring over the ages, and petroleum hydrocarbons might have been formed by many processes and each contributing its own share, may be small.

In conclusion, regarding the biologic origin of petroleum there need not be any doubt, because of the oil association with sediments containing a relatively large amount of organic matter, the presence of optically active compounds and complex substances of obvious biological origin. Oil is also absent from formations having adequate traps and porous strata, but without any organic material. The formation is at low temperature, usually less than 200°C or even 100°C. The thermophilic bacteria (bacteria that can survive at high temperature) plays the major role in conversion of this organic mass into liquid hydrocarbons.

1.2 RESERVES AND DEPOSITS OF WORLD

There are very few parts of the world which are self-sufficient in energy. Russia and Eastern Europe are largely self-sufficient in energy and they never look for outside supplies of oil. Middle East, Africa, the major oil producers are exporters of huge energy, while Japan is a leading importer of energy. USA is also energy producing country, but its commitments have grown to such an extent, that it is now a major importer of energy.

Petroleum deposits mainly occur in some elevated sections of porous sandy strata.[1b] Sedimentary rocks accumulate in sea bed at a very slow rate. These layers over millions of years stratify under pressure and temperature and are transformed into metamorphic rocks, sand stones, marbles, etc. These are the basic reservoirs of gas and oil. Whenever fissures and dislocations in zones of earth occur, oil and gas reach surface and get burnt continuously by accidental fires like lightning, etc. and sometimes oil drains to nearby streams too. A false understanding prevails that volcanic eruptions are followed by huge quantities of hydrocarbons.

It is seen coal is more uniformly distributed throughout the world rather than oil; oil is scattered randomly and 80% of oil found to date occurs in what has been called oil-axis-pole.[5b] Gulf Caribbean, Mesopotamian and Persian Gulf are such areas confined to depressions on earth's crest. Initial migration of oil takes place during compaction of dense shales.[15] Although the migration mechanism is not fully understood, evidently it is not effective, with the result, much oil is left in the formations. At present oil from such source is not economically recoverable.

Bacteria in marine sediments may also contribute to the liberation of oil from oil bearing materials in various ways. According to Stone and Zobell obvious import is the bacterial decomposition of organic complex in which oil is trapped.[16] Secondary by dissolving carbonates or sulfates on which oil is absorbed by action of bacterial acids. Thirdly bacterial gases reduce the viscosity of oil.

In erstwhile Soviet Union, the South Caspian Basin, Ural-Volga Basin, and Western Siberian Basins are famous for hydrocarbon deposits which rank next only to Iran-Arabian Basin, the richest in the world.

Once the richest basin, Aspheron-Peninsula has been exploited since last hundred years. It is found through approximate records that oil extracted throughout the world till 1975 was about 43,000 MMT, a considerable amount of which was produced during 70's only.

Kuwait, a small country with 18% of world reserves held its 4th position for a long time in the production of oil after USA, Russia and Venezuela, but now Iran and Saudi Arabia have surpassed Kuwait. Saudi Arabia is about 100 times bigger than Kuwait and is ranking next to Iran with 10 MMbbls. per day. Kuwait is producing approximately 1.25 MMbbls. per day.

US crudes are relatively less sulfur ones and make up about 7% of world reserves.[17] Venezuela and African crudes of relatively low sulfur content constitute about 10% world oil reserves. African countries like Libya, Algeria and Nigeria account to about 300 million tones per annum. A characteristic feature of the oil wells near Middle East is high yields per well. Iraq and Iran are famous for such types of wells, which can produce even 10,000-20,000 tons per day per well. Thus near and Middle East countries with highest proven reserves of 70% of world's petroleum stock; holds the control in the energy export.

At present the trade is governed by OPEC*; fortythree percent of world crude produced in 1972 was shared among the group members, which is very much true today also.

1.2.1 World Reserves

Several estimates of world hydrocarbon reserves were made by almost all leading institutions and companies. As furnished by Weeks the estimates of proven reserves are $1,900 \times 10^9$ bbls of oil against the estimates of Warman (British Petroleum Company) 1,200 to 2,000 $\times$ 10^9 bbls. There is no congruency of expression in these reserves; one can judge by the following ultimate reserves as presented by Richard G. Sieidl[n], how the estimates once cited regularly change with better methods of surveys and techniques.

1965	Hendricks	$2,480 \times 10^9$ bbls
1968	Shell Company	$1,800 \times 10^9$ bbls
1970	Moody	$1,800 \times 10^9$ bbls
1971	US National Petroleum Council	$2,670 \times 10^9$ bbls
1972	Weeks (8th WPC, Proceedings 2, p. 99–106)	$3,650 \times 10^9$ bbls

*OPEC: Organisation of Petroleum Exporting Countries A 13 members body consists of: (1) Algeria, (2) Iran, (3) Iraq, (4) Saudi Arabia, (5) Gabon, (6) Kuwait, (7) Ecquador, (8) Libya, (9) Indonesia, (10) Nigeria, (11) Qatar, (12) United Arab Emirats, (13) Venezula.

1975	Whiting (recoverable)	556.2 $\times 10^9$ bbls and
		80 $\times 10^{12}$ Cu.M. gas.
1977	Harry Warmen[18] (recoverable oil reserves)	270 $\times 10^9$ tons of oil
		Gas equivalent to 55 $\times 10^9$ ton of oil

Total proven reserves as per Whiting goes upto 2,200 $\times 10^9$ bbls of oil and 250 $\times 10^{12}$ Cu.M. of gas; of which 30% will be a Off-Shore product and rest onshore. Cumulative oil production of world till January 1974, was put around 270 $\times 10^9$ bbl oil and 20 $\times 10^{12}$ Cu.M. of gas.[19]

Next to coal, natural gas reserves of 60 billion tons oil equivalent are highest in the world; with a proven to potential capacity of 30% for gas and 60% for oil, the chance of finding more amount of gas is always bright.[20]

John J. McKetta[21] *et al.*, have presented the following data on the available energy of today (1980) and estimates of reserves of energy sources:

Available energy	(1980)
Fossil Fuels	
Oil	570 $\times 10^9$ bbls
Natural gas	2,500 Exajoules
Coal	637 $\times 10^9$ tons
Shale	30 Gigatons
Tar sands	15.30 Gigatons
Nuclear source	MW
Renewable energy	
Geothermal	132 MW
Hydroelectric	5.7 Exajoules
Solar	negligible
World oil estimates	240-260 Gigatons
Natural gas	10,500 Exajoules
Of which undiscovered	8,150 Exajoules
Coal	10,125 Billion tons
Of which known reserves	637 Billion tons
Nuclear	
Uranium	3 Million tons
Thorium	630 Thousand tons
Additional undiscovered Uranium	80–280 Million tons

Exajoule = 160 Million barrels
Gigaton = Billion barrels

At present the production of oil in the world is about 3300 MMta and gas over one trillion cubic meters.

Fears of impending oil shortage have raised hopes of finding huge deposits of oil even under deep oceans. The activity now-a-days is more confined to such areas.

According to Klemme there are 334 giant fields in 66 basins that contains 70-75% of oil. Indian subcontinent is placed in Type 4 basin (India and Assam) and the coastal belt in pulled apart basins.

Estimated sedimentary rocks of the world, by D. Ion and Hendricks is about 62×10^{17} tons, which contain an organic matter of 3.5×10^{15} tons, which in turn can yield a hydrocarbon content of 7.7×10^{13} tons.[22]

World oil reserves estimates 2004 : 1188.6 billion barrels (Department of Energy USA)

Of which OPEC 890 and ME 733.8 MMT

NG consumption world 2420 MILLIONTOE 3.3% over 2003 consumption (N. America 705.9)

Reserves of NG World   179.3 Trillion Cubic meters

Reserves of N. America.   7.32 America Central 7.01 ME 72.83 Asia Pacific 14.06

Nuclear energy (World) 382 Gw_e, India capacity (14 plants) 4013 MW_e By the end of X1 Plan 9935 Mwe capacity

Uranium Deposit India (Recoverable at 130 $/kg) 52.7 Thousand tons Probability of going upto 80 thousand tons

Uranium production decreased from 36,700 tons in 1996 to 3,281 tons thousand tons in 2000

Coal reserves (Sub bituminous + Bituminous + lignite + anthracite) World 985 Billion tons (Recoverable 2000)

India Recoverable (Sub bituminous + Bituminous + Lignite) 84.453 Billion tons,

World production 4.342 Billion tons, India 323 MMT

Hydel — India potential capacity 150000 MWe only 20% of which is tapped against this World Hydel energy 692,420 with an additional of 110,183 MWe

Geothermal is enormous World feasible presently 14,400TWH/year; Technically exploitable 40,704 TWh/year

Direct use installed world 18,000 MWe (USA 5366, India 80).

1.2.1.2 The Difference Between Oil Shale and Shale Oil

There is a huge difference between oil shale and oil produced from shale reservoirs, often called shale oil. The former remains a promising, yet expensive-to-produce, also a resource that may eventually require more development. The latter generates significant, real production growth for a host of independent North American E&P firms; with crude around $70 to $80 a barrel, many shale oil projects are generating an after-tax return on investment of as much as 100%.

1.2.1.3 Oil Shale Today

Oil shale is an inorganic rock that contains a solid organic compound known as kerogen. The term "oil shale" is a misnomer because kerogen isn't crude oil, and the rock holding the kerogen often isn't even shale. Conventional liquid crude oil is organic material-plant and animal remains-exposed to heat and pressure in the absence of oxygen over millions of years within the earth. Kerogen is among the first stages in the process of petroleum generation from organic matter; bitumen-the hydrocarbon targeted in oil sands projects-is formed from kerogen and represents a later stage in the process. Kerogen to oil is done by retorting. To generate liquid oil synthetically from oil shale, the kerogen-rich rock is heated to 500 degrees Celsius in the absence of oxygen, a process known as retorting. Shale can be heated under-ground known as in situ retorting or can be mined like coal and retorted on the surface.

There are several competing technologies for producing oil shale. Exxon has developed a process for creating underground fractures in oil shale, filling it with a material that conducts electricity and then supplying power through the shale to heat it . Because of heat the kerogen, gradually converts into recoverable oil. Shell uses electric heaters that it buries underground to heat the kerogen slowly. Although estimates of the cost to produce oil shale vary widely, it's more expensive and energy-intensive and is economical if oil price goes higher than $100 a barrel and feasible for production

There are a handful of projects in the works, but none are likely to produce significant quantities of oil for well over a decade. Brazil's national oil company (NOC) Petrobras (PBR), Shell, Exxon and Japan's Mitsui (MITSY) are among the companies involved in US various oil-shale projects. Clearly, none of these companies represent pure plays. The potential size of the resource is huge. Oil shale naturally occurs in more than 20 countries around the world, and Brazil, Estonia and China all have small commercial projects producing a total of 14,000 barrels of oil per day. However, by far the world's largest oil shale resource is located in the Green River formation of Colorado, Wyoming and Utah. Colorado (1980, Exxon Mobil (XOM) acquired Atlantic Richfield's 60% interest in the Colony) Oil Shale contains the richest shale. Total recoverable resources could be as high as 1 trillion barrels, roughly four times Saudi Arabia's total proved reserves. If crude prices remain elevated for a prolonged period, prompting an increase in oil shale development, it's possible that production will hit 150,000 barrels per day by the late 2020s. The Energy Information Administration (EIA) projects that US oil shale will produce 144,000 barrels per day by 2030.

1.2.1.4 Shale oil

Shale oil plays such as the Bakken have far more in common with shale gas plays like the Marcellus Shale of Appalachia and Haynesville Shale of Louisiana than they do with oil shale of Colorado. Shale oil plays are known as unconventional fields. Truly, natural gas and oil do not exist underground in some giant caverns like oil lakes. Rather, hydrocarbons are found trapped in the pores and cracks of a reservoir rock. A typical conventional reservoir rock is sandstone; which is very pores and capable of holding hydrocarbons. Typically, those pores are also well connected like pipelines (channels) so that oil and gas can easily travel through sandstone reservoir rock because of having a high permeability. geologic pressures help in pushing these hydrocarbons However, Shale fields and other unconventional fields aren't particularly permeable (no channels). This prevents free travel of oil and gas even though there is plenty of oil and/or gas in the rock.. Thus even in shale fields under high geologic pressure, the hydrocarbons are essentially locked up in place.. To free that oil and gas puncturing/fracturing these rocks is done. This is known as "fracking". Producers presently have developed and refined two major technologies to unlock shale: Horizontal drilling and fracturing.: Horizontal wells are drilled down and sideways to expose more of the well to productive reservoir layers.

Fracturing is a process that describes water (with chemicals) is pumped into a shale reservoir under such tremendous pressure that it cracks the reservoir rock. . The chemicals help in increasing pore cavity, reduction of viscosity of oil and increase the permeability of the bed This creates channels through which hydrocarbons can travel. This process is called fracking Fracturing and horizontal drilling are now common in the US and producers have perfected their use on gas shale plays Shale oil, unlike oil shale, does not have to be heated over a period of months to flow into a well. And the oil produced from these plays is; in fact crude oil not differing in any manner from conventional crudes, many producers say that it's even better quality on average than West Texas Intermediate (WT!), the US standard crude The Bakken Shale is an unconventional oil play located in North Dakota, Montana and across the Canadian border in Saskatchewan. The US side of the play offers thicker deposits of oil; however, the Bakken looks like a commercial play in Saskatchewan as well. Fracking has become a World wide tendency which is showing that world can meet energy needs through gas and oil demands through 250 years! Exxon estimates that it would process about 66,000 tons of raw shale per day to produce around 47,000 barrels per day of shale oil (2009). And while Exxon was the largest company to invest in oil shale, it wasn't the only player in the industry. Unocal, Royal Dutch Shell (RDS.A), Amoco and Ashland Oil, among others, all had projects in the region. But almost two years later Exxon abruptly pulled out of the Colony Shale project, writing off more than $1 billion of its investment.

1.3 PETRO GLIMPSES AND PETROLEUM INDUSTRY IN INDIA

Many economists believe that economic rate is directly linked to the consumption of oil. It is seen that the World economy growth rate declining to 3.1% from 4.1% in the previous year 2004 while China's economy is kept at 8.1%. Indian Economic Survey says India's economic growth rate is expected around 6-8% as per the GOI, but in reality it may not be even more than 5% during 2005. Many Planners believe that oil consumption is directly related growth rate, though marginally varies.

The world oil demand during 2005 is expected to be about 90 MMBPD (the major growth in consumption will be from China and India) i.e equivalent to 4.5 billion tons of oil. At this rate of consumption the world reserves are expected to last another 40 years. Oil consumption by USA alone is nearly around .900 million tons per annum. While the margins in refining in many countries are coming down as low as 3 $ per barrel (Europe 2.37$) in India it is very high more than world average, about 4.05 $. Due to low marginals many US refineries of smaller sizes less than 3 MMTPA were closed *i.e.* More than 50% refineries about 250 refineries were closed our of 650 refineries operating at peak time. By 2010 the demand of oil in India will be about 140 MMTPA of which 100 MMT will have to be imported.

Petronet the first importer of LNG into the country is making arrangements to bring 5 MMT PA of LNG. Petronet commissioned its terminal at Dahej in Gujarat and successfully marketed 2.5 MT. From Qatar 20 MMT LNG is to be imported. After liberalisation the Multi Nationals have entered the oil field and are giving good competition in all petroleum products to native companies. Royal Dutch Shell returned to India and setting up retail outlets to market the products. Specially in marketing LPG Petronet is a competitor for ONGC and GAIL as they are entering market along with IOC, HPCL, BPCL and IBP.

RIL will step up its production from KG basin to 40 MMCMPD from 2007 with an investment of 10,757 crores (Dhurubhai 1 and 3) RIL has till date made 12 gas finds in deep sea block KGDWN. The estimates show about 8.3 trillion cft in the fields. The discoveries in D6 block has estimates of 14 trillion cft gas. The growth in consumption of petroleum products, mainly in fuels, is surpassing 6%, which is very difficult to be managed by Indian industries hence foreign companies are invited again correcting the fault of nationalization of oil companies.

A second LNG terminal is being setup at Hazira by Shell to sell 2.5 MMT of LNG from 2005. Hazira is being designed to handle 5 million ton LNG and the capacity will be expanded to 10 milion ton. IOC is setting up its terminal on the East coast near Chennai to handle the gas from Iran. This will boost the total sales of IOC to 4.5 MMTPA of gas. In addition Cairn Energy India operates with 5 licences Ravva, KG-OS/6, KG- DWN - 98/2 all in KG basin and CB/OS-2 and RJ-ON-90/1 in West Coast. Again from Qatar 20 million ton LNG is to be imported in the next few years.

IOC's Sri Lanka subsidiary, GAIL's, OVL entry into Egypt and Iran and ONGC's Videsh Nigam acquisitions in Sudan and Libya are some of the ventures Indian companies are doing. OVL's first fruits of investment by producing 3.3 MMT in Sudan. The consumption of petroleum by the whole World is higher than 4200 million tons, OPEC presently produces 27 MMBPD. The demand is much more from China, India and Korea. In addition to the demand often, hurricanes and storms disrupted US oil fields and refining. This marked the abrupt increase in crude price to 60 $ per barrel in 2005. Indian scene in production of oil is same as it was a decade ago, OIL produces around 3 MMTPA while small private players produced 2.5 MMTPA. Only consolation is OVL and ONGC are having reserves of 199 and 1128 MMT. By inviting private parties and multinationals the pricing of petroleum products may change, although Government will exercise its share of opinion, hence there may not be any benefit for consumer except perhaps more availability.

The challenges before the oil companies will be to manage the product prices and keep the products of high quality as per Euro 11 norms. In fact though pollution may be regarded as the highest priority there is no need to go for different high grades of petrol like 91 octane, high speed, Xtra-premium, Xtra-mile. In the name of these brands consumer can be easily cheated as there is no strict method of testing as is followed in Western countries. Further the conditions of the roads, the waiting time at signal points over rule the quality of speed and quality oils thus the Euro 11 normas will have no meaning. Even in USA different grades of gas are sold, the major consumption is of regular quality *i.e* the cheapest one as is in India

Indian situation is clear in having poor resources in oil and gas. Hence the alternative fuels and blending play vital role. In this direction the hope is in alcohol, vegetable oils. Any vegetable oil is suitable for running Diesel engines, however in our country where there is huge shortage of edible oils, and with limited land the program of cultivation of oil seeds will face tremendous difficulties. In other countries, in Malaysia palm oil is plenty, in USA soya bean oil largely available, sunflower or rape seed oil is available in Europe. Some studies in US have shown that these bio and ethanol fuels burn more energy than they produce. Researches at Cornell and California University contend 29% more fossil energy is required to turn corn into ethanol. Also the various forms of grass/wood require 45% more energy and soybeans/sunflowers require more than double the energy than the bio diesels. However the arguments and counter arguments are many and have to be studied carefully. Further the cost difference between Diesel and Petrol in those countries is zero or marginal and the edible oil is not very costly when compared to fuels oils, hence the bio diesels is a working proposition, similarly blending gasoline with alcohol not only to reduce the consumption of hydrocarbon but to reduce pollution is a necessity as it is available in plenty including India.

Strategic considerations are not given much importance in India unlike many western and European countries. USA produces only 45% of oil of its needs. Earlier in 1970-80 there were more than 600 refineries in USA, presently this has been reduced to less than 300. The reasons for shutting the refineries may be more, conspicuously environmental concerns, quality enhancement of motor fuels are on the top of the list. Thus the refining activity is shifting towards Africa and Asia. This obviously requires strategic oil reserves, which the president of USA operates whenever the supply is shrinking due to various reasons. The merger of various national oil companies in to two integrated outfit, i.e ONGC, Bharat Petroleum and HP to form an integrated company with exploration, production to down-stream refining and marketing, while the second merger is OIL and IOC. This consolidation is to reduce the operating and overhead costs, which is very high in India. The recent understanding agreement between Iran and India to Supply 7.5 MMT of LNG from 2009 is a significant thrust in the direction of procurement of petroleum. By the end of the decade India may have to import 120 MMTPA assuming a yearly 5.5% growth in consumption of petroleum fuels. This is mainly due to growth in automobile and aviation industries. The present production of gas is about 29 billion cubic meters. Though the earlier predictions of gas deposits were to a limited extent giving an active life of 25 years, the recent discoveries by RIL and ONGC have raised the hopes that gas demands can be met to substantially large extent. Thus the gas accounting to 6% of commercial energy of the country can go up to 7-8% in two/three years. Presently the primary energy demand increased from 1716 MMTOE (million ton oil equivalent) in 1992 to 2421 MTOE in 2001 i.e. commercial energy increased from 199 to 225 during that period. The gas consumption in fertilizer industry and power units is more than 80% thus leaving only small % for petrochemical and domestic needs The total current requirement is around 67 MMSCMD of gas.

The World's petroleum industry dates back to 1855 when Samul Kier started the first refinery of 5 bbl capacity at Pittsburg, incidentally first oil well was dug by Colonel Drake in August 1859 at Titusville. Indian oil industry was not lagging at its inception as the first oil well was dug in Captain Goodenhow in 1866 and the first refinery was started in 1893 by Assam Railways and Burma Oil company. Later on in the name of Assam Oil Company the refinery was shifted to Digboi and named as Assam Oil Company and it was the only Refinery in India till 1954. People were under the firm impression that there was no possibility of finding oil resources in

India. International Oil companies as a group written off India as barren land where there was no possibility of finding oil except in some parts of Assam. The Government of India started the refineries on sea coasts to easily procure oil from international market. Providence helped India to invite the Russian experts; accordingly in 1955 a team of top petroleum geologists arrived. Discreet surveys made by the team shoed promising areas. The discovery of Cambay oil field (1958) followed by Ankaleswar, strengthened the hopes of India and the collaboration between Russia and India started.

The real foundation of oil industry took place during 2nd Five-Year-Plan (1956-61) when the Government of India launched a planned program of exploration, production, refining and distribution of oil and products. Accordingly ONGC took birth in August 1956 with assigned duties of exploration and production of oil on On-Shore and Off-Shore. Next to come in this field is Oil India Limited, a joint venture of GOI and Burmah Oil Company (1956), which was given lease to explore and produce oil in Eastern Region. Indian Oil Corporation was floated in 1958 with objectives of procuring, processing, and distribution of oil. It has been divide into two wings, Refinery wing and Marketing wing. Refinery wing procures crude for the refineries, establishment of new refineries and new techniques and developments in quality specifications etc. Marketing division takes care of distribution and transportation imports and exports etc.,

The birth of ONGC, the discovery of new oil fields in Eastern and Western parts of the country had laid the major path for the country's progress on the oil front. Since 1949 to 1960 Standard Vacuum Oil Company alone explored the West Bengal basin and later jointly with Government of India. Since many decades the major oil producing organizations are ONGC, OIL, Assam Oil Company (Now amalgamated with Indian Oil) additional since last 8-9 years Reliance Petroleum, the Government's nationlisation and denationalisation processes to some extent destabilized the industry initially although now open market competition had vitalized the industry.

1.3.1 Oil and Gas Scene

The sedimentary basins of India, onland and Off-Shore up to the 200m isobath, have an areal extent of about 1.79 million sq. km. So far, 26 basins have been recognized and they have been divided into four categories based on their degree of prospectivity as presently known. In the deep waters beyond the 200m isobath, the sedimentary area has been estimated to be about 1.35 millon sq. km. The total thus works out to 3.14 million sq. km. of which about 30% is unexplored. Although this appears to be a sizeable proportion, it is a marked improvement over the corresponding figure of 50% in 1995-96. Also, exploration has been initiated in about 33% of unexplored areas 1995-96. Thus, across eight years, there have been significant forward steps in exploring the hydrocarbon potential of the sedimentary basins of India. (Director General of Hydrocarbons).

Total hydrocarbon resources, inclusive of deep waters, are estimated at around 28 billion tonnes oil and oil-equivalent of gas (O+OEG). As on 01.04.2004, initial in-place oil of 7.89 billion tonnes and ultimate reserves of 2.94 billion tonnes have been established. The resources estimated by DGH for its 'internal use', for the country, are 32 billion tonnes (O+OEG). India's sedimentary area (up to economic zone 200 miles) is 3.14 million square kilometers, of which 0.42 million sq. km area is Off-Shore area, spread over 6000 km coastal line and 0.4 million sq. km area is available in the form of continental slope, while the remaining area is land based[23]. India's Off-Shore basins are divided into 8 regions out of total 26 basins, as shown in Figure 1.1b. The hydrocarbon potentiality of the entire region is 23 billion tons. The biggest Off-Shore activity started with discovery of Bombay High in 1973. Oil and oil equivalent (OEG) gas available in the country as per survey's of April 2004 is 1650 MMT. According to estimates the reserves of gas are 763 billion cu. Meters while the production has not gone up beyond 90 MMSCD falling short of demand.

7 Gas hydrate structures with free gas accumulation have been mapped in Andaman Off-Shore areas. Perhaps the recent tsunami effects may lead to migration of oil from Indonesian coast to coastal Andamans. The gas reserves are about 763 billion cu. meters while the oil proven reserves are put at 732 million tons.

Some countries and the reserves are listed here:

Iran	130.8	Billion barrels
Iraq	112.5	......
UAE	87.8	......
Kuwait	96.5	......
Venezuela	78.0	......
Saudi Arabia	262	......
USA	22.5	
Mexico	18	......
Brazil	14	......
Russia	69	......
UK	25.4	......
Norway	9.8	......

Canada is having 178 billion barrels, however most of the reserves are oil-sand deposits which is difficult to extract. Usually oil-sand deposits are mined and oil is extracted by steam heating or extraction methods.

1.3.2 HBJ Gas-Grid (Figure 1.1a)

Associated gas comes out along with oil. The quantity of gas dissolved depends upon the saturated pressure of the reservoir. Commonly about 500 SCf dissolves in a barrel of oil under a pressure of 2500 psi. when the production of oil has to be stepped up naturally the associated gas quantum also increases. During the intial stages of Bombay High development there were no infrastructure facilities to consume the gas hence it was flared. During 80-90 the quantum of gas flared was enormous.

About two decades back gas was ranked second to oil, but today the thinking has changed. Right from a boiler fuel to petrochemical feed stock and a clean fuel for automobiles the utility and serviceability of gas has brought increase its creditability to No. 1 fuel. Today GAS Refinery is a reality. The enhanced gas consumption during last decade and ever increasing demand is a real meter to show how it is being regarded as most important material in daily life. GOI has started looking for Hydrocarbon deposits through out the country. The results of which were significant. Rajasthan, Gujarat, Tripura, Palk straits etc. are some of the examples. KG basin discovery in the late 80s brought significant change in the hydrocarbon activity in this region.

To successfully utilise the enaromous amount of associated gas from Bombay High, at Hazira a gas processing plant was commisioned and this sweetned gas is transported through a pipe line known as Hazira-Bijapur-Jagdishpur Pipe line (HBJ). The length of this pipe line initially was 1700 km, (presently it is 2300 km long) and in the second phase was able to transport 20 MMCMD gas. The gas is supplied to six fertiliser plants, and power house for generation of 400 MW besides to a gas cracker plant at Gandhar and IPCL. With branch line drawn up to Kanpur, Faridabad and Salimpur Petrochemical complex (originally planned). In full capacity the line can transport 33 MMMD of gas.

KG basin is the next biggest Off-Shore activity in India, approximate reserves are 7 trillion CM. With the entrance of Reliance Industries with the partnership of Niko Resources of Canada into this basin remarkable achievements were attained. With the success in Gas discovery in M-l and H-l wells it has plans to produce 45

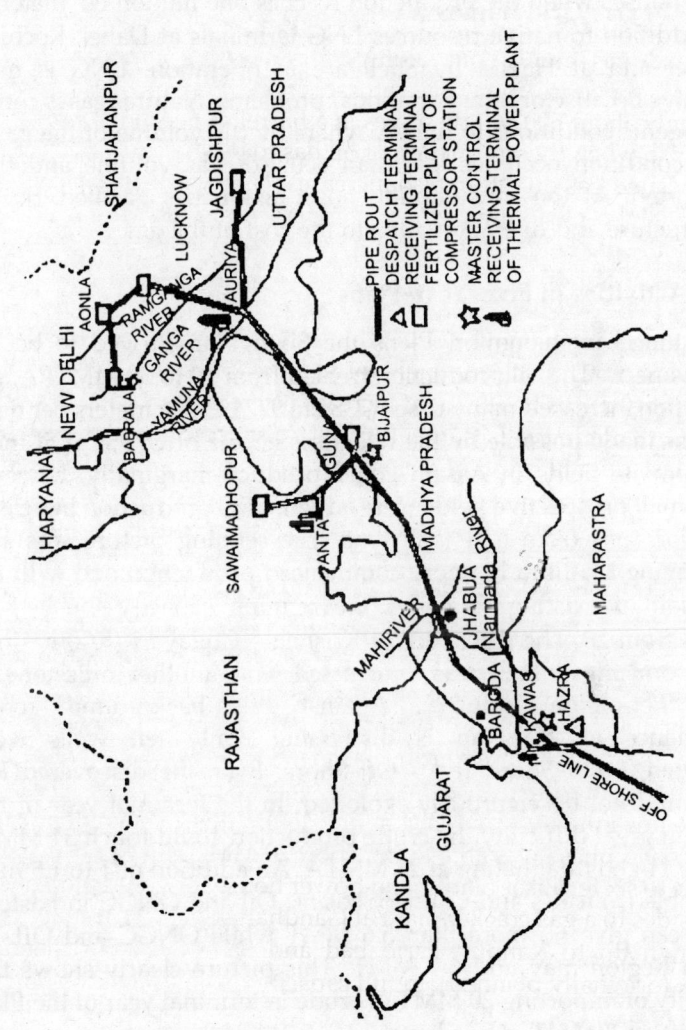

Figure 1.1a GAIL, HBJ Gas Pipe Line.

MMCMD of gas from its structure D6 (The reserves of which are 1.45 trillion cubic meters gas). Presently 60,000 to one lakh cu. meters of gas is being supplied to Power house in Kovvur 80 Km away from KG basin and other gas consumers. The future of gas is secure in KG basin and will provide gas to Nagarjuna Fertilisers and other gas based power houses when the production reaches one million cu. meters.

In addition to native resources LNG terminals at Dahej, Kochi by Petronet and at Hazira by Shell are in operation. LNG is more expensive because of transportational problems. Natural gas is cooled in cryogenic conditions to liquefy where by the volume of the gas in liquid condition occupies less than 600th of gas volume and then transported. At the consumption point again it is gasified. So the infrastructure and operation adds to the cost of the gas.

1.3.3 Activities in Five–Year–Plans

Of all the quinequennium Plans the Sixth Plan Proved to be the fruitful one [25]. The oil production trebled from 10 to 30 MMTPA, gas production increased many times (2385 to 9774 Cubic meters per day). This was made possible by the relentless efforts of ONGC. Oil India Ltd. from its fields in Assam could produce marginally, however many small prospective fields were added. LPG production increased to million tons from 6.91 lakh tons. The refining picture was also encouraging Mathura Refinery commenced production and with the expansion of existing refineries the refining capacity touched 45 M^2TPA from 26. The discovery of Krishna – Godavari (KG) basin is most promising than it was anticipated, was another milestone in Indian Petroleum Industry. Against this background actual performance of 7th Plan is distressing. Only ten wells were discovered, 3 On–Shore and 7 Off Shore. Even the discovered KG basin could not be vigorously exploited. In the terminal year of the Plan with great difficulty the crude production could touch 34 MMT. Bombay High is stagnating at 21 MMTA. An addition of 1 to 1.5 may be possible from KG and Cauvery basins. Oil and ONGC in Eastern Region can give no more than 5.6 MMT while ONGC and OIL in western Region may add 6-7 MMT. This picture clearly shows the possibility of importing 20 MMT of crude in terminal year of the Plan and addition 3 MMT of products.

In 8th Plan, in KG basin a mini refinery was established. The biggest Refinery at Jamnagar by Reliance of 27 MMTPA was also commenced. The refining capacity was gradually raised to more than 90 MMTPA. The 3 MMTPA refining capacity expansion and modernisation of Chennai Petroleum Corporation Ltd. (CPCL) has resulted in improving the distillate yield and produce cleaner fuels to meet Euro 111. It is also unique in handling Hydrocracker bottoms in Fluid Catalytic Cracker (FCC). Thus the yield of LPG has gone up to 4 lakh ton from 1.8 tons, further the expansion increased the yield of gasoline to 7 lakh tons. Thus the overall capacity of the refinery gone up to 9.5 million tons. Many changes have been brought in refinery administration. Nagarjuna Oil Corporation Ltd. is going to start a refinery at Cuddalore in Tamilnadu (6 MMTPA) 1976 nationalisation was the talk in 1997 deregulation of all the oil refineries and private participation was again allowed. Complete deregulation was allowed since 2002. Similarly APM (Administered price mechanism) has been dismantled since 2001-2002 in favour of market driven price mechanism (MPM).

During 2001-2002 IOC commissioned the following projects:
- FCC and Vacuum distillation unit at Haldia Refinery
- Production of unleaded petrol at Guwahati Refinery
- Gas turbine at Panipat Refinery
- Bulk storage terminals and depots at eight localities
- LPG bottling plants at six locations

The seven IOC refineries achieved crude throughput of 33.76 million tons, with Chennai and Bongaigon Refineries combined, the total handling of crude gone up to 42 million tons. The LPG bottling capacity had also gone up to 3.24 million tons. It could handle 47 million tons of petroleum products including low sulfur diesel (0.05% S). In Ninth–Five–Year–Plan, the small refinery at Thatipaka, KG Basin is able to supply about ten thousand tons of high speed diesel just by flashing the crude as the crude is highly paraffinic. Nagarjuna Oil Corporation Ltd. is going to start a refinery at Cuddalore in Tamilnadu (6 MMTPA). There is no rational out look in this regard as the automobile industry is allowed to expand freely. Even the surveys show that the Diesel consumption and petrochemicals consumption is down by 1 to 1.5%. Similarly naphtha consumption has come down due to alternate fuels.

Total demand for the petroleum products in the year 2005-06 is estimated to be 111.063 MMT registering a negative growth of 0.5% over 2004-05. Depleted growth was mainly due to less demand for products such as Naphtha (9.4%), LDO (34.8%) and HSD (0.2%). However, demand for other major products *viz.*, LPG,

MS, ATF, SKO and Lubes are likely to go up by 1.7%, 4.6%, 14.6%, 1.8% and 11.3% respectively during the year 2005-06 over 2004-05.

Indigenous crude production including condensate for the year 2005-06 is estimated at 32.427 MMT. The crude intake of Indian refineries during the year is 127.334 MMT. Crude imports for refineries were estimated at 98.471 MMT. (67.605 MMT for PSU refineries 30.866 MMT for private refinery).

Against the gas demands, the production of 75 MMCMPD is almost half. The Tenth–Plan shows a short fall of 94 MMCMPD of gas. If the power sector expansion by 25000 MWs really takes place in coming Five–Year–Plan then the gas demand alone for the plants would be 150 MMCMPD. Thus relatively, even if better availability of gas is envisaged, presently even 50% of the demand could not be met due to lack of infrastructure and the Pipe Line project is under the authority of GAIL. However private producers are allowed to develop their own net work to serve consumers within 100 km from the point of production. Coal Bed Methane (CBM) is also being considered. The Reliance has plans to produce CBM from Sohagpur mine showing about 3.65 GBM.

The total current requirement of gas is around 67 MMSCMD. The present allocation is 30.5 MMSCMD from domestic sources against which supply is 23 MMSCMD. In addition to 23 MMSCMD, there is an additional requirement of 44.50 MMSCMD for conversion of naphtha and FO/LSHS units, expansion units, de-bottlenecking projects and revamping of closed units. Of the 23 MMSCMD being supplied to the fertilizer sector, 17.3 MMSCMD is APM gas and 5.8 MMSCMD is RLNG. An additional 6 MMSCMD of gas is expected from Panna-Mukta-Tapti (PMT) in next two years.

Similarly the lubes demand is increasing, just in three years time 1997-2000 the increase in imports shows 25-66 thousand tons. With strict laws enforced by environmentalists the production of lubes has become more complicated. There are many varieties of additives for specific purposes and the purpose and uses are many. The disposal of wastes is the biggest problem. The quality of the crude is first important for the production of lubes. The present capacity of Indian refineries is about 8.4 lakh ton of base stock against a demand of 1.2 million tons, hence the imports are increasing.

1.3.4.1 Growth of World Oil Industry

150 years ago the first petroleum refinery of 5bbl capacity was started by Samuel Kiers in Pittsberg in 1855-56. The first oil well (79 ft deep)

was dug by Col. Edwin L Drake in Titusville, Pennsylvania in August 1859. These incidents were the characteristic landmarks of American Petroleum Industry as well as to some extent the world's petroleum industry. George Bissell (1821 - 1884) is often considered the father of the American oil industry Against this the world's first oil refinery opened at PloiestL Romania in 1856. Later Several other refineries were built at that location with the investment from the US companies which were occupied by Nazi Germany during World War II.. Till the middle of 19th century there was no growth of petroleum industry and the products were waxes and burning oil. The advent of IC engines changed the industrialisation like electric motor. With the invention of automobile and airplane transportation and other industrial and defense needs changed displacing all other industries to second place and petroleum refining became the order of the day. These engines sought better quality and quantity of fuel hence a new wave of design and practice of refinery operations were pouring in from the European and western countries. I.

1.3.4.2 Russian Oil Industry

Oil extraction on the Aspheron Peninsula where Baku is located , dates back to 7th and 8th centuries. In the tenth century Arabian traveler. Marudee, reported that both white and black oil were extracted from Baku . And the area shot into fame as "oil Belt Of Baku" known by the name Black city and also became the 'Black Gold Capital'. Reports say that First oil well was dug in 1847 itself in Baku at BibiEyat oil field by Russian Engineer F.N. Semyonov

1.3.4.3 Indian Oil Industry

Indian petroleum Industry cannot be treated in isolation, as API and Indian petroleum industry are closely accompanying in the historic voyage of 150 years. IPI is just seven years to API at its inception. Infact this is the second oldest industry next to Sugar/Gur/Textile industry in India. In March 1867 the first oil well in the Asian continent was dug using mechanical means in the Makum Namdang area in Upper Assam.

India's oil industry is discernible into three distinct phases

    (a) Pre independence
    (b) Post independence
    (c) Private Sector (economic Liberlisation era)

1.3.4.4   Pre Independence Oil Industry:

The history of Assam since 19th century is closely linked with the discovery of oil and its subsequent and continuing exploitation firstly, by the British, and later in the post-Independence period by different organizations of India Even though the sub surface oil explorations started in Assam in 1860, the beginning of Indian petroleum industry was registered with digging of an oil well (102 ft) by Cap. Goodenough of McKillop, Stewart & Co., Calcutta in November 1866, at Nahorpung about 30 miles south east of Digboi, just seven years after the world's first commercial oil well operated in USA . After finding no oil , the mission was given up. In March 1867 the first oil well in the Asian continent was dug using mechanical means in the Makum Namdang area in Upper Assam. The hit of oil at 118 feet and over a tone of crude oil production was the real foundation of Oil industry in India In 1889 the Assam Railway & Trading Company began massive oil exploration and production in Digboi. 1893 saw the formation of Assam Oil Syndicate to handle oil production in Assam and a complex sprung up in north of Digboi (1889). In 1901 with the establishing of the Assam Oil Company that started producing 500 barrels of crude oil per day and established a refinery to refine this crude in Digboi itself.

And the first oil refinery was started in 1893 at Margherita (3000 bbl capacity), which was later shifted to then commissioned Digboi refinery in 1901. by AOC. This followed by the second well at Makum near Margherita, about 8 miles from Digboi. In 1911 the Burma Oil Company came to Assam with the intention of oil exploration and production and soon they discovered massive oil reserves in the so called Burma Valley.

1.3.4.5   Post independence

Waman Bapuji Metre, (1906–1970), admiringly referred Metre in the Indian oil industry circles, was the doyen of Indian petroleum geologists. W. B. Metre joined Assam Oil Company Ltd. (AOC), a wholly owned subsidiary of Burmah Oil Company Ltd. (BOC), at Digboi in upper Assam as a geologist in 1930 It was under his leadership in 1953, that Assam Oil Company made the discovery of the first new oil field in post free- India, striking oil at No. 1 Nahorkatiya exploration well, at a location based on a pre-war seismic survey The Independent India's Oil industry was begun with

Second Five Year Plan.(1956-61) Cambay oil field discovered by Russian Experts is the start of indigenous industry .Born were the organizations for surveys & exploration and development of oil fields and erection of refineries and the sustainable programs were drawn up. This resulted in the birth of ONGC,OIL,IOC initially and many off-shoot organisations like GAIL,OVL,DGH etc

1.3.4.6 Oil And Natural Gas Commission (Corporation) ONGC: (August 1956)

. With its head quarters at Dehra Dun, it was the premier institution for surveys and production and transportation of crude. Until IOC was born it was even taking care of refining activity. In fact it is the kick starter of refining and petrochemical activity in India. Established aromatic complex that became Koyali refinery olefin complex became prodigious IPCL in IPCL. It is the major crude producing organization in India. The biggest discovery of Bombay High later KG Basin were the triumphs of this institution. It has over sea organization (OVL) for all exploration works and extended its presence from Russia to Iran, Nigeria and Vietnam. Import of gas through Bangladesh from Burma and Gas from Iran are main ventures pending. Flagship explorer ONGC has discovered four new oil fields in Assam.

The new fields discovered in 2006-07 include Mekeypore, Kalyanpur,Panidihing and the latest, Disangmukh. At present, ONGC's Assam Asset has a recoverable reserve base of 72.8 million metric tonnes (MMT) of oil.

Under its cap are Ankaleswar oil filed, Jaisalmeer gas fields Kalolo.Balol fields of heavy crudes, Bombay High, KG, Kaveri Basins, Andmans, Some parts in Eastern region.

The gas cracker plants MGCC, GANDHAR, PATA are real gems in its crown. It has over seas venture in the form ONGC (OVL) The Assam Asset has a production target of 1.55 MMT for the year 2007-08. However, with a sustained annual increase in production, the asset now plans to achieve a target of 1.95 MMT in the year 2011-12 in Assam ONGC has single-handedly scripted India's hydrocarbon saga by: Establishing 6.42 billion tonnes of In-place hydrocarbon reserves with more than 300 discoveries of oil and gas; in fact, 6 out of the 7 producing basins have been discovered by ONGC: out of these In-place hydrocarbons in domestic acreages, Recoverable Reserves are 2.29 (BMT) of Oil Plus Oil Equivalent Gas. Cumulative production of

762.3 Million Metric Tonnes (MMT) of crude and 440.7 Billion Cubic Meters (BCM) of Natural Gas, from 115 fields. ONGC produces about 22 MMTPA through Off-Shore fields. The KG basin has been a potential zone for extraction of oil and gas for a long time. The ONGC started exploration for oil and gas in the KG Basin in April 1977, and drilled its first well near Narasapuram in 1978, and discovered gas. As part of exploration for oil and gas, the ONGC has drilled 622 wells, including 483 on land, and 136 offshore. At present 55 oil wells including 34 on land, 21 offshore and 176 gas wells, including 151 on land, and 25 offshore are in operation in the KG basin.

ts onshore oil production witnessed a gradual change over the period, from 0.207 MMT in 2005-06, to 0.284 million metric tonnes in 2008-09. ONGC has set up a mini-refinery at Tatipaka to distill crude oil into naphtha, high-speed diesel, superior kerosene oil and low sulphur high stock. It has rigs at Penikilpadu, South Mahadevapatnam, Kalla, Kesudaspalem, Malapuram and at Vadali. Currently, the ONGC produces 840 tonnes of oil and 44 lakh cubic metres of gas daily.

1.3.4.7 Oil India Limited (February 18, 1959,)

Established with its Head quarters at Duliajan,Oil India Private Limited was incorporated to to look after the prospects of oil industry in Eastern sector and expand - develop the newly discovered oil fields of Naharkatiya and Moran in the Indian North East. In 1961, it became a joint venture company between the Indian Government and Burmah Oil Company Limited, UK. In 1981, OIL became a wholly-owned Government of India enterprise.

Its discoveries Gelki oil fields and Tripura gas fields are note worthy. The organization had developed indigenous technology to transport waxy crudes and established an LPG plant based on Turbo Expansion Technology,(50,000tpa) first in Asia.(1982) It is also a partner in Brahmputra gas cracker . The Company produces around 2.7MMMSCM (06-07) Natural gas and has a dedicated pipeline network for collection/supply of gas as fuel and feedstock to many nearby industries such as Refinery, Fertilizer & Petrochemical Plant, Power generation Plant and 200 Tea Gardens. Additionally, OIL's exploration activities are spread over onshore areas of Ganga Valley and Mahanadi. OIL also has participating interest in NELP exploration blocks in Mahanadi Offshore, Mumbai Deepwater, Krishna Godavari Deepwater, etc. as well as various overseas projects in Libya, Gabon, Iran, Nigeria and Sudan. OIL, operates 1,432 km of cross-country crude oil pipelines. Commissioned in 1962, OIL's crude oil pipeline traverses 79 river crossings,. The state-of-the-art pipeline can transport over 8.0 MTPA of crude oil, feeding 4 Public Sector Refineries in North-east India. It produces 2264.57 MMSCUM natural gas and oil 5MMTPA during 2006-2007.

1.3.4.8 *Indian Oil Corporation Ltd. :(1959)*

Beginning in 1959 as Indian Oil Company Ltd., Indian Oil Corporation Ltd. was formed in 1964 with the merger of Indian Refineries Ltd. (Estd. 1958) Head Quarters at Delhi, Indian Oil Group of companies owns and operates 10 of India's 23 refineries with a combined refining capacity of 70 million tonnes per annum - the largest share among refining companies in India Two refineries of subsidiary Chennai Petroleum Corporation Ltd. (CPCL) and one of Bongaigaon Refinery and Petrochemicals Limited (BRPL) amalgamated in to IOC. The Corporation's cross-country crude oil and product pipeline network spanning about 9,300 km meets the vital energy needs of the country.

IOC achievements include

Largest and the widest network of petrol & diesel stations in the country, numbering about 21,000. Including Auto LPG Dispensing Stations. It supplies Indane cooking gas to over 66.8 million. In addition an R&D center at Faridabad supports, develops and provides the necessary technology solutions to the operating divisions of the corporation. They are backed for supplies by 170 bulk storage terminals and depots, 101 aviation fuel stations and 89 Indane LPG bottling plants. Indian refining major IOC had listed out its areas of interest in twelve different African countries -- namely Algeria, Angola, Congo Brazzaville, Egypt, Equatorial Guinea, Gabon, Libya, Mauritius, Nigeria, South Africa, Sudan and Tunisia.

Take over of Bongaigaon Refinery & Petrochemicals Ltd. INDMAX technology for the 4 MMTPA Fluidised Catalytic Cracking (FCC) unit at the Corporation's upcoming 15 MMTPA refinery-cum-petrochemicals complex at Paradip in Orissa, as well as for the FCC unit coming up at BRPL and HPCL: The Corporation's cross-country crude oil and product pipeline network spanning about 9,300 km meets the vital energy needs of the country.

1.3.4.9 DGH

To promote hydrocarbon activity in the country t, DGH has carried out several surveys to upgrade information covering a total area of 2 million sq kms of which, 84% is in offshore and 16% in on land. It has done pioneering work for initiating gas hydrate exploration in the country. East Coast and Andaman Deepwater areas is found to be promising areas for Gas Hydrates.. The total prognosticated gas resource from the gas hydrates in the country is placed at 1894 TCM.

Area Opened Up for CBM Exploration	13600 Sq Km
Blocks Awarded	26 Nos.
CBM Resources in Awarded Blocks	1374 BCM
Production Potential in Awarded Blocks	38 MMSCMD

1.3.4.10 GAIL(Gas Authority of India Est. 1984 Aug, New Delhi

It is India's flagship Natural Gas company, integrating all aspects of the Natural Gas value chain (including Exploration & Production, Processing, Transmission, Distribution and Marketing) and its related services. In a rapidly changing scenario, It is spearheading the move to a new era of clean fuel industrialisation, creating a quadrilateral of green energy corridors that connect major consumption centres in India

GAIL's Business Portfolio includes

6,700 km of Natural Gas high pressure trunk pipeline with a capacity to carry 130 MMSCMD of natural gas across the country

7 LPG Gas Processing Units to produce 1.2 MMTPA of LPG and other liquid hydrocarbons

1,922 km of LPG Transmission pipeline network with a capacity to transport 3.8 MMTPA of LPG, 30 oil and gas Exploration blocks and 3 Coal Bed Methane Blocks ,13,000 km of OFC network offering highly dependable band - width for telecom service providers (TELECOM), Joint venture companies in Delhi, Mumbai, Hyderabad, Kanpur, Agra, Lucknow, Bhopal, Agartala and Pune, for supplying Piped Natural Gas (PNG) to households and commercial users, and Compressed Natural Gas (CNG) to the transport sector 6,700 km of Natural Gas high pressure trunk pipeline with a capacity to carry 130 MMSCMD of natural gas across the country . GAIL began its city gas distribution in Delhi in 1997 by setting up nine CNG stations, catering to the city's vast public transport fleet. GAIL established North India's only gas based integrated petrochemical complex in Pata using natural gas as the feedstock and has a capacity of 4,40,000 TPA of Ethylene and 3,10,000 TPA of Polymers (HDPE 1,00,000 TPA & LLDPE). In 2001, GAIL commissioned world's longest and India's first Cross Country LPG Transmission Pipeline from Jamnagar to Loni. The Brahmaputra Cracker and Polymer Limited (BCPL) is a joint venture promoted by GAIL (India), and Oil India Limited (OIL),

The company has also extended its presence in Power, Liquefied Natural Gas re-gasification, A Rs 5,640 crore, 2,80,000 TPA Petrochemical Complex in Assam, A 6,200 crore, 4,00,000 TPA gas cracker complex plant in Kochi, Kerala - the feedstock for this plant, will be imported LNG, which will be regassified at the terminal of Petronet LNG Ltd. at Kochi.

GAIL has signed a Production Development Agreement with the National Petrochemicals Company, Iran, for jointly developing a polyolefin plant with a capacity of 1 MMTPA ethylene. Present day gas supply position shows:

Sector wise supply and demand for NG MMSCMD

	2007–08	2009–10
Power	79.7	102. 7
Fertilizers	41.0	55.9
City gas	12.1	13,8
Petrochemicals	25.4	29.1
Sponge iron	6.0	6.9
Total demand	179.2	225.5
Supply	80.5	120.0

GAIL's LPG plants Capacity (MTA)(from NG)

(a) Bijaipur (2 Nos), Madhya Pradesh 4,06,000

(b) Auraiya Pata, UP2,58,250

(c) Gandhar (2.2 lack tons) & Vaghodia, Hazira in Gujarat. 2,07,000 & 73,000

(d) Ussar, Maharastra 1,39,500

(e) Lakwa, Assam 85,000

(f) Duliajan assam 50,000

(g) Nagapattanm 50000

Domestic Gas Supply Outlook(MMSCMD)

Sources	07–08	08–09	09–10	10–11	11–12
ONGC	47.28	48.42	45.69	44.67	41.08
Pvt./JV	23.26	61.56	60.28	58.42	57.22
Anticipated D6 (RIL)	20.00	30.00	40.00		
GSPC	54	54	54		
Total	70.54	109.98	179.97	187.09	192.30

1.3.4.11 GSPC Gujarat State Petroleum corporation: (1979)

One of the leading oil and gas exploration, development and production companies in India. KG BASIN OFF-SHORE BASIN, primary asset is the Deen Dayal field in the Krishna-Godavari basin (the "KG basin"), which has significant gas reserves, part of which, Deen Dayal West ("DDW"), It t operates of the offshore KG-OSN-2001/3 block (the "KG block"), which includes the Deen Dayal field, and hold an 80.0% Working Interest in the block. Gujarat State Petronet Limited (GSPL) was set up to complement the efforts of GSPC. While GSPC harnesses and procures natural gas, GSPL is building the infrastructure that transmits the gas across the state of Gujarat and ultimately allows last-mile linkage to the end-user. This holds Working Interests in 15 producing fields in the Cambay basin further working Interests in 64 onshore and offshore exploration and production blocks. 53 of these blocks are located in India and 11 are located in Australia, Egypt, Indonesia and Yemen.also engaged in other activities in the energy sector as well like wholly owned subsidiary, GSPC LNG Limited "GSPC LNG"), is developing an LNG terminal at Mundra in Gujarat. Associated company Gujarat State Energy Generation Limited ("GSEG"), owns and operates a gas based power plant at Hazira in Gujarat. Another wholly owned subsidiary, GSPC Pipavav Power Company Limited ("GPPC"), is setting up a gas-fired combined cycle power plant at Pipavav in Gujarat. GSPL had also set up a wind farm at Jakhau in Gujarat.The company acquired several discovered oil and gas fields in the first and second rounds of bidding initiated by the Government of India during 1994-95. . State-owned (GSPC) has struck gas in the Krishna Godavari basin, off Andhra Pradesh . The reserves are estimated at 20 trillion cubic feet (tcf). In 1976, ONGC found gas in the Vasai offshore fields (in the Arabian Sea) with an estimated reserve of 24-27 tcf.

KG BASIN GSPC's Deendayal block, Reliance's Dhirubhai blocks, ONGC's G1 deepwater blocks and Cairn Energy's Annapurna,

Padmavati and Kanaka Durga blocks may tilt the energy matrix for India as it goes about globe trotting in search of energy security. Total output at 100 mcm per day IT dwarfs the recent gas discoveries in the region by Reliance and ONGC. While Reliance was the first to strike gas off the Andhra coast in November 2002, estimated at 14 tcf, ONGC announced its find in March this year with estimated reserves of 4 tcf.

There are four major offshore platforms in the KG Basin – Oil & Natural Gas Corporation's GS-15 and GS-23, Reliance Group's D-6, one of the biggest offshore platforms in the country, Cairn Energy's Ravva and Gujarat State Petroleum Corporation's platform near S. Yanam. All the four have been placed on high alert. Exploration in deepwaters of Krishna Godavari Basin, off the Andhra cost, has given seven gas discoveries with reserves of 10.5 trillion cubic feet. "Another 10 tcf reserves are estimated in another shallow water block (awarded to GSPC) in the same basin," he said adding the total potential of in KG Basin was 40-50 tcf RIL started its production of crude oil at the KG-D6 block of the KG basin in 2008 with a production of 5,000 barrels a day. They achieved hydrocarbon production of 5,50,000 barrels. The block is located in the Bay of Bengal, 50 km off the Kakinada coast, a ONGC's natural gas reserves increased from 540.7 BCM in 2006-2007 to 550.27 BCM in 2007-2008. However, private companies and joint ventures walked away with all the accolades as they were able to increase their natural gas reserves from 459.4 BCM in 2006-2007 to 613.01 BCM in 2007-2008.t a depth of 8,000

ONGC's natural gas reserves increased from 540.7 BCM in 2006-2007 to 550.27 BCM in 2007-2008. However, private companies and joint ventures walked away with all the accolades as they were able to increase their natural gas reserves from 459.4 BCM in 2006-2007 to 613.01 BCM in 2007-2008.

1.3.4.12 Reliance Era: Jamnagar (1999)

Started with a refinery (the biggest in India 29MMTPA It has spread its wings to all fields of petroleum. With in 6 years it established another refinery of 33 MMTPA at Jamnagar. Its business comprises of Petroleum refining, marketing products and Petrochemicals production. Oil & gas exploration production- transportation and marketing are its activvtues. It has producing blocks in KG Basin Supply of D-6 gas to fertilizer industry-: RIL's pipeline 1.5 mmscmd of gas.Reliance produces 550,000BPA, Cairon produces 40,000BPA

Reliance Industries has struck 9.46 tcf gas in D6 block while Cairn Energy of UK has found 1.2 tcf in the adjacent D5 block, he said. While reserves of state-owned companies ONGC and OIL increased marginally by 1.5%, reserves of private companies and joint ventures,led by RJL, increased by a whopping 33.4% in 2007-2008

Manufacturing Facilities: Reliance Industries Limited operates world-class manufacturing facilities across the country at Allahabad, Barabanki, Dahej, Dhenkanal, Hazira, Hoshiarpur, Jamnagar, Kurkumbh, Nagothane, Nagpur, Naroda, Patalganga, Silvassa and Vadodara.

1.3.4.13   Petronet LNG Limited (PLL)

It is government of India 's a joint venture company promoted by the Gas Authority of India Limited (GAIL), Oil and Natural Gas Corporation Limited (ONGC), Indian Oil Corporation Limited (IOC) and Bharat Petroleum Corporation Limited (BPCL) The country's first LNG receiving and regasification terminal at Dahej, Gujarat, and is in the process of building another terminal at Kochi, Kerala. While the Dahej terminal has a nominal capacity of 10 MMTPA [equivalent to 40 MMSCMD of natural gas], the Kochi terminal will have a capacity of 5 MMTPA [equivalent to 20 MMSCMD of natural gas]., a JV promoted by GAIL, IOCL, BPCL and ONGC was formed for import of LNG to meet the growing demand of natural gas. PLL has constructed a 5 MMTPA capacity LNG terminal at Dahej in Gujarat and expanded to 10 MMTPA capacity. is likely to. expand this terminal to 10 MMTPA capacity by 2008-09 Shell's 2.5 MMTPA capacity LNG terminal at Hazira has been commissioned. Dabhol LNG terminal (total 5 MMTPA capacity,) is functioning. PLL has been formed for setting up of LNG import and regasification facilities. PLL has a long term LNG supply contract with RasGas, Qatar, for import of 7.5 MMTPA of LNG. PLL has successfully implemented a pilot project for supplying LNG through cryogenic road tankers. PLL is also coming up with a LNG terminal at Kochi, Kerala, with an initial capacity of 2.5 MMTPA, expandable up to 5 MMTPA and it is scheduled to be operational by end of 2011. GAIL has 12.5% equity stake in PLL, along with BPCL, ONGC and IOCL as equal partners.

1.3.4.14   Ratnagiri Gas and Power Pvt. Ltd. (RGPPL)

RGPPL is a joint venture company between GAIL, NTPC, Financial Institutions and MSEB. The capacity of the Ratnagiri Gas & Power

Station is 2,150 MW, which is the largest gas based power generation facility in the country and is currently producing 1,850 MW of power. RGPPL is in the process of commissioning an LNG import terminal of 5 MMTPA capacity. GAIL has 32.88% stake in the Company along with NTPC as equal partner.

1.3.4.15 Tripura Natural Gas Company Limited (TNGCL)

TNGCL is presently supplying gas to around 7,500 domestic, 170 commercial and industrial consumers and has set up one CNG station in Agartala, which is catering to more than 1,400 vehicles. TNGCL has received authorisation from MoPNG for CGD in Agartala. GAIL has 29% stake in the Company. "Dominion is marketing 4.6 million tonnes per annum and Gail has booked 50% of such capacity for 20 years," It has signed a deal with US energy firm Dominion for using capacity at its Cove Point terminal at Lusby in Maryland.

1.3.4.16 Indian Petroleum (At the End of Tenth Plan): Capacity 134.MMT 2012 Refinery through put is 180MMTPA

(IOC Refineries 10 at the End Of 11th Plan)

EXISTING	CAPACITY	EXPANSION (MMTPA)
Koyali	12.5	15-18
Barauni	4.2	6
Guwahati	1.0	
Haldia	3.75	6.5
Mathura	9.0	
Numaligarh	3.0	
AOC	1.065	
Panipat	6.0	12-15
Bongaigon	2.35	
Paradip	15	
[Chennai		
Refineries ltd	6.5	9.5
Manali 3rd phase	3	4
Narimanam Chennai	0.5	
HPCL		
Vizag	7.5	8.33
Mumbai	5.5	
+ Swing Refinery	2.0	
Bhatinda	9.0	

BPCL		
Mumbai	8	12
Kochi	7.5	
PRIVATE		
Vadinar (Essar)	14	20
Reliance	27	33/30
Nagarjuna	6.0	15 (2015)
GAIL		
Bina	9.0	
ONGC		
MRPL complex	9.69	
KG Basin	23000Bbl	
Kakinada ??	30?	
	159	+ 50 by the end of 12th Plan

1.3.5.1   Boost in Refining Refineries (IOC Refineries -10)

1. Guwahati Refinery, IOCL (Assam)

Guwahati Refinery is the country's first Public Sector Refinery as well as Indian Oil's first Refinery since 1962. Built with Romanian assistance the crude processing capacity at the time of commissioning of Refinery was 0.75 MMTPA and the capacity was subsequently enhanced to 1.0 MMTPA. Refinery was designed to process a mix of OIL and ONGC crude. Quality LPG, Motor Spirit, Aviation Turbine Fuel, Superior Kerosene Oil, High Speed Diesel, Light Diesel Oil and Raw Petroleum Coke are the products of this Refinery. It is fitted with a secondary Delayed Coking Unit (DCU).

2. Barauni Refinery, IOCL (Bihar)

Barauni Refinery was built in collaboration with the Soviet Union at a cost of Rs.49.4 crores and went on stream in July, 1964. The initial capacity of 2 MMTPA was expanded to 3 MMTPA by 1969. The present capacity of this refinery is 6.00 MMTPA. A Catalytic Reformer Unit (CRU) was also added to the refinery in 1997. Barauni Refinery was initially designed to process low sulphur crude oil (sweet crude) of Assam. After establishment of other refineries in the Northeast, Assam crude is unavailable for Barauni. Hence, sweet crude is being sourced from African, South East Asian and Middle East countries like Nigeria, Iraq and Malaysia. The refinery receives crude oil by pipeline from Paradip on the east coast via Haldia.

3. Koyali Refinery - IOCL(Gujarat)

This refinery was built with Soviet assistance at a cost of Rs.26 crores and went on stream in October 1965. When commissioned, the refinery had an installed capacity of 2 mmtpa and was designed to process crude from Ankleshwar, Kalol and Nawagam oilfields of ONGC in Gujarat and capable to process indigenous and imported, both low sulphur and high sulphur grades of crude oil. The product slate includes besides fuels, petrochemical products such as Linear Alkyl Benzene (LAB), Polypropylene Feed Stock, Food & Polymer Grade Hexane Its facilities include five atmospheric crude distillation units. The major units include CRU, FCCU and the first Hydrocracking unit of the country, 1.2 MMTPA for conversion of heavier ends of crude oil to high value superior products.. Its mega project worth around Rs.7000 crore to comply with the road map for supplying eco-friendly Bharat Stage-III and IV compliant MS and HSD and to upgrade the bottom of the barrel to improve the gross margin of the Refinery India's first diesel hydrodesulfurisation. By September 1999 ,ts capacity reached 13.7MMTPA. The refinery's facilities include five atmospheric crude distillation units.. company plans to go for the expansion and take the capacity up to 16 MMTPA as part of its long-term plan The refinery was modified to handle imported and Bombay High crude. The refinery also produces a wide range of specialty products such as benzene, toluene, MTO, food grade hexane, solvents and LABFS

4. Mathura Refinery, IOCL(Uttar Pradesh).

The refinery, was built at Rs. 253.92 crores It was commissioned in January, 1982 to processes 6.0MMTPA. The FCCU and Sulphur Recovery Units were commissioned in January, 1983. The refining capacity of this refinery was expanded to 7.5 MMTPA in 1989 with further increase ,the present capacity is 8MMTPA. The refinery processes low sulphur crude from Bombay High, imported low sulphur crude from Nigeria, and high sulphur crude from the Middle East. A DHDS Unit was commissioned in 1989 for production of HSD with low sulphur content of 0.25% wt. (max.). The major secondary processing units provided were Fluidised Catalytic Cracking Unit (FCCU), Vis-breaker Unit (VBU) and Bitumen Blowing Unit (BBU). a

Continuous Catalytic Reforming Unit (CCRU) Through Hydrocracker Unit (from Chevron, USA)

5. Digboi Refinery (Assam)

This is the oldest working Refinery of the World, established in 1901 by Assam Oil Company. Earlier this was at Margheretta processing 3000bpd established in 1893. IOC took over this refinery in 1981(October.14th) A new delayed cracking unit of 1,70KTPA was commissioned in 1999,A new solvent dewaxing Plant for maximum production of micro crystalline wax was commissioned in 2003.Also a hydrotreater for diesel installed. The capacity is almost same around 0.7MMTPA , having most of the units in this oldest refinery still working..

6. Haldia Refinery (IOCL)(West Bengal)

This Refinery was commissioned in January 1975 , to process Middle East crude, capable of producing lube oils. The fuel sector was built with French collaboration while Rumanian collaboration launched lube stocks. Presently it is capable of refining MM. 7.5 MMTPA Products include Jute Batching Oil. Diesel Hydro Desulphurisation (DHDS) Unit (1999,) for production of low Sulphur content (0.25% wt) High Speed Diesel (HSD). for producing BS-II and Euro-III equivalent HSD. Residue Fluidised Catalytic Cracking Unit (RFCCU) was commissioned in 2001 in order to increase the distillate yield of the refinery as well as to meet the growing demand of LPG, MS and HSD. Refinery also produces eco friendly Bitumen emulsion and Microcrystalline Wax. A Catalytic Dewaxing Unit (CIDWU) was installed and commissioned in the year 2003 for production of high quality Lube Oil Base Stocks (LOBS

7. Panipat Refinery, IOCL, (Haryana)

This refinery was set up in the IOC chain in 1988 costing Rs 3868 crores Installed to process 7MMTPA expanded to 13MMTPA. A Naphtha Cracker at Panipat, built at a cost of Rs 14,400 crore, is the largest operating cracker in India. It was expanded to process 15MMTPA in 2006. The major secondary processing units of the Refinery include Catalytic Reforming Unit, Once Through Hydrocracker unit, Residue Fluidised Catalytic Cracking unit, Visbreaker unit, Bitumen blowing unit, Sulphur block and associated Auxiliary facilities. In order to improve diesel quality, a Diesel Hydro

Desulphurisation Unit (DHDS) was subsequently commissioned in 1999. Referred as one of India's most modern refineries, Panipat Refinery was built using global technologies from IFP France; Haldor-Topsoe, Denmark; UNOCAL/UOP, USA; and Stone &Webster, USA The feed for the unit is sourced internally from Indian Oil's Koyali, Panipat and Mathura refineries. Panipat Refinery has also developed new products like 96 RON petrol, and sub-Zero diesel for the Indian army The Naphtha Cracker produces - Polypropylene (capacity: 600,000 tonnes), High Density Polyethylene (HDPE) (capacity: 300,000 tonnes) and Linear Low Density Poly Ethylene (LLDPE) (350,000 tonnes Swing unit with HDPE), Mono Ethylene glycol unit.Another technologically advanced plant is manufacturing Paraxylene (PX) from captive Naphtha and thereafter, converting it into Purified Terephthalic . It receives crude from Vadinar through the 1370 km long Salaya-Mathura Pipeline which also supplies crude to Koyali and Mathura Refineries of IndianOil. Panipat refinery with an expected consumption of 15430 TMT of crude worth Rs.64,517 crore will be the single biggest crude consuming refinery

Consumption of crude at eight IOC operated refineries is expected to be 54,750 TMT, worth about Rs.231,996 crore.

Panipat refinery with an expected consumption of 15430 TMT of crude worth Rs.64,517 crore will be the single biggest crude consuming refinery

8. Paradip (IOCL) (Odisha)

After delays Paradip refinery took the shape when Indian Oil Corporation Limited announced in March, 2006 that it is going to set up a refinery and petrochemical complex in the state of Odisha. Initially the production capacity was stated to be 9 MMTPA which was augmented to 15 MMTPA in due course. A 300,000bpd crude and vacuum distillation unit, a 104,000bpd vacuum gas oil hydrotreater, and 80,000bpd delayed coker unit, a 78,000bpd fluidised catalytic cracker (FCC) and a 53,000bpd crude catalytic reformer are established .Chemicals like paraxylene (24,000 bpd), polypropylene and styrene (10,000 bpd alkylation unit),are to be produced. It also commissions a pipeline from refinery to Ranchi. The refinery will produce 5.97 million tonnes of diesel, 3.4 million tonnes of petrol, 1.45 million tonnes of kerosene/ATF, 536,000 tons of LPG, 124,000 tons of naphtha and 335,000 tonnes of sulphur. The fluidised catalytic cracker will use the Indmax process developed by IOC. The process will convert heavy distillate and residue into LPG and light distillate products.Consumption of crude at eight IOC operated refineries is expected to be 54,750 TMT, worth about Rs. 231,996 crore

Consumption of crude at eight IOC operated refineries is expected to be 54,750 TMT, worth about Rs.231,996 crore. Panipat refinery with an expected consumption of 15430 TMT of crude worth Rs.64,517 crore will be the single biggest crude consuming refinery the first indigenous grass root Refinery in the country integrated with a Petrochemical complex at one location.

9. Chennai Refinery, (CPCL) (earlier MRL) Manali (Tamil Nadu)

Under the name of Chennai Petroleum Corporation Ltd was established in joint sector with GOI and AMOCO- National Iranian Oil Company in 1965 to process 2.5 MMTPA and now its capacity 9.5. CPCL became a group company of Indian Oil Corporation Limited (IOCL). After AMOCO and NIOC withdrew from the company. CPCL has two refineries with a combined refining capacity of 11.5 Million Tonnes Per Annum (MMTPA). The Manali Refinery is one of the most complex refineries in India with Fuel, Lube,Wax and Petrochemical feed stocks production facilities. CPCL's second refinery is located at Cauvery Basin at Nagapattinam. This unit was set up in Nagapattinam with a capacity of 0.5 MMTPA in 1993 and later enhanced to 1.0 MMTPA. The main products of the company are LPG, Motor Spirit, Superior Kerosene, Aviation Turbine Fuel, High Speed Diesel, Naphtha, Bitumen, Lube Base Stocks, Paraffin Wax, Fuel Oil, Hexane and Petrochemical feed stocks. Propylene Plant with a capacity of 17,000 tonnes per annum was commissioned in 1988 to supply petrochemical feedstock to neighbouring downstream industries. The unit was revamped to enhance the propylene production capacity to 30,000 tonnes per annum in 2004. CPCL also supplies LABFS to a downstream unit for manufacture of Liner Alkyl Benzene. The crude throughput for the year 2011-12 was 10.557 million metric tonnes (MMT) Delayed Coker Unit produces 50,000 bpd. A hydrocracker revamp, and sulfur recovery unit are also installed.

10. Bongaigon Refinery(BRPL) (Assam){subsidiary of IOCL)

The one million ton Bongaigaon Refinery & Petrochemicals Ltd. was incorporated on the 20th of February in 1974 under public sector of the government of India. BRPL became a subsidiary company of the Indian Oil Corporation Ltd., which is a Union Government undertaking because of its 76% equity. This was established under political considerations as there was no sufficient crude to run this refinery. The refinery presently processes crude oil produced in the Assam oil fields, as well as, Ravva oil fields in Andhra Coast through Pardip-Haldia- Bongaigon. The best crude of KG basin is lost to Assam while HPCL at Vizag has to import crude from outside (high sulfur) The production facilities of the company consists of a refinery with a crude processing capacity of 2.35 million tones per year While commercial production of the refinery started in 1979, the petrochemicals complex is designed to have a Xylene plant, dimethyle terephthalate (DMT) plant and a polyester staple fibre (PSF) plant which were commissioned in stages, The major products from the refinery are LPG, MS, Naptha, ATF, SKO, HSD, LDO, LSHS, LVFO, RPC & CPC. DMT and PSF, which are the most important products of the Petrochemicals sector. While the petroleum products (except RPC & CPC) are marketed by the Indian Oil Corporation Ltd., (Marketing Division), the Petrochemical products and RPC & CPC are marketed by BRPL itself through its own marketing network. Two Crude Distillation Units (CDU), two Delayed Coker Units (DCU) and a Coke Calcination Unit (CCU) with a processing capacity of 2.35 MMTPA of crude oil.. An LPG Bottling Plant with a capacity of 44,000 MTPA was also commissioned in the year 2003... BRPL has the unique distinction of being the first integrated refinery of India.

HPCL-2Refineries

11. Mumbai Refinery (HPCL), (Maharastra- Thane)

The refinery was established in Independent India by ESSO in 1954 with a capacity of 3MMTPA increased to 6.5 MMTPA. After Nationalisation under the nameof Hindustan Petroleum Corporation Ltd. (1974/7/04) existing. Plans to expand the capcity is 7.9 MMTPA (2012). The Lubricating Oils Refinery set up at Mumbai is largest lube refinery in India(3.35 lakh tons). Crude through put from two refineries is 16.19MMT in 2012 (Vizag 8.68 MMT) The refinery is installing a new FCCU of 1.4 MMTPA capacity, which will increase the FCCU processing capacity from the existing level of 1 MMTPA to 2.4 MMTPA. It has adopted cogeneration principle of steam FCCU CO boiler at Mumbai refinery HPCL is going to set up a Guru Gobind Refinery in Punjab.

12. Visakh Refinery (HPCL) (Andhra Pradesh)

Established as Caltex Refinery in1957,with a capacity1.0 MMTPA was amalgamated into HPCL in May 1978. The capacity was changing periodically presently standing at 7.5 MMTPA(1999) and likely to 8.33 MMTPA. effective since April 2010. Diesel Hydro desulphurization (DHDS) project was commissioned in the year 2000 to meet BS-I/II specification of diesel. The facilities were further augmented in 2005 by addition of 2nd Reactor in DHDS unit for supplying BS-III grade diesel. Visakh Refinery is executing the Mounded storage system for LPG and Propylene in place of existing LPG /Propylene Horton sphere

BPCL - 2Refineries + Joint Ventures: 2

13. Mumbai (BPCL) (Mahaul)

Established by Burma-Shell in 1955 to process imported crude at a rate of 5.25 MMTPA, expanded to 6.9 MMTPA. Further, it was expanded to 12 MMTPA in a Refinery Modernization Project undertaken by the company. The project added CDU/VDU, HCU, LOBS, HGU units After nationalization it was christened as Bharat Petroleum Corporation Ltd(1976). The crude throughput at BPCL's Mumbai Refinery, during, 2011-12 was 13,355.4 KT, which included 4,359.8 TMT of indigenous crude and 8,995.6 TMT of imported crude. High sulphur crude made up 47.5% of the throughput. During the period, 3,568.9 TMT of light distillates were processed by the refinery against a target of 3,224.5 TMT. 7,057.9 TMT of middle distillates were processed against a target of 6,175.4 TMT and 2,099.6 TMT of heavy distillates were processed against the target of 2,298.4 TMT. BPCL's bitumen and LPG business during April-November, 2011-12, grew at 30.17% and 8.87% as against the industry's growth of 6.33% and 7.51%. The total amount of high sulphur imported crude processed at the Mumbai refinery during June 2012, stood at 540,338 MT. This comprised of 139,716 MT of Kuwait export crude, 108,423 MT of Arab extra light, 85,775 MT of lower zakum crude, In addition, the refinery also processed 140,223 MT of imported low sulphur crude during June. This was made up of 91,332 MT of Mellitah crude, 34,579 MT of Miri light crude, 9,952 MT of AKPO and 4,360 MT of EL Sahara? Refinery currently processes about 12 Million Metric Tons of crude oil per annum. BPMR has processed 61 different types of crude in five decades of its operations, making it one of the most flexible Refineries in the country. Bharat Petroleum Corporation Limited only PSU Oil Company in top category of "BRICS Carbon Ranking" amongst BRICS nations comprising of companies from Brazil, Russia, India, China & South Africa.

14   Kochi Refinery, Cochin, Kerala

Kochi Refinery, a unit of Bharat Petroleum Corporation Limited (BPCL), embarked on its journey in 1966 with a capacity of 50,000 barrels per day. Formerly known as Cochin Refineries Limited and now named as Kochi Refineries Limited it was originally established as a joint venture in collaboration with Phillips Petroleum Corporation, USA. It is one of the two Refineries of BPCL that has refining capacity of 9.5 MMTPA. Its fuels wing produces Liquefied Petroleum Gas, Naphtha, Motor Spirit, Kerosene, Aviation Turbine Fuel, High Speed Diesel, Fuel Oils and Asphalt. Specialty products include Benzene, Toluene, Propylene, Special Boiling Point Spirit, Poly Iso Butene and Sulphur. High sulphur crude made up 49.7% of the crude throughput at the refinery during the period with 10.5 MMTPA (2012) capacity, The Kochi refinery's delayed coking unit will have the capacity to process 3.84 tpd of vacuum resid and FCC products based on Lummus Technology. The refinery is under expansion from the present capacity of 9.5 MMTPA to 15.5 MMTPA and LPG production is expected to go up from 0.519 MMTPA to 1.26 MMTPA. 0.5MMTPA shall be consumed at Kochi whereas the balance approximately 0.76 MMTPA will be required to be exported Upgrading Refinery to produce clean automotive fuels; Installation and commissioning of Hydrocracker unit under Refinery Modernization Project (RMP) for production of Euro lll Grade Auto Fuel product (High Speed Diesel): Installation of Diesel Hydro-desulphurization facility for production of Euro lll Grade Auto Fuel product (High Speed Diesel): Revamp of catalytic Reformer 'Unit for production of Euro III Grade Motor spirit. Installation Methyl Tertiary Butyl Ether (MTBE) Unit to replace Tetra Ethyl Lead (TEL) from motor spirit. BPCL is upgrading their refinery units for producing Euro-III/IV quality auto fuels from 2010. Production of High Quality Group II + Lube Oil Base Stock (LOBS) for manufacturing environment friendly (ultra low sulphur, long life) lube oils.

15. Numaligarh Refinery,(1985) Numaligarh, Assam

Popularly known as Assam Accord Refinery BPCL has major shares followed by OIL, has been commissioned in 1999 having a capacity of 3MMTPA. and a VDU 1.3 MMTPA. It has Delayed coker 0.3 MMTPA, with coke calcination plant, hydrocracker 1.1 MMTPA and sulfur recovery unit. It produces MS.LPG and naphtha of 313 k tons and middle distllates ATF132, SKO270, HSD 1856 k tons respectively.

16. The Bharat Oman Refineries Limited (BORL) (Bina MP)

Conceived in the early 1990s as the Central India refinery, Bina Refinery, iocated at Bina, was built as a joint venture between India's Bharat Petroleum Corporation (BPCL) and Oman's Oman Oil Company, and the 6 MTPA refinery was commissioned in May 2011During the next two-three years, capacity of the refinery will be expanded from six to nine million tonnes with the ultimate objective being 15 MMTPA by 2015-16.

The project faced significant delays on account of environmental clearances and poor infrastructure and suffered high escalation of a budgeted ₹6,300 crore to ₹9,100 crore. The refinery also consists of a 1 MTPA naphtha hydrotreater, a 0.5 MTPA catalytic reformer to produce gasoline, a 1.95-million-tonne hydrocracker, a 1.63-million-tonne diesel hydrotreater and a 1.36-million-tonne delayed coker. Exports of naphtha from the refinery began in 2012. The plant is equipped to produce Euro III and Euro IV petroleum products and is capable of producing Euro V petroleum products with minimal additional investment. The crude is transported through a 935-km long pipeline from Vadinar to Bina. The products of the refinery are then transported through the 257- km Bina-Kota pipeline where it joins the Mumbai-Manmad-Bijwasan pipeline to reach the markets of North India.

Mumbai Refinery: The crude throughput at BPCL's Mumbai Refinery, during April to March, 2011-12 was 13,355.4 KT, which included 4,359.8 TMT of indigenous crude and 8,995.6 TMT of imported crude. High sulphur crude made up 47.5% of the throughput. During the period, 3,568.9 TMT of light distillates were processed by the refinery against a target of 3,224.5 TMT. 7,057.9 TMT of middle distillates were processed against a target of 6,175.4 TMT and 2,099.6 TMT of heavy distillates were processed against the target of 2,298.4 TMT..

BPCL's all-product sales during April to March, 2011-12 stood at 31.2 MMT, registering a growth of 6.6%. .BPCL's bitumen and LPG business during April-November, 2011-12, grew at 30.17% and 8.87% as against the industry's growth of 6.33% and 7.51%.

17. Bhatinda Refinery (2008) Punjab

Known as Gobind Singh Refinery (GGSR) is a refinery owned by Hindustan Mittal Energy Limited (HMEL) a joint venture between HPCL and Mittal Energy Limited he work for refinery started in and the refinery became operational in March 2012. The crude from ME is transported via Kandla - Bhatinda pipe line. Its annual capacity is nine million tons (180,000 barrels per day). It was built at a cost of $4 billion. The refinery will get its crude oil supply from Mundra a coastal town in Gujarat through a 1,014 km pipeline? It is a refineru of highest yields of propylene,. One of the highest Nelson Complexity Indices in the region with the ability to process Heavier and Sour crudes. Also, Petronet India planned a 2,290km cross-country pipeline to cater to north and central India. ExxonMobil believed there was no market and withdrew in February 1999.

Private sector -4

18. Essar Oil Refinery, Vadinar,

Though started in 1996, it could not be completed till 2006. Thegrass roots refinery in Gujarat was refinery has a capacity of 300,000 barrels per day . which started commercial production on May 1, 2008. It is configured to produce Euro II and Euro III grades of petrol and diesel fuel After its expansion it runs at a Nelson complexity index of 12.8. This means it will be able to refine all varieties of crude, producing Euro 5 grade fuels. The refinery recently expanded to a 20 million tons per annum. Essar has a global portfolio of onshore and offshore oil and gas blocks, with about 35,000 sq km available for exploration. It has over 750,000 bpsd of global crude-refining capacity (Vadinar + Stanlow U.K + Kenya refinery 50% stake). Essar's exploration and production business has 2.1 billion barrels of oil equivalent of reserves and resources. It is capable of producing LPG, Naphtha, light diesel oil, Aviation Turbine Fuel (ATF) and kerosene and it is designed to handle sweet to sour and light to heavy crudes. Vadinar Refinery (Essar Oil) is capable of processing over 80% of ultra heavy and heavy crude in its crude mix and produce higher grade products like Euro-

IV and Euro-V compliant petrol and diesel to cater to the domestic and international markets. The capacity of the refinery raised to 20 MMTPA (June 2012).This changed its complexity from 6.1 currently to 11.8 on the Nelson index, making it India's second largest single-location refinery to produce fuels compliant with the latest Euro IV and Euro V emission standards. Essar's exploration and production business has 2.1 billion barrels of oil equivalent of reserves and resources. Of this, approximately 150 million barrels are 2P and 2C resources, 1 billion barrels are prospective resources and 1 billion barrels are unrisked, in-place resources. Essar Oil Refinery will be expanded in two phases to achieve a capacity of 36mmtpa.

19. **Nagarjuna Oil Corporation Ltd, (NOCL) Cuddalore, Tamil Nadu ,(2013)**

The Refinery is one of the ventures of Hyderabad-based Nagarjuna Group, which was founded in 1973, by KVK Raju In the first phase, the implementation of six million tons a year capacity unit will be completed and operations are expected to commence by April 2014, It will be expanded to 9 million tons per annum and bringing the total crude processing capacity to 15 million tons per annum by 2015-16 The refinery can process some of the world's sour crudes and is said to have Nelson Complexity Factor (NCF) of around 8.74. The Crudes that are processed in the refinery are Bonny Light and Arab Mix crude.25 The Refinery's primary units include Crude Distillation Unit (CDU) and Vacuum Distillation Unit (VDU) which is said to be the mother units, and secondary units include Catalytic Hydrodesulphurization Unit, Methyl Tertiary Butyl Ether Unit, Cold Box Unit, Fluid Catalytic Cracking Unit, Delayed Coker Unit, Diesel Hydrodesulphurization Products include. High Speed Diesel, naphtha and Aviation Turbine Fuel. The refinery is designed for producing feedstock of EURO III and EURO IV standards Apart from regular petroleum fuels that are expected from this expansion, NOCL plans to set up a Xylene production facility, a Purified Terephthalic Acid (PTA) plant and a Propylene Recovery Unit.

20. **Tatipaka Refinery(Mini) (2001) Andhra Pradesh**

ONGC'S mini refinery at Thatipaka, approximately handles 0.1MMtPA light crude produced in KG basin. It removes lights suitable for farming application.

21. Mangalore Refinery: (April 2012) Mangalore (Karnataka) Phase-III

ONGC, set up in 1993 MRPL oil refinery at Mangalore . The refinery has a versatile design with high flexibility to process crudes of various API gravity and with high degree of automation. MRPL has a design capacity to process 9.69 million metric tonnes per annum and is the only refinery in India to have two hydrocrackers producing premium diesel (high cetane). It is also the only refinery in India to have two CCRs producing unleaded petrol of high octane. Currently, the refinery is processing about 12.5 million tonnes of crude per year. It is setting up a 3 MMTPA CDU/VDU, a 2.2 MMTPA Petrochemical Fluidised Catalytic Cracking Unit (PFCCU), a 3 MMTPA Delayed Coker Unit (DCU), a 3.7 MMTPA Diesel Hydrotreating Unit (DHU), a 0.65 MMTPA Coker Heavy Gas Oil Hydrotreating Unit (CHTU), a 70 KTPA Hydrogen Generation Unit and a 440 TMTPA Poly Propylene Unit,. Before acquisition by ONGC in March 2003, MRPL was a joint venture oil refinery promoted by M/s Hindustan Petroleum Corporation Limited (HPCL), a public sector company MRPL's Refining Capacity is around 8 % of India's total Refining capacity as on 1-04-2012. Also with enhanced capacity of 15 MMTPA with a Refinery complexity of around 9.5, MRPL is well placed globally, to compete with the peers in the Refining Industry.

21 & 22. Reliance Refineries , Jamnagar, Gujarat

The refinery was commissioned on 14 July 1999 with an installed capacity of 668,000 barrels per day (106, 200 m^3/d). It is currently the largest refinery in the world. The Jamnagar Complex is the first manufacturing complex of its kind, having a fully integrated petroleum refinery, petrochemicals complex, The Jamnagar Refinery with a refining capacity of over 33 million tons per year and paraxylene production of 1.5 millions tons per year, Reliance Jamnagar is the world's largest grassroots refinery and aromatics complex. Reliance Petroleum's parent company, Reliance Industries Ltd., is the largest private sector company in India.In 2008, a second refinery was built adjacent to the first, with an investment of over $6bn to double the company's Jamnagar facility's capacity to approximately 1.24 million barrels per day. Construction on the refinery was started in 2005 and took 36 months to complete in 2008.

On 25 December 2008, Reliance Petroleum Limited (RPL) commissioned its refinery into a Special Economic Zone in Jamnagar, Gujarat, from where only exports are permissible The completion of the RPL refinery has enabled Jamnagar to emerge as a 'Refinery Hub', housing the world's largest refining complex with an aggregate refining capacity of 1.24 million barrels oil per day, more than any other single location in the world. It is among the top ten refineries in the world. The second refining unit can process approximately 29 million tons. The combined capacity of the two refineries is 62 million tons, ranking the Reliance's Jamnagar refineries With its commissioning, the country's refining capacity jumped to 178 million tonne per annum (MTPA) compared with its annual demand of 129 MTPA. (2008)

23. Kakinada Refinery and Petrochemicals Ltd (KRPL). Andhra Pradesh

A16 MMTPA refinery is planned in 2006 with investors. ONGC being the major player Still the materialization is not complete. It is marked in the petro corridor of Kakinada- Rajahmundry route.

1. 3.5.2 *Reserves and Potential Hydrocarbon locations*

The Earth's total original endowment of oil amounts to 2.1 -2.8 trillion bls. As of 1998 we consumed 800billion bls. World energy consumption was growing about 2.3% per year. While in India it is up to 3% World Oil consumption is likely go up to 121 million bbls/day by 2020. It may be around 200 MMTPA in India by 2015. The oil shock of 1973 had unfurled the most intricate avenues of alternate energies. Almost same time the Scientists of MIT forwarded a hypothesis of LIMITS TO GROWTH. (Under club of Rome) And the lamentable Peak Oil Theory by Hubbert had added additional fuel to the burning prices of oil. Today if we look at those theories they are notjhing but Oil industry/ Rich capitalistic countries theories to extract more resources from third countries at far lower rates than prevailing or expected to reach. Since the oil exploitation started there was no smoothsailing for industry, like pollution, environmental disorders, safety, carcinogenic fear all were adversaly affecting still the oil became a sincere slave to humanity.

Proved Reserves of Natural Gas (Trillion Cubic Feet)

	2007	2008	2009	2010	2011
North America	310.229	316.706	343.577	NA	NA
United States	237.726	244.656	272.509	NA	NA
Central & South America	240.745	261.795	266.541	266.803	268.541
Brazil	10.820	12.280	12.890	12.862	12.940
Venezuela	152.380	166.260	170.920	175.970	178.860
India	37.960	37.960	37.960	37.960	37.928
Middle East	2,566.038	2,548.900	2,591.653	2,658.273	2,686.373
Africa	484.433	489.630	494.078	495.250	517.706
Canada	57.946	58.200	57.906	61.950	61.950
Mexico	14.557	13.850	13.162	12.702	11.966
World	6,216.033	6,219.265	6,289.147	NA	NA

See the figure 1.1b3 World oil consumption it is evident no where Peak consumption is shown but continues with increasing consumption.

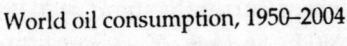

World oil consumption, 1950–2004

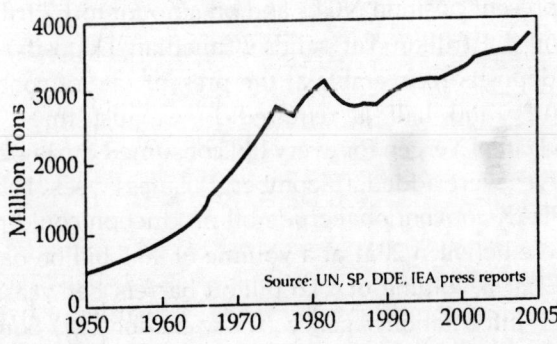

Source: UN, SP, DDE, IEA press reports

Figure 1.1b2 Oil consumption in the World (UN Source)

2000 USGS survey shows 2,300 billion barrels ($370×10^9$ m3) reserves proved are available and unconventional other types are many more times. Unconventional sources, such as heavy crude oil, oil sands, and oil shale are not counted as part of oil reserves. In 2009, the USGS updated the Orinoco tar sands (Venezuela) recoverable "mean value" to 513 billion barrels ($8.16×10^{10}$ m3), with a 90% chance of being within the range of 380-652 billion barrels ($103.7×10^9$ m3), making this area "one of the world's largest recoverable oil accumulations" The Earth's total original endowment of oil amounts to 2.1 -2.8 trillion bls. As of 1998 we consumed 800billion bls. World energy consumption was growing about 2.3% per year. While in India it is up to 3% World Oil consumption is likely go up to 121 million bbls/day by 2020. It may be around 200 MMTPA in India by 2015.

Oil sands &bitumen	30%
Conventional oil	30%
Extra heavy oil	25%
Heavy oil	10%

Present World gas reserves are 300,000 billion CM (higher than shown above) while Russia leads with 55,000, Iran 33,500 billion CM, while India is having just 1,200BCM (Central Intelligence Agency: The World Fact Book). According to current estimates, more than 80% of the world's proven oil reserves are located in OPEC Member Countries, with the bulk of OPEC oil reserves in the Middle East, amounting to 66% of the OPEC total. OPEC Member Countries have made significant additions to their oil reserves in recent years. As a result, OPEC's proven oil reserves currently stand at 1,199.71 billion barrels. The present position NGLs and oil amount to 1.2 trillion bbls, (2010)Shale oil 4.8 trillion Tar sands (Canadian Deposits) 6.trillion bbls known deposits recoverable at the present rate of consumption another century and half is required to exhaust these reserves. According to Daniel Yergen for every bbl consumed during 2007-9 1.6 bbl new reserves were added, (BloombergBusiness week, Feb 5, 2012)

WORLD PEAK conventional crude oil production could plausibly occur anywhere between 2021 at a volume of 48.5 billion barrels per year and 2112 at a volume of 24.6 billion barrels per year product usage" Energy Information Agency (www.eia.doc.gov) Some Great World energy consumption was growing about 2.3% per year. While in India it is up to 3% World Oil consumption is likely go up to 121 million bbls/day by 2020. It may be around 200 MMTPA in India by 2015. World Oil Reserves (2004) 1188.6 billion barrels US Dept of Energy) or 928 billion ton hydrocarbons of which OPEC has 890 billion barrels (32 billion tons probable Indian reserves for internal use by DGH India) 1,161 billion barrels of oil equivalent .(Natural gas 1150 billion cubic meters & gas hydrates 1900 trillion cubic meters. in India) Oil Reserves estimation show 778 Million tons recoverable 2011-12)

1.3.5.3 Incredible Discoveries:

In 1976 Offshore discovery Cantarell oil field off Mexico became a minting machine for Petroleos(Pemex) Today it is producing 2.3 mmbd and has a reserve for 25 mmb in deep waters. New technologies are helping to probe more clearly beneath the Earth's surface. These advances allow to maximize the production of existing oil and natural gas fields and to drill deeper than ever before to locate and recover resources that were once considered too difficult to develop and exploit. The Tahiti Field in the deepwater U.S. Gulf of Mexico demonstrates Chevron's capabilities. Tahiti's deepest producing well is more than 26,700 feet (8,140 m), a record for the Gulf of Mexico. (Production began in May 2009). Yamal peninsula holds the largest gas reserves of the Russian Federation 32.84 BCM. In 2009 alone the Natural gas production was registered at 33.846 million cubic metres.

1.3.5.4 Future Scenario

Energy consumption in the developed countries may increase by 3-3.5%a year while developing economies like China, Brazil, India, Turkey may reach unexpected consumption rates. China's strategy of tapping in and around may produces 70% of its oil demands by 2035 while looking for another 150 MMTPA. Brasil with 4 or more million tons of alcohol can go for cleaner energy. Its inorganic oil fields along with one of the biggest discoveries in the world can export about 100MTPA by 2035. India on the other hand is in a deplorable condition.. No Systematic policies to find alternative energy sources have been implemented, Solar panels/solar vehicles should have been the order of the day, but no hope even for another ten years. Stunning growth in automobile sector would drain of continuously exchange reserves, making the country more dependent on oil from other countries. Even nuclear energy is not progressing, facing hurdles due to local resistance hence energy sector's future is dim. Aided by no strict policy on growth of population as followed in China, India's population will grow at much faster rate than any other country adding more perils to energy sector.

Royal Dutch Shell, when major oil resources getting harder to find, and the need for cleaner, reliable alternative energies to share growing needs of population (June 11, Fortune: 2012) Shell's slogan for 250 year supply of NG to the World(www.shell.us/letsgo) is well known, however the effect of it in Market price is not good.

US domestic oil production peaked in 1970. Global production of oil fell from a high point in 2005 at 74 mb/d, but has since rebounded, and 2011 figures show slightly higher levels of production than in 2005. Most of the remaining oil is from unconventional sources. Rough estimates indicate that out of an available 2 trillion barrels of oil, about half has been consumed. Between 1995 and 2005, U.S. consumption grew from 17,700,000 barrels per day (2,810,000 m^3/d) to 20,700,000 barrels per day (3,290,000 m^3/d), a 3,000,000 barrels per day (480,000 m^3/d) increase. China, by comparison, increased consumption from 3,400,000 barrels per day (540,000 m^3/d) to 7,000,000 barrels per day (1,100,000 m^3/d)

The World has witnessed ups and downs in oil consumption after Oil shock of 1973. Oil production per capita has declined from 5.26 barrels per year (0.836 m^3/a)in 1980 to 4.44 barrels per year (0.706 m^3/a) in 1993, but then increased to 4.79 barrels per year (0.762 m^3/a) in 2005 In 2006, the world oil production took a downturn from 84.631 to 84.597 million barrels per day (13.4553 × 10^6 to 13.4498 × 10^6 m^3/d) although population has continued to increase. This has caused the oil production per capita to drop again to 4.73 barrels per year. Canada has been looked upon as a favorable supplier given its gigantic reserves of hydrocarbons. The country is also the third largest natural gas producer in the world which makes it a lucrative long term supplier to the needed. Its average annual gas production totals 6.4 TCF with reserves over 1.72 TCM. U.S. petroleum consumption reached an estimated 18.87 million barrels per day in 2011, and is expected to increase to 18.96 million barrels per day in 2012. Demand for 2012 is projected to continue to decline to 8.74 million barrels per day.. This equates to an average of 33 miles per vehicle per day. The cost of crude oil accounted for 72% of the cost of a gallon of gasoline in the United State while refining accounted for just 12%. Taxes and distribution/marketing accounted for 11% and 5% respectively.

World non energy refinery products (2009) include: Naphtha & spirits 56% bitumen 21%, asphalt, wax 1% , lubes 9% petroleum coke 23%. Important fuels and refining in the world and India are shown below

Production of energy fuels '000 tons

		Total, oil fuels	Aviation gasoline	Motor	jet fuel	kerosene	Diesel	fuel oil	LPG gas	Refinery gas
World	2006	3230197	1618	889670	228016	92420	1178057	588396	112278	129741
	2009	3224425	1430	913897	221817	83453	1230897	518459	118749	125618
India	2006	107299	–	12539	7805	8491	54268	15697	6315	2184
	2009	144440	–	22554	9304	8545	77605	17535	6515	*2382

1.3.5.6 Indian Hydrocarbon Activity

Indian crude production during April-February 2013, was 34.639 MMT. The targeted production for the cumulative period was 36.517 MMT. Gas production April-February 2012-13 period came to 37.503 MMCM, which was also lower then the planned target of 37,716 MCM. Cumulatively, in the April-February period, Indian refineries processed 165.637MMT of crude, which was 1.8% above the target of 162.658 MMT. Against planned the crude oil production target of 206.73 MMT in the eleventh plan, the actual achievement is only 177 MMT, that is, 14 per cent below the target. .and natural gas production was 212.54 BCM as against the production target of 255.76 BCM,. Refinery capacity addition too suffered: Just 88.42 per cent of the target was scaled and major projects, including the MRPL expansion and Paradip refinery projects have slipped into the Twelfth Plan.

1.3.5.7 Indian refinery throughput million ton

1960-61	6.130	95-96	58.74
65-66	10.233	99-2k	85.96
70-71	18.38	2k8	156.11
75-76	22.28	2k9	160.67
78-79	25.97	2011	206.121
80-81	26	2012	211.42
90-91	51.77		

1.	Domestic Gas Supply Outlook(MMSCMD)					
2.	Sources	07-08	08-09	09-10	10-11	11-12
3.	ONGC	47.28	48.42	45.69	44.67	41.08
4.	Pvt./JV	23.26	61.56	60.28	58.42	57.22
5.	Anticipated D6 (RIL)		20.00	30.00	40.00	
	GSPC			54	54	54

6. Total   70.54   109.98   179.97   187.09   192.30
7. KG BASIN OFF-SHORE BASIN
8. Reliance produces 550,000BPA
9. Cairon produces 40,000BPA

ONGC's 3rd LPG plant at Uran on hold affected GAIL's plant at Usar, NG production is 47.56 billion cubic meteres less than 2011 by 8.92% (53Billion c.m) Oil imported was 171.73 million tons 4.5%increase over previous year. 9.7 Million ton LNG) Refinery products were 196.8Million tons during this period additionally 2.13 million tons of LPG form NG were available

1.3.5.8 LNG

Progress in LNG re-gasification capacity of 13.6 MMTPA,(2012) is expected to reach 53.5 MMTPA by 2016-17 with the commissioning of new terminals at:

10 MMTPA at PLL's terminal	12.5 MMTPA by 2013
Dahej	3.6 MMTPA
Shell's terminal at Hazira.	5 MMTPA in 2014-15.
RGPPL's Dabhol terminal	5 MMTPA in 2014-15.
GSPC-Adani's terminal at Mundra	5-MMTPA
IOC's terminal Ennore	5 MMTPA

With the production of naphtha rising from 12.6 MMT in 2003-04 to 19.4 MMT in 2010-11. In contrast, consumption of naphtha declined from 11.9 MMT to 10.7 MMT during the same period. the export of naphtha has increased from 2.2 MMT in 2003-04 to 10.7 MMT in 2010-11. In comparison, naphtha imports, after going up to 6 MMT in 2007-08, fell to 2.1 MMT in 2010-11

1.3.5.9 Reserves (See Figure 1.1b)

The reserves of India are going up day by day, due to the improved technological skills and massive investments. The 7th Plan additions itself constitute nearly 1000 M^2T of hydrocarbon deposits. The following are some known reserves.

KG offshore	13 M^2T oil and 3 BCUM gas
KG Basin	570 M^2T oil

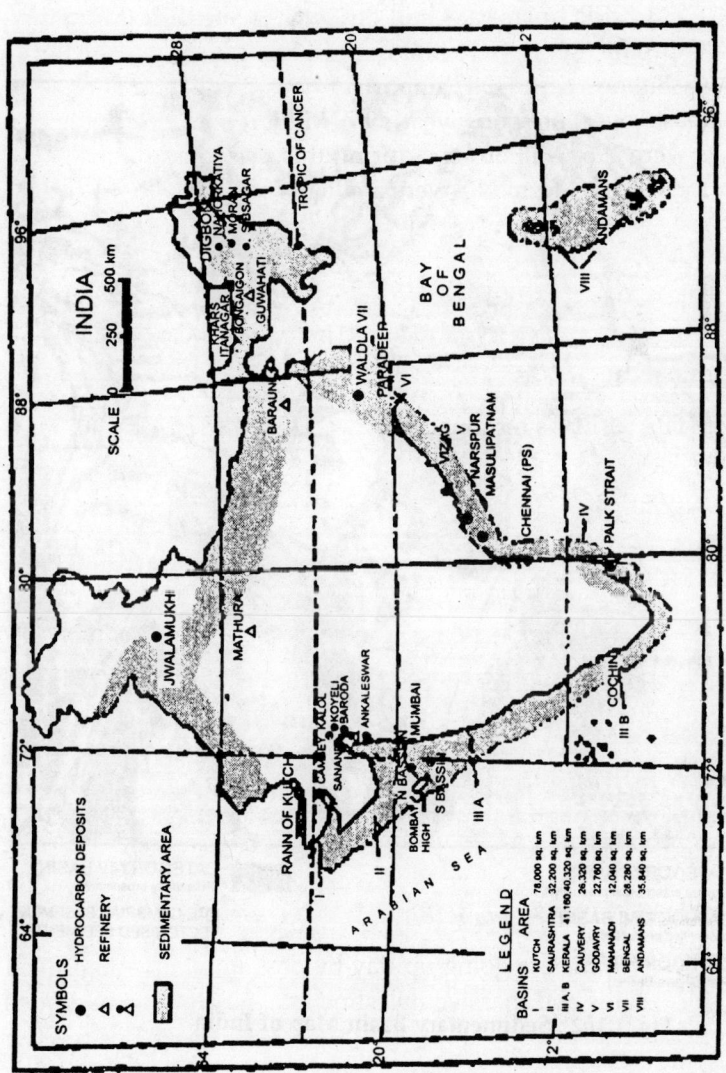

Figure 1.1*b* Prospective and Existing Hydrocarbon Deposits and Other Petroleum Activities.

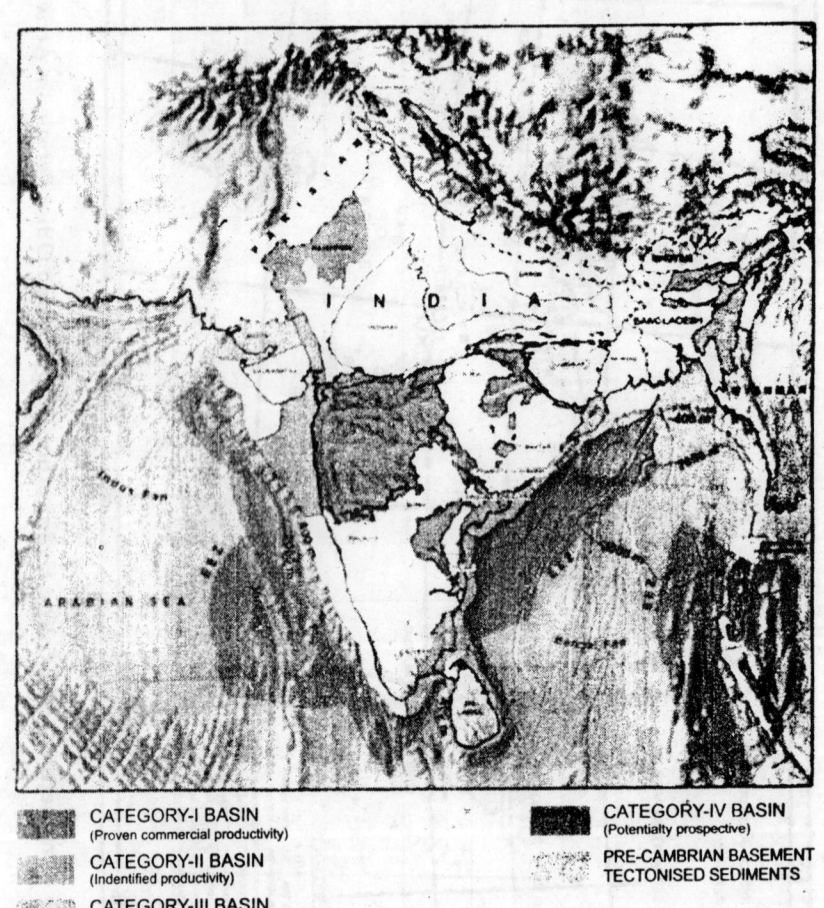

CATEGORY-I BASIN
(Proven commercial productivity)

CATEGORY-II BASIN
(Indentified productivity)

CATEGORY-III BASIN
(Prospective Basins)

CATEGORY-IV BASIN
(Potentialty prospective)

PRE-CAMBRIAN BASEMENT
TECTONISED SEDIMENTS

Fig. 1.1b2 : Sedimentary Basin Map of India

TABLE 3b : (Contd.) Data From DGH.

BREAK-UP OF DOMESTIC CRUDE OIL & NATURAL GAS

Company	2003-04		2004-05		2005-06 (Till Nov. 05)	
	Crude (MMT)	N Gas (MCM)	Crude (MMT)	N Gas (MCM)	Crude (MMT)	N Gas (MCM)
ONGC	26.057	23.584	26.484	22.971	16.241	15.000
OIL	3.002	1.887	3.196	2.005	2.192	1.513
Pvt/JV	4.314	6.491	4.300	6.77	2.962	4.832
Total	33.373	31.962	33.980	31.746	21.395	21.345

KG basin wells are producing initially 600 bbls/day. The production can be increased to 2000 bbls/day per well by introducing early production system (EPS) and the gas flow rates touch as high as 52,000 CUM per day.

Amalapuram wells	1,300 bbls/day
Mandapeta wells	52,000 CuM/day
Razole, Bhiminipally,	
Tatipaka, Pasarlapudi	$3 M^2CuM/day$
Cauvery Basin	$370 M^2T$
Vedanarayanapuram	
wells yield rich quality oil of	1500 bbls/day
	40% gasoline
	25% kerosene
	11% Diesel
Bhuvangiri wells	150-200 bbls/day
Gas	52,000 CuM/day
Nagapattanam wells	150-200 bbls/day
Madanam coast wells	4,300 bbls/day
Assam free gas	34 BCu M
Associate gas	27 BCu M
Assam/Arunachal	$59 M^2T$
Assam-Arkan Basin	3 Billion ton Hydro-Carbon
(One lakh sq km)	reserves
Tripura gas fields	881 BCu M
(Gas yields	1.3 lakh Cu M per day)
Rukia	4.5 lakh Cu M per day
Rajasthan :	
Bikaneer	$25 M^2T$
Jaisalmeer	$1 BCu M; 75 M^2T$ (Reserves)
	1.4 lakh Cu M/day (Production)
Tapati Structure	$14.2 M^2 Cu M$
Bombay High (1 RS)	$1000 M^2T$
BHN	$100 M^2T$

1.3.5.10 *Energy Reserves*

India is relatively well endowed with both renewable and exhaustible energy resources. Coal, oil, and natural gas are the primary commercial sources of energy. While the Indian coal reserves have increased from 62.54 BT (billion tonnes) in 1991 to 69.9 BT in 1997, the world coal reserves have declined from 1040.52 BT to 1031.6 BT for the same period. However, on a per capita basis, coal resources are well below the world average. The R/P (reserve-production) ratio for natural gas is higher than that of crude oil. This clearly brings out the possibility of a higher rate of production of natural gas than at present. India's R/P ratio for crude oil has fallen from 25.6 in 1991 to 15.6 in 1997. Table 1.3a1 shows proven reserves of fossil fuels and R/P (reserve-production) ratio. Gas hydrates to an extent of 1900 trillion cu m is another valuable addition to the hydrocarbon activity.

TABLE 1.3a1   Proven Reserves and R/P Ratios.

| Fuel | End 1987 | | | | End 1997 | | | |
| | Reserves | | R/P ratio | | Reserves | | R/P ratio | |
	India	World	India	World	India	World	India	World
Coal (billion tonnes)	62.54	1040.52	195.00	239.00	69.90	1031.60	212.00	219.00
Crude oil (billion tonnes)	0.80	135.40	25.60	43.40	0.60	140.90	15.60	40.90
Natural gas (trillion cubic metres)	0.70	124.00	48.80	58.70	0.49	144.80	22.90	64.01

TABLE 1.3a2   Energy Accommodation During Ninth–Plan.

Fuel	1996/97	1997/98	1998/99	1999/20	2000/01	2001/02
Coal (MMT)	311.00	323.40	334.00	359.90	392.80	405.70
Petroleum Products (MMT)	81.2	86.2	94.2	100.6	106.8	112.8
Natural Gas (mscmd)		76.64	79.70	82.68	82.68	82.68
Power (GW)	68.373	73.458	78.936	84.466	90.093	95.757

1.3.6   Estimation of Reserves

The estimation of hydrocarbon content in a reservoir under operation is usually done by material balance. However the estimation in a fresh finding is done by knowing the physical conditions of the reservoir. Usually the deposit occurs in the porous sedimentary region. The porosity of the reservoir ($\in$) multiplied by the volume of the reservoir is the maximum amount of the hydrocarbon that can be present in a reservoir. However it is accompanied by formation water which is known as connate water (S_w). Hence the total deposit will be exclusive of the water and is given by

The volume of the deposit = vol. of the reservoir $\times \in \times$ (1-S_w).

This is the volume of the deposit under reservoir conditions. Usually the deposit is several metres deep which experiences more than 100 atmospheres pressure. This keeps the deposit under compression. As the oil or gas comes out of the well the volume increases because of decrease in pressure. Thus when the amount of deposit is calculated it is usually expressed at the standard conditions of pressure and temperature. A correction coefficient known as liquid volume factor (B_0) or gas volume factor (B_g) is incorporated in the above calculation.

Liquid/gas volume factor is defined as the volume of barrel under reservoir conditions to surface conditions. In case of liquid the coefficient compression and coefficient of thermal expansion are small; for gases they are very high.

Thus the volume of deposit at surface conditions = Vol. of the reservoir $\times \in \times$ (1-S_w) B_g.

Problem 1.1

A reservoir is having a volume of 1.35 kilometers cube. The porosity of the rock is given as 22%, the connate water is estimated to be of 16.5%. Find the volume of the oil deposit. Given the pressure of the reservoir 204.8 bars and temperature of the reservoir 39.8°C. B_0 given as 1.12 STB/bbl

Solution

Volume of deposit = 1.35 $\times 10^9 \times$ 0.22 $\times$ (1–0.165) $\times$ 1.12
$\times$ 5.043* = 1.4007 billion barrels

Problem 1.2

In a gas reservoir, the gas is flowing out at a rate of 2500 SCF per hour. The pressure of the reservoir is constant at 1575 psia. The residual gas saturation is assumed to be 23% and the gas volume factor at the reservoir pressure is 55.23 SCF/cft. and at surface condition it is 176.8 SCF/cft. Find the time taken for recovering 55%

---

*bbls per cubic meter

of the reservoir gas? Reservoir volume is 3.2 ac-ft. Porosity is 22%
Connate water 18.6%.

Solution

Volume of the gas in the reservoir $= 3.2 \times 43560 \times 0.22 \times$
$$(1 - 18.6) \times 176.8 \text{ SCF}$$
$$= 441.334 \times 10^6$$

Volume of the gas, retained in the
reservoir $\qquad = 3.2 \times 43560 \times 0.22 \times 0.23$
$$\times 55.23 \text{ SCF}$$
$$= 389, 550$$

Net available gas $= 440.944 \times 10^6$ at surface

Time taken $\qquad = 440.944 \times 10^6 \times 0.55/2500 \times 24 \times 360$
$$= 11.2 \text{ years}$$

1.4 COMPOSITION OF PETROLEUM

Petroleum occurs in nature in all three possible states solid, liquid and
gas. The liquid petroleum is usually coloured from dark brown to
bluish black or black, exhibiting sometimes bloom or fluorescence.
The semi-solid or solid petroleum is well known by the name pitch,
usually black in colour. The famous pitch lake of Trinidad is an
example of such vast deposits of petroleum in solid state. Such kind of
deposits are assumed to form after the evaporation or migration of
lighter fractions. The gaseous deposits of petroleum are known as
natural gas deposits, where sometimes wild gasolines are also
accompanied. Gas from condensate reservoirs contain a good portion
of lighter fractions of a boiling point upto 30°C. Associated reservoirs
contain gas mainly in the dissolved form in liquid petroleum.

Although the composition of petroleum depends not very much on
the origin of formation, but certainly change with the time of
formation, storage and different stratas through which it migrated. It
is a homogeneous mixture of various hydrocarbons of saturates and
ring-structures. The average ultimate composition of petroleum is
mainly given in terms of constituents of hydrocarbons, namely carbon
and hydrogen as follows:

Carbon   :   84-86%
Hydrogen :   11-14%

The other major elements of importance are sulfur, oxygen and
nitrogen. These elements in hydrocarbons are usually treated as
impurities because of their inherent properties like odour, colour

corrosiveness etc. Generally these three elements combined do not exceed 5% on an average. Exception to this statement can be traced in some Gulf crudes, Russian crudes and Mexican crudes. Ratwi (Neutral zone) contain as much as 5% of sulfur alone. Middle East and Gulf crudes contain upto 3%, compared to these, crudes from East possess very less amount of sulfur, examples being Indonesian, Indian, Nigerian, and Libyan crudes.[26] The crudes of U.K., like Beryl, contain upto 0.5% sulfur.

The bulk of petroleum is made up of hydrocarbons of saturated compounds like paraffins, naphthenes and unsaturated cyclic compounds mainly aromatics.

The highest carbon atom present in the crude is C_{70}. Further, except first few hydrocarbons, all other hydrocarbons exhibit isomerism. The general properties of these homologous series are discussed below:

1.4.1 Paraffins

C_nH_{2n+2} is the general formula of paraffins. First three compounds are gases while compounds upto C_{16} are liquids and beyond that, they assume semisolid consistancy. Well beyond C_{30} assume the shape of solid blocks, sometimes even crystalline forms. There are number of isomeric compounds for each compound, profoundly differing in properties. For example, upto C_3 no isomers are possible, C_4 exhibits only two isomers, as shown here:

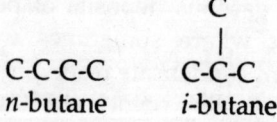

$$C\text{-}C\text{-}C\text{-}C \qquad\qquad C\text{-}C\text{-}C$$
$$n\text{-butane} \qquad\qquad i\text{-butane}$$

And C_5 exhibits three isomers. The number of isomers increases as the number of carbon atoms increase. $C_{13}H_{28}$ exhibits 802 isomeric forms.

1.4.1.1 General Properties of Paraffins

Paraffins are stable, not attacked by sulphuric acid or other oxidising agents. However, paraffins of higher order > C_{30} are prone to oxidation. Even usual oxidising agents like potassium permanganate can cause good amount of oxidation. The aptitude to contribute the substituted products with halogens has magnified the petrochemical industry. Higher paraffins are very much insoluble in water; though the lower ones are soluble in ethers and alcohols (Figure 1.2).

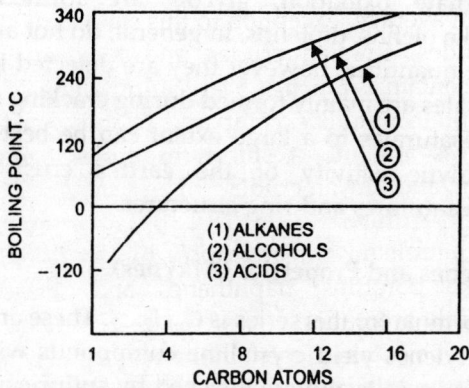

Figure 1.2 Boiling Points of First Few Alkanes, Alcohol and Acids.

Paraffins upto 3 carbon atoms have inclination to form hydrates such as $CH_4\ 7H_2O$, $C_2H_6\ 7H_2O$ and these hydrates offer clogging and corrosion difficulties. Hence drying is essential before usage.

The specific gravity of the series increases with molecular weight, still paraffins have less specific gravity and boiling point than aromatics. Viscosity of paraffins is less but viscosity index is high in contrast to aromatics. The smoke point of the paraffins is very high, with poor illuminating characteristics. The pour point of paraffins is usually high; due to this paraffin rich crudes and products bring difficulties in transportation and storage.

Isomers differ from n-paraffins by having slightly low boiling points, low pour points, high viscosity and viscosity index. Usually i-paraffins, are more reactive than n-paraffins but burn like n-paraffins without much illumination and smoking.

High molecular compounds ($>/C_{20}$) may be of saturated or unsaturated nature and decompose if exposed to a temperature of above 370°C. Vacuum distillation is essential for distilling such boiling stocks to prevent them from thermal degradation.

1.4.2 Unsaturates (Olefins and Properties)

Olefins are represented by the general formula C_nH_{2n}. The first four are gases and upto C_{15} are liquids and beyond C_{15} are solids. The boiling points of olefins are generally lower by few degrees than the saturated compounds of the same carbon number. Chemically these differ very much from paraffins. They are easily attacked by sulfuric acid and some of them even polymerize. Treatment with sulfuric acid and subsequent hydrolysis yields alcohols (e.g. isopropyl alcohol) and with permagnate oxidation, glycols are formed. Unsaturated compounds like olefins, diolefins, in general, do not appear in crudes to measurable quantities, however they are detected in some crudes. These unsaturates are mainly formed during cracking operations. The absence of unsaturates to a large extent can be best judged by the probable catalytic activity of the earth's crust in converting unsaturates to saturates and ring structures.

1.4.3 Acetylenes and Properties (Alkynes)

The general formula for this series is C_nH_{2n-2}. These are isomeric with diolefins. Acetylenes yield crystalline compounds with ammoniacal solution of copper salts and are attacked by sulfuric acid. Acetylenes can be readily hydrogenated to give stable compounds.

1.4.4 Diolefins

These are represented by the formula C_nH_{2n-2}. Like other unsaturates, these are produced during cracking reactions. These can be distinguished from acetylenes as they do not form salts with ammoniacal solutions of copper salts. But with mercuric chloride these form precipitates and sulfuric acid polymerises these unsaturates.

1.4.5 Naphthenes

(These are saturated ring compounds bearing the general formula C_nH_{2n}. The prominency of ring structure starts with five carbon atoms. Although C_3 and C_4 ring structures,[27] are in existence, their stability is decreased because of excessive strain (Bayer's Strain Theory). Naphthenes are isomeric with olefins but differ profoundly in properties. Naphthenes exhibit both the properties of saturated paraffins and unsaturated aromatics, the result of which, all the properties like sp. gravity, viscosity, pour point, thermal characteristics lie in between the two mentioned homologues. Usually, all the ring structures are having branched chains, where the isomeric character predominantly occurs, followed by positional isomerism in rings (Figure 1.3).

1.4.6 Aromatics

The first and smallest of the aromatics is benzene; other simple aromatics to follow are toluene, xylene, cumene etc. Even though benzene is unsaturated, yet it follows the principles of substitution with halogens rather than addition. This is mainly due to symmetric grouping of closed ring structure and resonance.

Aromatics are usually having high boiling points, low pour points (freezing points), high octane numbers, high viscosity and low viscosity index and these burn characteristically with a red flame with much soot. As these behave like saturates, they resist oxidation. In petroleum fractions aromatics beyond 3-ring structure (Anthracenes) are probably non-existent. Aromatics usually extend their presence from a temperature of 80°C onwards and well dominate in lower middle cuts and heavy cuts. Actually the light aromatics (BTX) do not exceed even 5% of crudes of general nature. Bulk of the aromatics are with side chains together with naphthenes exist in heavier portion of crudes.

1.4.7 Inorganics

Sulfur compounds: Sulfur is found in most of the crudes in variable amounts. Generally sulfur compounds are present in more quantities in higher molecular weight stocks. Usually the sulfur content does not exceed 5%, however rare exemptions are: Venezuela (5.25%), California (USA 5.21%), Qaiyarah (Iraq—7%) etc. crudes.

Sulfur in crude occurs in different forms like free sulfur, hydrogen sulfide, mercaptans and thiophenes etc. These are frequently occurring compounds in almost all fractions of the crude though to a different degree. Heavier fractions contain sulfides, polysulfides, sulphonates and sulphates.

Sulfur occupies prominent position in refining due to its ominous problems of corrosion and odour. Pollution problems and following cost of waste treatment is punitive for all refiners with high sulfur-stocks. However, refiners habitually remove more detrimental sulfur compounds and leave the less harmful ones into the products, as seen in the case of sulfides converted to disulfides in gasolines. Some of the sulphonates are regarded as good emulsifiers and detergents, hence promptly extracted for use in cutting oils. Conspicuous effect of sulfur is reflected in increasing the density of crude.

A correlation presented by Obolentsev[28] shows the influence of sulfur on gravity

$$(\delta \text{at } 20°C)\ \rho = 0.0087\ (S\%)^2 + 0.0607\ (S\%) + 0.7857$$

ALKANES (C_nH_{2n+2})

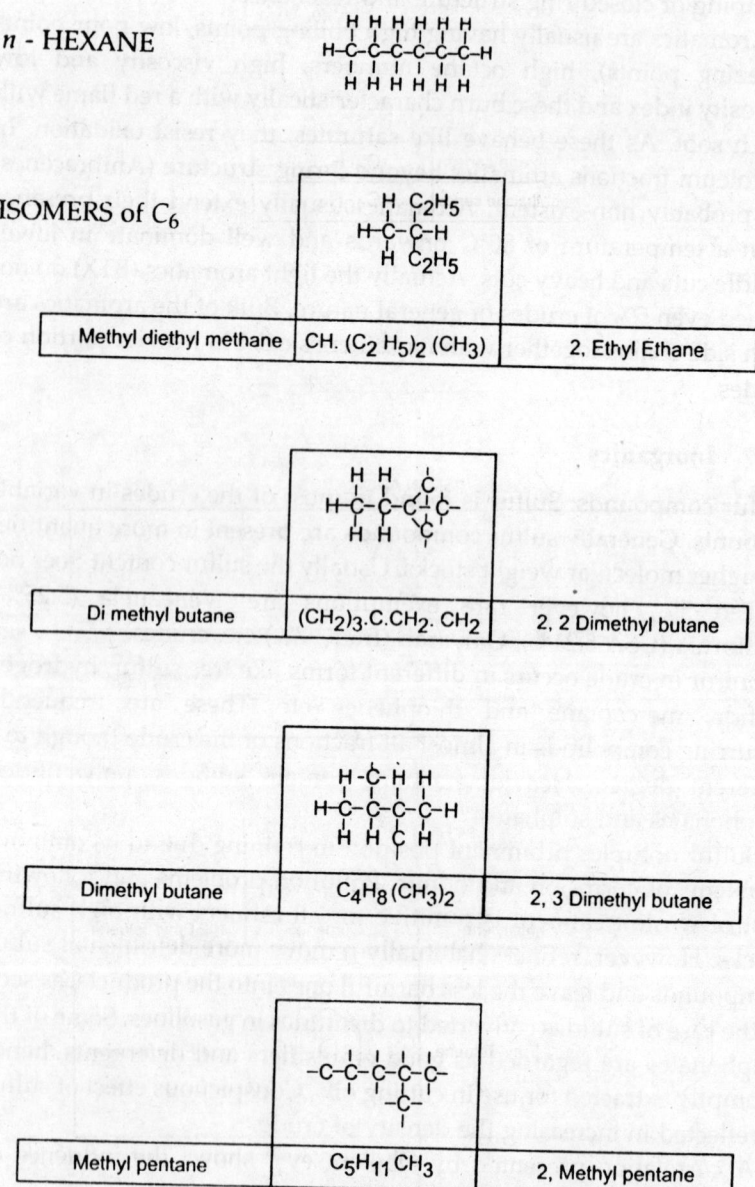

Figure 1.3 **Structures of Some Organic Compounds.**

OLEFIN (TERPENS) C_nH_{2n-4}

Triotafins	C_nH_{2n-6}
Di-acetylenes	C_nH_{2n-6}

AROMATICS

Benzene series	C_nH_{2n-6}
Nephthalene series	$C_{10}H_8$

Diacetylenes
$H - C \equiv C - C \equiv C - H$

ISOMERISM IN AROMATICS

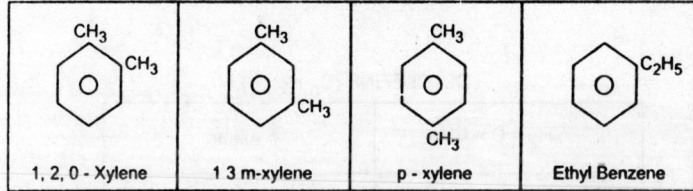

Mesitylene C_9H_{12}

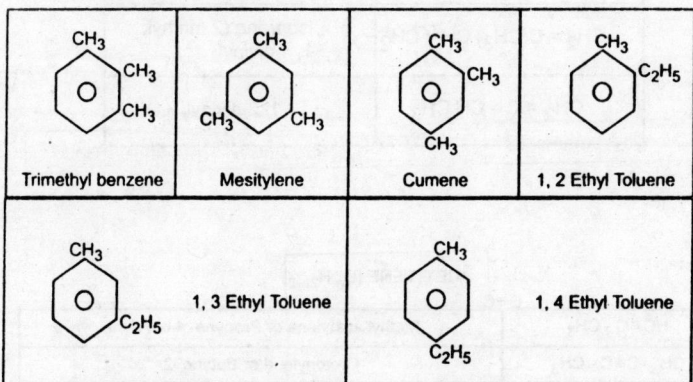

Figure 1.3 : (Contd.)

OLEFINS (C_nH_{2n})

C = C – C – C	α - Butylene
C – C = C – C	β - Butylene

C — C = C
    |
    C
	γ - Butylene
	(iso - butylene)

CYCLO - PARAFFINS ($C_n H_{2n}$)

⬡	Cyclo - hexane
▢	Cyclo - butane
⬠	Cyclo - pentane

DI - OLEFINS ($C_n H_{2n-2}$)

$CH_2 = C = CH_2$	Allene
$CH_2 = CH.CH = CH_2$	Divinyl (butadiene)
$CH_2 = C(CH_3).CH = CH_2$	Isoprene (2 methyl, 1, 4 divinyl)
$CH_2 = C = CH.CH_3$	1, 2 divinyl

ACETYLENE ($C_n H_{2n-2}$)

$C \equiv C$

$HC \equiv C - CH_3$	Methyl acetylene or Propyne- 1 or Allyne
$CH_3 - C \equiv C - CH_3$	Crotonylene or Butyne -2
$C_2H_5 - C \equiv C - CH_3$	Ethyl methyl acetylene or Pentyne -2

Figure 1.3 : (*Contd.*)

SULPHUR COMPOUNDS

Mercaptans	$C_nH_{n+1}SH$	
	C_2H_5SH	Ethyl mercaptan
Sulphones	RSO_2	
	$C_7H_{15}SO_2$	Heptyl Sulphone
Sulphides	R_2S	
	$(CH_3)_2S$	Dimethyl sulphide
Di Sulphides	R_2S_2	
	$CH_3\,S.S\,CH_3$	Dimethyl disulphide
Sulfoxides	$\begin{matrix} R \\ R \end{matrix}\!\!>\!SO$	
	$\begin{matrix} CH_3 \\ CH_3 \end{matrix}\!\!>\!SO$	Dimethyl sulfoxide
Thio phenes		
	 CH_3-S-C_6	Thio deeyl thiophene
Thio phanes	$C_nH_{2n}S$	
	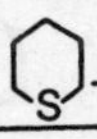 $-C_4$	Butyl thio cyclohexane
Sulphates	$(C_nH_{2n+1})SO_4,\ R_2SO_4$	
	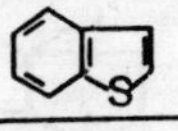	Thio benzols
Sulphonates	 $SO_3\ \ R\,SO_3$	
Carbonyl Sulphide	COS	
	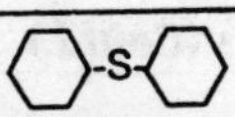 $-S-$	Cyclohexyl sulphide
	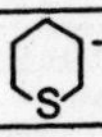 $-C$	Thio cyclo heptane

Figure 1.3 : (*Contd.*)

NITROGEN COMPOUNDS

Carbozole	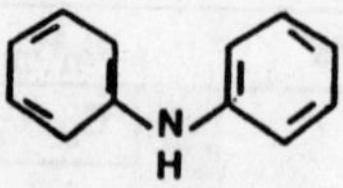
Pyrole	
Methyl Quinoline	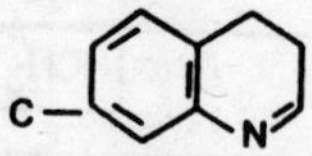
Indoles	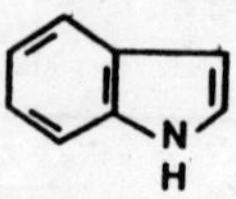
Quinoline	$C_9 H_7 N$ $C_{12} H_{17} N$ $C_{13} H_{18} N$ $C_{14} H_{19} N$ $C_{15} H_{19} N$
Pyridine	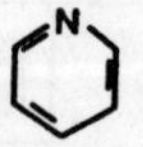
Pyrroles	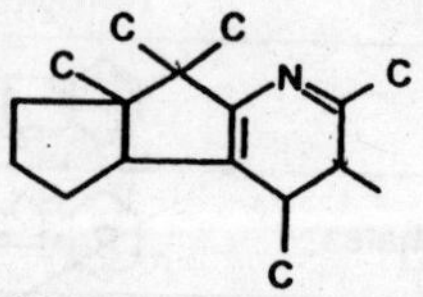

Figure 1.3 : (*Contd.*)

OXYGEN (Naphthenic Acids) Compounds

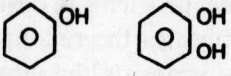

$C_6H_{11}COOH$	Hepta naphthenic
$C_7H_{13}COOH$	Octa naphthenic
CH—CH HC CH (Furan ring)	Furan
(benzofuran structure)	Benzo Furan

Porphyrins

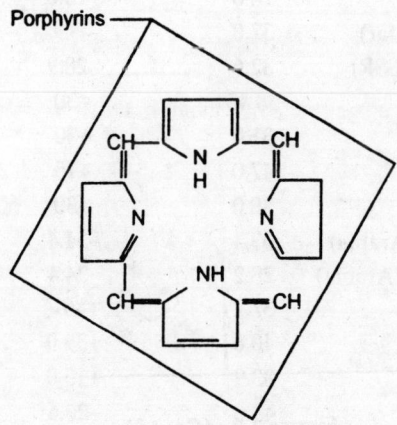

Figure 1.3 : (Contd.)

Different crudes are presented in Table 1.1. It clearly shows the effect of sulfur on API gravity of crude and pour point of crude. All sulfur crudes mysteriously exhibit low pour points.

Further, sulfur containing residuums when cracked leaves cross linked structures, resembling the phenomenon of valcanization of rubber and offer perennial problems in desulfurisation. Its presence in different fractions complicates the refining and treatment methods. Yet another problem is, it desists the effects of additives. Sulfur in gasoline inevitably depresses the effect of lead and demands more amount of additive. When crude contains more than 0.5% S, it is denoted as high sulfur crude. A terse distinction, at this juncture between sour crudes and sulfur crudes is desirable. Free hydrogen sulfide is available in some crudes, which naturally fosters corrosion.

TABLE 1.1   Effect of Sulfur on Gravity and Pour Point.[29]

Crude	API	Pour Point °C	Sulfur %
Cyrus (Iran)	19.0	–23.3	3.48
Iranian Heavy	30.8	–20.6	1.6
Kuwait	31.2	–17.8	2.50
North Slope (USA)	26.8	–20.6	1.04
Quatar Marine (Quatar)	37.0	–3.9	1.50
Romashkinskaya (USSR)	32.6	–28.9	1.61
Bassein (India)	38.45	+30	0.15
Nahorkatiya	31.0	+30	0.16
Ankleshwar	47.0	+15	0.05
Bombay High	38.0	+30	very low
Arabian (Light) (S. Arabia)	33.4	–34.4	1.80
Arabian (Heavy) (S. Arabia)	28.2	–34.4	2.84
Arjuna (Indonesia)	37.7	+26.7	0.12
Bu Attifel (Libya)	40.6	+39.0	0.10
Basrah (Iraq)	33.9	+15.0	2.05
Brass (Nigeria)	43.0	–20.6	0.08
Darius (Iran)	33.9	–17.9	2.45

Such crudes are classified as sour crudes; other sulfur bearing compounds are not taken into this account. The crudes containing sulfur compounds other than hydrogen sulfide and exceeding 0.5% are denoted as high sulfur crudes.

1.4.8 Oxygen and Nitrogen

Oxygen and nitrogen do not occur in free state either in crudes or in fractions. Nitrogen presence in free form is well known in natural gas only. Oxygen occurs as oxygenated compounds like phenols, cresols, naphthenic acids, sulphonates, sulphates and sulfoxides.

Nitrogen exists in the form of indoles, pyridines, quinolines and amines, usually well below 2%. Nitrogen compounds exasperate problems in processing and stability of products. Catalyst deactivation or poisoning, gum formation are some of the offshoots of nitrogen. Nitrogen is present in two forms, basic and non-basic. Basic[30] nitrogen is characterised by its titratability with perchloric acids, whereas nonbasic nitrogen is not titratable hence no possibility of extraction. Most of the nitrogen[31] pigments impart color to crude and fractions. The most interesting compounds of nitrogen are porphyrins. These are obtained from living organisms and preserved in petroleum. It stands to reason that aneorobic conditions were prevailing during petroleum formation; otherwise oxidation would have destroyed them. Chlorophyll[32] is also a complex of porphyrin, where central atom is magnesium instead of nickel or vanadium or iron. Iron porphyrins are also known as heme, the constituents of red cell in the blood.

Porphyrin[33,34] pigments are usually associated in complex form with metals like, copper, iron, vanadium and nickel. The proper understanding of these pigments[35,36,37], may augment the knowledge of origin and formation of petroleum. The following is an example of nitrogen complex:

Gravity API	—	38.8
Sulfur wt. %	—	0.2-0.0%
Vanadium ppm.	—	0.5-2.5%
Nickel ppm.	—	1.0-170
Vanadyl porphyrin ppm.		0.7-1130
Nickel porphyrin ppm.		1.0-390

1.4.9 Asphalts, Resins and Bitumen

Asphalts are high molecular weight complex molecules, black in color, soluble preferably in aromatic solvents and carbon disulphide.

TABLE 1.2a World Oil Well Completions During 1975.

Country	Oil	Gas	Dry	Others	Average Depth Metres
N. America (Total)	17,449	9,615	14,404	1,243	1,407
U.S.A	16,626	7,437	13,203	1,121	1,451
Venezuela	246	6	35	96	1,806
Africa (Total)	273	14	332	151	1,852
Algeria	65	10	8	20	2,417
Nigeria	57	–	178	111	1,621
Middle East (Total)	409	20	140	191	2,330
Iran 95	11	29	18	2,775	
Iraq 36	9	–	2,699		
Saudi Arabia	101	–	16	95	2,135
Asia (Total)	576	81	358	51	1,691
India	90	6	19	8	2,171
Indonesia	410	38	145	37	1,283
World (Total)	20,089	9,995	16,047	1,807	1,472

Resins are mostly compounds of highly condensed ring structures, containing oxygen, sulfur and nitrogen, sometimes inorganics too. Though bitumen is a manufactured product, it is essentially made up of three components, asphalts, resins and mineral oil. These three components comprise a colloidal system; asphalts are suspended in oil and resins contribute to the stability of the system.

1.4.10 Less Inorganics

The other elements present are nickel, vanadium, iron, silica, sodium, magnesium, and halogens etc. Even though the analysis is not desirable as these metals hardly exceed 0.01%, yet may be analysed for sensitive purposes. The ash formation is mainly due to these metals and inorganics. Sometimes organometallic compounds are available in colloidal form. Inorganics always leave a marked influence on fractions, for example halides may give off halogens during hydrolysis or thermal decomposition. Corrosion, pollution, ash etc. are mainly contributed by inorganics and the quality of crude and fractions are always debased by these small amounts.

Table 1.2a shows some statistics on completed oil wells during 1975 in the world.

TABLE 1.2b World Primary Energy Consumption During 1985.

	Liquid fuels	(MillionTons) Gas	Solid fuels	Hydro	(Power MW) Nuclear	Thermal
USA	442.714	391.714	464.249	84,986	81,516	533,611
Canada	120.800	108.973	50.738	57,458	10,889	30,937
L/S America	256.251	59.268	13.319	57,853	1,675	33,601
U.K.	182.670	56.761	76.809	4,190	7,064	56,356
W. Germany	5.861	19.926	117.327	6,668	16,095	69,938
M.E.	760.971	40.129	0.900	2,973	–	43,784
Japan	0.786	3.245	14.580	34,337	24,686	110,291
USSR	851.665	755.758	514.284	61,259	28,100	229,936
China	178.421	17.129	591.468	26,500	–	55,700
World	4224.358	2051.537	2978.168	545,463	250,697	1,607,712

TABLE 1.2c Utilisation of Natural Gas and Oil by Different Countries.

Country	Utilisation of Natural Gas (10⁶ Cu.M.) (1974)	Growth of World wide resources (10³ Barrels) (1977)	Oil produced (10³ Barrels) (1975)
North America	697,673	37,199	3,570,926
U.S.A.	611,662		
Latin America	45,938	29,609	1,603,695
Western Europe	163,305 ⎱	24,539	3,955,917
Eastern Europe	309,773 ⎰		
U.S.S.R.	261,000		7,143,494
Middle East	41,934	367,681	1,821,954
Africa	16,617	60,570	
Asia	63,380	19,391 ⎱	1,377,919
India	1,010	⎰	
China	33,980		
World (Total)	1,338,602	640,081	19,473,903

(Source for Tables 1.2a, & c 1.3a to 1.3c Oil Statistics New Delhi) 1975 & 1976; Petroleum Information Service.

TABLE 1.3a Oil and Gas Reserves (India).

Area	1966	1970	1975	1977	1978*	1979*	1980*
1. Crude Oil @ (MillionTonnes)							
On Shore	153.00	127.84	130.13	127.90	125.99	128.15	135.38
Off Shore	–	–	13.77	175.28	221.04	226.29	230.95
Total	153.00	127.84	143.90	303.18	347.03	354.44	366.33
2. Natural Gas @ (Billion Cubic Metres)							
On Shore	63.15	62.48	81.39	80.39	79.71	79.58	80.95
Off Shore	–	–	6.28	148.08	186.15	264.64	270.96
Total	63.15	62.67	87.67	228.47	265.86	344.22	351.91

*As on 1st January of each year.
@Proved and indicated balance recoverable reserves.

TABLE 1.3b Production of Crude Oil and Natural Gas.

Company	1960-61	1965-66	1970-71	1975-76	1978-79	1979-80	1980-81*
1. Crude Oil Production ('000 tones)							
(a) On Shore							
AOC	176	151	104	66	53	40	48
OIL	272	1895	3084	3103	2671	2215	1241
ONGC	–	1427	3634	5279	5599	5091	4229
Total (a)	448	3473	6822	8448	8323	7346	5518
Of which							
Gujarat	–	1427	3455	4148	4238	3768	3808
Assam	448	2049	3367	4300	4085	3578	1710
(b) Off Shore							
ONGC	–	–	–	–	3310	4422	4985
GRAND TOTAL (a+b)	448	3473	6822	8448	11633	11768	10503

(Contd.)

TABLE 1.3b : (Contd.)

Company	1960-61	1965-66	1970-71	1975-76	1978-79	1987-88	1980-81*
2. Natural Gas Production (Million Cubic Metres)							
(a) On Shore-							
AOC	–	–	65.24	40.33	44.33	1.390	38.14
OIL	–	–	905.25	1404.88	1256.79	1.390	709.19
ONGC	–	–	474.41	822.06	1124.45	3.311	933.16
Total (a)	–	–	1444.90	2367.27	2425.57	4.701	1680.49
of which							
Gujarat	–	–	464.94	773.08	907.60		838.91
Assam	–	–	979.96	1594.49	1517.86		841.58
(b) Off Shore							
ONGC	–	–	–	–	385.87	10.760 365**	673.23
Grand Total (a + b)	–	–	1444.90	2367.27	2811.44	15.826	2353.72

* Provisional
** K.G. & Cauvery Basins
– Not Available

TABLE 1.3c Refinery Crude Throughput.

('000 Tonnes)

Refinery	1960-61	1965-66	1970-71	1975-76	1977-78	1978-79	1979-80	1980-81*
(a) PUBLIC SECTOR	–	1966	10820	17045	24364	25450	27066	25334
BPCL, Bombay	–	–	–	–	4512	4693	4821	4901
CORIL/HPCL/(VMU)/(4)	–	–	–	–	1304	1329	1100	1319
CRL, Cochin	–	–	2498	2294	2933	2862	2866	2912
HPCL, Bombay	–	–	–	2830	2897	2800	3130	3113
IOC, Barauni	–	753	2191	2949	3060	2661	2285	505
IOC, Gauhati	–	804	686	827	817	825	646	639
IOC, Gujarat	–	409	3463	4107	4129	5251	6714	6974
IOC, Haldia	–	–	–	1348	2096	2213	2492	2308
MRL, Madras	–	–	1982	2690	2616	2759	2822	2611
BRPL, Assam @	–	–	–	–	–	57	190	52
(b) PRIVATE SECTOR	6130	8267	7559	5238	534	524	408	503
AOC, Digboi	446	488	476	527	534	524	408	503
BSR, Bombay (1)	2879	3989	3541	3630	–	–	–	–
CORIL, Vizag (2)	901	1118	1175	1081	–	–	–	–
ESRC, Bombay (3)	1904	2672	2367	–	–	–	–	–
Total Crude								
THROUGHPUT (a+b)	6130	10233	18379	22283	24898	25974	27474	25837

(1) BSR was taken over by the Govt. in Jan. 1976 and renamed as Bharat Petroleum Corporation Ltd.

(2) CORIL was taken over by the Govt. in Dec. 1976.

(3) ESRC was taken over by the Govt. in March 1974 and renamed as Hindustan Petroleum Corporation Ltd. in July 1974.

(4) CORIL was merged with HPCL on 9th May 1978.

@ BRPL came in production on 6-2-1979.

*Provisional

TABLE 1.3d   Refinery Production.

('000 Tonnes)

Products	1960-61	1965-66	1970-71	1975-76	1984-85	1988-89	1979-80	1980-'81
TOTAL PRODUCTION	5777	9561	17110	20829	35.600	48,000	25794	24124
1. Light Distillates	1050	1770	3021	3630			4459	4098
of which								
LPG	8	45	169	331	596.3	10,222	406	367
Mogas	1037	1421	1526	1275	2153.5	2,000	1512	1519
Naphtha	–	181	1205	1910	–		2415	2111
2. Middle Distillates	2593	4543	8562	10769		22,274	13080	12119
of which								
Kerosene	929	1463	2896	2439	3351.3		2539	2396
ATF	–	128	710	925	1336.9		1104	1000
HSD	1063	1816	3840	6285	11086.2		7975	7371
LDO	525	789	986	946	1251.6		1230	1110
3. Heavy Ends	2134	3448	5527	6430		10,294	8255	7907
of which FO	1342	2455	2987	3595			4086	4040
Fuel Oil (Total)	1641	2709	4090	5083	7953.5	8,300	6351	6124
Lube Oils	22	46	231	342	414.1	614	487	426
Bitumen	421	562	805	697	421	500	1103	1081
Petroleum coke	13	88	151	160	100	110	99	85

* Provisional

(contd.)

TABLE 1.3d : *(contd.)*

Products	1980-81	1990-91	1995-96	1996-97	1997-98	1998-99	'000 Tonnes 1999-00
(a) From Crude Oil							
Light Distillates of which	4101	10023	12433	12883	13032	13776	18314
LPG	366	1221	1539	1598	1666	1724	2487
Mogas	1519	3552	4462	4704	4849	5573	6232
Naphtha	2115	4859	5975	6123	6103	6081	8170
Others	101	391	457	458	414	398	1425
Middle Distillates of which	12115	26344	29941	32423	33933	36168	44995
Kerosene	2396	5471	5267	6236	6701	5341	5735
ATF/RTF/Jet A-1	1001	1801	2127	2119	2147	2289	2292
HSD	7371	17185	20661	22202	23354	26716	34793
LDO	1108	1509	1351	1286	1246	1336	1624
Others	239	378	535	580	485	486	551
Heavy Ends of which	7907	12195	12707	13698	14343	14600	16102
Furnace Oil	4041	4879	5351	5980	6771	6407	6559
LSHS/HHS							
RFO	2079	4550	4228	4318	4309	4623	4793
Lube Oil	426	561	633	619	593	586	728
Bitumen	1082	1603	2032	2283	2158	2419	2485
Petroleum Coke	86	229	256	246	282	286	465
Paraffin Wax	–	49	43	31	27	40	47
Other Waxes	–	46	63	56	45	63	70
Others	193	278	101	165	158	167	955
Total (1+2+3)	24123	48562	55081	59004	61308	64544	79411
(a) From Natural Gas							
LPG	929		1715	1780	1787	2207	1989

TABLE 1.3e Installed Capacity and Refinery Crude Throughput.

'000 Tonnes

Refinery	Installed Capacity as on 1.4.2000	1980-81	1990-91	1995-96	1996-97	1997-98	1998-99	1999-00
			Refinery Crude Throughput					
(A) Public/joint Sector	85540	25333	51772	58702	59958	61313	64469	74052
IOC, Guwahati	1000	639	783	839	848	856	836	914
IOC, Barauni	4200	504	2416	2322	1895	2181	2204	3411
IOC, Gujarat	12500	6974	9334	10167	10352	10694	10935	11109
IOC, Haldia	3750	2308	2335	3416	3451	4706	4714	4105
IOC, Mathura	7500	-	7808	8332	8113	8565	8909	8125
IOC, Digboi	650	-	566	559	477	502	553	603
IOC, Panipat @ @	6000	-	-	-	-	-	2208	4153
BPCL, Mumbai	6900	4901	6957	7460	7640	7996	8878	8907
HPCL, Mumbai	5500	3113	5766	5965	6534	6378	5203	6007
HPCL, Visakha	7500	1319	3464	5037	4847	2467	3861	4555
CRL, Cochin	7500	2912	6006	7421	7293	7729	7770	7830
MRL, Chennai	6500	2611	5698	5599	6621	6965	6101	6377
MRL, Narimanam	600	-	-	370	345	556	644	636
BRPL, Assam	2350	52	1139	1215	1542	1718	1653	1905
NRL, Assam #	3000	-	-	-	-	-	-	215
MRPL, Mangalore @	9690	-	-	39	2912	3853	4069	5200
(B) Private Sector	27000	503	-	39	2912	3853	4069	11912
AOC, Digboi (1)	-	503	-	-	-	-	-	-
RPL, Jamnagar ##	27000	-	-	-	-	-	-	11912
Total (a + b)	112040	25836	51772	58741	62870	65166	68538	85964

(1): AOC refinery was taken over by the Govt. in Oct. 1981 & merged with IOC

@ : Commenced production from 25.3.1999

Commenced production from April 1999

@ @: Commenced production from May 1998

: Commenced production from July 1999

Source: Ministry of Petroleum & Natural Gas

TABLE 1.3*f* World Oil Production Capacity by Region and Country, Reference Case, 1990-2020

(Million Barrels per Day)

Region/Country	History (Estimates)			Projections		
	1990	1999	2005	2010	2015	2020
OPEC						
Persian Gulf						
Iran	3.2	3.9	4.0	4.3	4.6	4.8
Iraq	2.2	2.8	3.1	3.8	4.7	5.8
Kuwait	1.7	2.6	2.8	3.5	4.1	5.0
Qatar	0.5	0.6	0.5	0.6	0.7	0.7
Saudi Arabia	8.6	11.4	12.6	14.7	18.4	23.1
United Arab Emitates	2.5	2.7	3.0	3.5	4.4	5.1
Total Persian Gulf	18.7	24.0	26.0	30.4	36.9	44.5
Other OPEC						
Algeria	1.3	1.4	1.9	2.1	2.3	2.5
Indonesia	1.5	1.7	1.5	1.5	1.5	1.5
Libya	1.5	1.5	2.1	2.5	2.8	3.2
Nigeria	1.8	2.2	2.8	3.2	4.0	4.7
Venezuela	2.4	3.4	4.2	4.6	5.0	6.0
Total Other OPEC	8.5	10.2	12.5	13.9	15.6	17.9
Total OPEC	27.2	34.2	38.5	44.3	52.5	62.4
Non-OPEC						
Industrialized						
United States	9.7	9.3	9.0	8.7	9.0	9.3
Canada	2.0	2.7	3.0	3.2	3.4	3.5
Mexico	3.0	3.5	4.1	4.2	4.4	4.4
Australia	0.7	0.8	0.8	0.8	0.8	0.8
North Sea	4.2	6.3	6.6	6.5	6.2	6.0
Other	0.5	0.7	0.8	0.8	0.8	0.7
Total Industrialized	20.1	23.3	24.3	24.2	24.6	24.7
Eurasia						
China	2.8	3.2	3.1	3.1	3.0	3.0
Former Soviet Union	11.4	7.2	9.6	11.9	13.6	14.8
Eastern Europe	0.3	0.3	0.3	0.3	0.3	0.4
Total Eurasia	14.5	10.7	13.0	15.3	16.9	18.2
Other Non-OPEC						
Central and South America	2.3	3.6	4.2	4.8	5.5	6.4
Middle East	1.4	1.9	2.2	2.4	2.5	2.4
Africa	2.2	2.8	3.1	3.8	4.6	5.8
Asia	1.7	2.2	2.6	2.6	2.6	2.5
Total Other Non-OPEC	7.6	10.5	12.1	13.6	15.2	17.1
Total Non-OPEC	42.2	44.5	49.4	53.1	56.7	60.0
Total World	69.4	78.7	87.9	97.4	109.2	122.4

Note: OPEC - Organization of Petroleum Exporting Countries.

TABLE 1.3g Growth of Indian Petroleum Industry at a Glance.

Parameters	Unit	1980-81	1990-91	1995-96	1996-97	1997-98	1998-99	1999-00
1. Reserves @ (Balance Recoverable)								
(i) Crude Oil	Mn. Tonnes	366	739	732	727	747	715	658
(ii) Natural Gas	Bn. Cub. Mtr.	352	686	660	640	692	675	628
2. Consumption								
(i) Crude Oil (in terms of refinery crude throughput)	Mn. Tonnes	25.84	51.77	58.74	62.82	65.17	68.54	85.96
(i) Petroleum Products (excl. RBF)		30.90	55.04	74.67	79.16	84.29	90.56	93.91
3. Production								
(i) Crude Oil	Mn. Tonnes	10.51	33.02	35.17	32.90	33.86	32.72	31.95
(ii) Petroleum Products		24.12	48.56	55.08	59.01	61.31	64.54	79.41
4. Imports & Exports								
(i) Gross imports:								
(a) Qty : Crude Oil	Mn. Tonnes	16.25	20.70	27.34	33.91	34.49	39.81	44.99
Pol. Products		7.29	8.66	20.34	20.26	19.53	18.78	13.07
Total (a) by IOC		23.54	29.36	47.68	54.17	54.02	58.59	58.06
(b) Value: Crude Oil	Rs. Crores	3349	6118	11517	18538	15897	14876	30695
Pol. Products		1917	4660	12578	15634	12432	9837	11119
Total (b) by IOC		5266	10778	24095	34172	28329	24713	41814
Pol. imports as per DGCI & C				25174	35629	30341	26919	45421
(ii) Exports : @@								
(a) Qty : Crude Oil	Mn. Tonnes	-	-	-	-	-	-	-
Pol. Products		0.04	2.65	3.44	3.16	2.93	1.40	.09
Total (a)		0.04	2.65	3.44	3.16	2.93	1.40	.09
Value : Crude Oil		-	-	-	-	-	-	-
(b) Oil Pol. Products	Rs. Crores	8	1004	1832	2085	1820	856	633
Total (b)		8	1004	1832	2085	1820	856	633

TABLE 1.4*a*   Indian Refineries and Cracking Practice.

	Types of Refinery Units	Capacity MMTA	Cracking BPSD	Refining
1. Assam Oil Company, Digboi (Bharat Petroleum)	$D_1D_2/D_C/L/B$	0.5	2300^x	
2. Burmah Shell, Bombay	$D_1D_2/C_C/R/B$	3.5 (5.0)	7000*FCC	7000*R
3. Hindustan Petroleum Corporation Ltd., (Formerly ESSO), Bombay	$D_1/C_C/B$	3.5	11000^{x}FCC	
4. -do- (Formerly CALTEX), Vizag	$D_1/C_C/B$	1.2 (3.5)	9600*FCC	
5. Cochin Refineries Ltd., Cochin	$D_1/Vis/R/B$	2.5 (3.5)	5950 (C)	6000*^{x}R 33,043* HDS
6. Madras Refineries Ltd., Madras	$D_1D_2/C/R/Vis/B/L$	2.8 (5.6)	6460(Vis)	2158* R 31,820 HDT

TABLE 1.4a : (Contd.)

	Type of Refinery Units	Capacity MMTA	Cracking BPSD	Refining
7. Indian Oil Corporation				
Gauhati	D_1/C	0.75	6200	
8. " Barauni	D_1D_2/C/B/L	2.5(3.4)	12000	
9. " Baroda	D_1/R/Vis	4.5(7.3)		7500*(R)
10. " Haldia	D_1D_2/Vis/R/B/L	2.5(5.5)	9000(Vis)	4900*(R) 4500 HDT
				13000 HDS
11. Mathura[1-A]	D/R/C_C/B/Vis	6	20000ˣ	5000ˣ
12. Bongaigon Refinery & Petro-chemicals (Complex)	D_1/R/isomerisation Unit 1			80000
150000 (isomerisation)				

(Nomenclature): D_1 :   Atmospheric column, D_2:VDC,C_c: Cat-Cracking, D_C: Dubbs cracking,
              C   :   Thermal cracking/coking, L: Lube oil, B: Bitumen, Vis: Visbreaking, R: Reforming, HDT:
                                          Hydrotreating, HDS: Hydrodesulfurisation

1 - A   :   Mathura Refinery Products: CEW, Oct. 1981, p. 22, HSD: over 2MMTA, Naphtha: 8,09,000 TA, Kerosene:
          6,58,000 TA, Petrol: 3,50,000 TA, LPG: 1,80,000 TA

(Fig. in brackets shows the expansions in progress)
     Ref: X: *Petroleum Times* 8th March 1974,* *Oil & Gas. J*, Dec. 27, 1976, p. 149

TABLE 1.4b Capacity of Refineries (M²T).

			85-86	86-87	89-90	(Actually Processed) 85-86	86-87	89
X_1	BPC	Bombay	6.00	6.00	6.00	6.389	5.580	6.500
	HPC	Bombay	3.50	3.50	3.50	4.375	5.013	5.365
X	CRL	Cochin	4.50	4.50	4.50	2.749	4.166	4.500
X_1	MRL	Madras	5.60	5.60	5.60	5.057	5.192	5.400
X_1	HPC	Vizag	4.50	4.50	4.50	2.659	3.715	4.500
X_1	10 C	Haldia	2.50	2.50	2.75	2.822	2.622	2.750
XX	"	Koyali	7.30	8.10	9.50	7.830	7.835	8.100
	"	Barauni	3.30	3.30	3.30	2.766	2.860	2.860
	"	Guwahati	0.85	0.85	0.85	0.766	0.811	0.810
X	"	Mathura	6.00	6.00	7.50	6.075	6.353	6.500
	"	Digboi	0.50	0.50	0.50	0.529	0.551	0.550
	BPRL	Bongaigon	1.00	1.00	1.35	0.890	1.011	1.100
		Karnal	–	–	–	–	–	
		Mangalore	–	–	–	–	–	
		Total	45.55	46.70	51.85	42.910	45.701	50.935
		Additional swing capacity	2.00	2.00	2.00			
		at HPC, Bombay	47.55	48.70	53.85			

X_1 0.6 M²T FCC units
X 1.0 M²T " "
XX 1M²T FCC & Hydrocraker.
These have increased the secondary processing facilities by 8 million tons.

TABLE 14c Process Used by Various Lube Units.

Refinery	Finished base stock capacity Tonnes/Year	1	2	3	4	5	6	7	8	9	10
1. Assam Oil Co.	66,150	x		x			x			–x	
2. Barauni Refinery	49,100	x				x		x		–x	
3. Madras Oil Refinery	200,000	x			x			x		x	x
4. Lube India Ltd.	162,000	x				x			x	x	x
5. Haldia Refinery	200,000	x	x		x			x	x		x

Key for the Table

1. Vacuum distillation; 2. Deasphalting; 3. Acid alkali-treatment; 4. Furfural extraction;
5. Phenol extraction; 6. Chilling and Pressing; 7. MEK dewaxing; 8. Propane dewaxing;
9. Clay contacting; 10. Hydro-finishing.

Note: (x) Units mentioned.   x* Acid some stock.

Ref:
P.K. Goel   : Chemical Age of India
Himmat Singh   : April 1971: p. 51
J.M. Sagar
K.K. Bhattacharya
B.S. Gulati

TABLE 1.4d Transformer Oils.

Apar Pvt. Ltd. Bombay	15,000 MT	Total licenced capacity: 65,000 MT
Nagpal Ambadi Petro-Chemical Ltd. Madras	15,000 MT	Production 35,000 MT
Savita Chemicals Ltd. Bombay	10,000 (licensed for 40,000 MT).	
Universal Petro-Chemical Ltd. Kolkata (Registered only)	10,000 MT	
		(Source: N.D. Desai 2nd LAWPSP Symposium January 1981, I.I.T. Bombay p. 4)
Lube India Ltd. (Base Stock for Transformer oils only)	17,000 MT	

TABLE 1.5 A Vista of Indian Petroleum Industry.

1866		First oil well, Nahorpung (Assam)
1890		Discovery of Digboi oil field
1893		Refinery at Mergherita
1899		A.O.C. Registration
1923		First geophysical survey by A.O.C.
1953		Discovery of Nahorkatiya field
1954		ESSO Establishment
1955		Burmah Shell
1956	(August)	ONGC was set up
1957		Caltex
1958	(18th February)	O.I.L India
1959		Lumej field
1960		Ankaleswar oil field 1960-11 P. Dehra Dun
1961		Discovery of Rudrasagar
1961		Gas field Kalol
1962		Noonmathti Refinery (Rumanian Collaboration)
1962		Gas field, Sonand
1963		Nawagam field
1963		Crude Oil produced by O.I.L
1964		Lakwa Oil field
1964		Barauni Refinery (Soviet Collaboration)
1964	(Sept.)	Establishment of I.O.C
1965		Koyali Refinery (Soviet Collaboration)
1966		Cochin Refinery (Phillips Petroleum)
1968		Gelki Oil field
1969		Madras Refinery (National Iranian and AMOCO)
1970	(Feb.)	Lube India Refinery foundation
1972	(Jan.)	Bongaigon Refinery & Petro-chemical Complex Foundation
1973		Haldia Refinery (French & Soviet)
1973		Acquisition of jack up rig Sagar Samrat
1973		ONGC entered into agreement with Iraq for exploration in southern parts of Iraq.
1974		Bombay High Discovery, Haldia Trial runs.
1974		Oil found in Bombay High.
1974		Taking over ESSO by G.O.I.
1976		Discovery of Bassein, near Bombay High
1976	(May)	Commercial production of oil from Bombay High two wells.
1976		Baramura gas field, Tripura
1978		Mathura Refinery foundation
1978		Mahim gas field at Bombay
1980	(Sept)	Oil struck in Godavari delta (Off Shore one well & On Shore two wells near Narsapur)
1981	(Sept)	Take over of Burmah Oil & A.O.C. by Government of India
1982		Kharsang, Oil field (Oil Ltd.) Arunachal
1982	(August)	First LNG plant Duliajan
1987		FCCs Introduction in refineries
1988		First Hydrocraker construction begins at Gujarat Refy.
1988		HBJ Line. GAIL
1988		Maharashtra Gas Cracker Plant
1989		S.B. Gas Structure
1994		ONGC Reliance Refinery Corporation

TABLE 1.6 Characteristics of Crudes Processed in India.

Crude Characteristics	Nahorkatia (Assam)	Ankleswar (Gujarat)	Light (Iranian)	Darius	Digboi
°API	31.0	47.0	34.4	34.1	
Sp. gravity 60/60°F	0.8710	0.80	0.8535	0.8280	0.84
Sulphur, wt%	0.16	0.05	1.35	2.47	0.20
Pour point, °C	30	15	−21	−40	29.8
Total C_4 minus wt.%	2.5	4.36	1.80	1.65	
C_5 to140°C wt.%	17.0	23.14	14.45	15.85	
RON clear	75.0	59.0	55.0	58	
140-250°C, wt.%	20.5	27.5	19.30	15.5	
Freezing point, °C	−60	−42	−56	−51	
Sulphur, wt.%	0.06	0.002	0.18	0.30	
Smoke point, mm	14	30	22	25	
250-371°C, wt%	24.0	23.2	21.80	20.5	
Sulphur, wt%	0.22	0.025	1.10	1.72	
Diesel index	39	66	53	55	
Pour point, °C	6	6	0	−3	
371°C + wt.%	36.0	21.8	42.65	46.5	
Sulphur, wt.%	0.38	0.12	0.53	4.2	
Viscosity cS @ 50°C	160	25	700	560	
Pour point °C	51	48	48	25	
Salt content NaCI wt.%	.0005				0.0005

(Contd.)

TABLE 1.6 : (Contd.)

	Bombay High	Gelki	Digboi
API	39.2	32.1	
Sp. gravity	0.8440	—	
Sulfur wt%	0.12	—	
Pour point °C	27 – 29	33	5 (CS)
Viscosity at 37.8°C	—	—	IBP : 72°C

Distillation Data, Volume % (Cumulative)

°C	Bombay High	Gelki (75°C)	Digboi (75°C)
50	4.5	3.0	0.5
100	11	9.7	4.0
150	23	17.5	19.0
175	29	21.7	25.0
200	33.5	25.6	30.0
225	36	29.5	35.0
250	43	33.1	41.0
275	54	38.9	45.0
300	60	44.6	54.5
325	65	49.2	
350	70	54.7	
370	74	57.0	

Mrs. N. Saikia, D. Chanda, et al.,
Proceedings of Symposium on Lubricants-Additives-Waxes & Petroleum Speciality products
Nov. 22-23, 1978, I.I.T., Bombay

94

Table 1.2*b* shows the world Primary Energy Consumption during 1985.

Table 1.2*c* shows utilisation of natural gas and oil by different countries.

Indian statistics on Oil and Gas reserves are shown in Table 1.3*a*.

Production statistics of crude oil and natural gas in India since 1960 to 1981 are shown in Table 1.3*b*.

Refining activity of Indian Refineries is shown in Table 1.3*c*.

Different products, produced by Indian refineries is shown in Table 1.3*d*.

Consumption pattern of petroleum products in India is presented in Table 1.3*e*.

Refining capacity and statistics on crude and gas reserves of some countries are presented in Table 1.3*f* Indian Refineries, Crude through put and various units are shown in Tables 1.4*a*, *b* & *c*.

Table 1.5 shows the milestones in Indian Petroleum Industry.

Table 1.6 shows the properties of different crudes processed in India.

REFERENCES

1. Meyer R.F. & Hocott C.R., The future supply of Man-made petroleum and Gas. Pergamon Press, Oxford, 1977, p. 1.
2. Hoyle F. Frontiers of Astronomy, 1955, p. 71.
3. Meyer R.F. & Hocott C.R., The future supply of Man-made petroleum and Gas. Pergamon Press, Oxford, 1977 p. 41
4. McConnel Sauders, J.; J. Int. Pet. 937, 23, p. 525.
5. (a) Graham B. Moody, Petroleum Exploration Handbook, McGraw Hill, 1st Ed. 1961, p. 5.
   (b) Ibid, p13.
6 (a) Ernest Beerstech. Petroleum Micrology, Eisevier, 1954, p. 74.
   (b) Ibid, p. 78.
7. Stone, R.W. & Zobell, C.E.: IEC, 1952, Vol. 44, No. 11, p. 252
8. Went, F. Nature, 20th Aug. 1960, p. 641
9. (a) Colombo, U. Fundamental Aspects of Petroleum Geochemistry (B. Naggy & U. Colombo) Eisevier, 1967, p. 331
   (b) ibid p. 79
10. Kenneth A. Kobe and John McKetta, Advances in Petroleum Chemistry and Refining, Vol. 1, p. 50, Inter Science, 1958.
11. Martins, R., Winters, J. & Williams, J.; Nature, 3th July, 1963, p. 11
12. Burton, D., Problems of Petroleum Geology, AAPG, 1934, p. 109
13. McNab, J.B., Smith, PV. & Betters, R.L.: IDC, Nov., 1952, Vol. 44, No. 11, p. 2558
14. Brooks, B.T., Kurtz, S.S., Board, C.E. & Schmerling, L. The Chemistry of Petroleum Hydrocarbons, Vol. 2. Reinhold 1954, p. 41.

15. Phillip, G.T.; Proce. 20th International Geology Congress, Mexico, 1956.
16. Stone, R.W. & Zobell, C.E.: IEC, 1952. Vol. 44, No. 11, p. 257
17. Shiraj: Proce. Of First Iranian Chemical-Indian Chemical Enginers Elsevier May 14-17. 1963 Vol l. p. 54
18. Derek Ezra: Coal and Energy Ernwst Bemm, London & Tonbridge 1978, p. 61-62
19. Daniel N. Lapedes, Encyclopedia of Energy, McGraw Hill, 1976, p. 508.
20. Ajit Loussine, N. HCP, April 1981, p. 131
21. John, J. Mcketta & Wall, J.D., HCP, July 1980, p. 59
22. Ion D. & Hendricks, 7th WPC. Vol. 9
23. Venugopal, RT; CEW, April 1981, p. 51
24. & 25 Hindu Survey of Indian Industry, 1987 & 88
26. Sokolov, V. Petroleum, Mire publications, Moscow, p. 257
27. Maitland Jones Jr. and Robert A. Moss; Carbenes, Vol. 1 John Willey & Sons, N. York, 1973, p. 2.
28. Obolenstev, R.D. Ibid., p. 109.
29. A guide to World Export Crudes, by Oil & Gas J.
30. Haines, W.E., Cook, G.L. & Dinneen, G.U. 7th WPC, Vol. 9, p. 84
31. James, G. Speight, The Chemistry and Technology of Petroleum, Marcel Dekker, N.Y. p. 75.
32. Charles, W. Keenan and Jesse. H. Wood, The General College Chemistry, Harper & Row Publications, N. York p. 655.
33. Hodgson, G.W., Barker, B.L. & Peake, E. 7th WPC, Vol. 9 p. 117.
34. David Dolphin. Porphyrins, Academic Press, London 1978, p. 59.
35. Berti V; IcPardi, H.M, & Nuzzi, M 7th WPC Vol. 9, p. 117
36. B.K.B. Rao, 3rd Law and PSP Symposium, I.I.T., Bombay, 1982.
37. Harold H. Schobert. The Chemistry of Hydrocarbon Fuels Butterworths-London, 1990: p. 34
38. Rao, S.P. Janamanchi, CEW, July 1978, p. 39.

2

Petroleum Processing Data

2.1 EVALUATION OF PETROLEUM

Petroleum is classified into mainly three congenial types as:

1. Paraffinic base
2. Mixed base or Intermediate and
3. Naphthenic base.

Some authors have referred the naphthenic base as asphaltic base. The name naphthenic is more convincing as it represents the homologous series whereas asphalt is a polymeric condensed material having very high carbon to hydrogen ratio of no equable nature of general formula. Although the catogerisation into these three types may not be very accurate and brief, it serves, without any reservations the purpose. Knowledge of base of a crude is very essential as it depicts, even though not fully the usual properties enmass and can adequately inform the refiners about the difficulties in processing too.

Irrespective of the base of the crude all the hydrocarbon series listed in Chapter 1 are present in all crudes, but to different extents.

Mallison classified the crude exclusively on the basis of residuum, a material left behind after distillation of fractions.

Accordingly:

Residue containing more than 5% paraffins is paraffinic crude
Residue containing less than 2% paraffins is naphthenic base
Residue containing 2%-5% paraffins is considered as mixed base

The proper method of evaluation of crude started with U.S. Bureau of Mines, which designated eight different bases of crude.

The basis of classification begins with two chosen fractions, Key fraction No. 1 and Key fraction No. 2 as given below:

Key fraction No. 1 (kerosene) has a boiling range of 250°C-275°C at normal pressure.

Key fraction No. 2 (Lube) has a boiling range of 275°C-300°C at a pressure of 40 mm Hg; i.e. 389°C-422°C at 760 mm pressure.

The API gravities of these fractions are found out and the base of crude then characterised as:

When key fraction No. 1 has: 40 or more API gravity it is paraffinic less than 33 API gravity, it is naphthenic and in the range of 33-40 API gravity, it is mixed base[1].

When Key fraction No. 2 has: 30 or more API gravity then it is paraffinic; less than 20 API gravity it is naphthenic and between 20-30 API gravity it is mixed base.

Above distinct classification dispels any ambiguity or dissension, however in the case of mixed base crudes, the nearness to extremities is mentioned for avoiding any moots. Say a key fraction No. 1 of a crude is having 39 API, this without any doubt, is mixed base but more oriented towards paraffinicity rather than naphthenicity. Similarly an API 33.3 is not only mixed base but tilted towards naphthenic nature rather than paraffinic nature. In such cases it is judicious to use (for the above fraction as) mixed base (paraffin), mixed base (naphthene) respectively.

In case of key fraction No. 2, often wax or wax free term is proposed. If the pour point is less than 3°C, for this fraction then the base is followed by the 'wax free' term; otherwise it is followed with subscript 'waxy'.

Earlier days, classification based on the location of crude was habitual such as Pensylvania crudes classified as paraffinic base and Gulf coast crudes as naphthenics. These ideas need not be any more entertained, as location or formation can never give the idea of crude. Some crudes rich in aromatic series were also classified as aromatic crudes or benzenoid base crudes. As all naphthenic crudes give plentiful aromatics such formal classifications need not be attended.

API (American Petroleum Institute) gravity: This is defined as:

$$\text{Deg. API} = \frac{141.5}{\rho} - 131.5 \qquad \text{... (2.1)}$$

where, ρ = sp. gravity of fraction at 15.6°C/15.6°C

API gravity magnifies the sp. gravity of fractions. As the sp. gravity of close boiling cuts usually lies very close to each other, this type of magnification is essential. For a blend, the API gravity is given as:

$$(API)_{mix} = (API)_a \times W_a + (API)_b \times W_b +$$
$$(API)_c \times W_c + ... \qquad .... (2.2)$$

Thus API gravity of a mixture is equal to the sum of the individual API gravity of a component multiplied by corresponding weight fraction in the mixture.

2.1.1 U.O.P. Characterisation Factor $(K)^{2a}$

Characterisation factor is of immense utility in refinery calculations. Besides it has got capacity to predict the qualities of crude, it can also provide with many valuable data on fractions. API gravity and molecular weight as other parameters, it can give almost all necessary information about the fraction, right from physical properties to tendency of cracking.

The relation is given by

$$K = \frac{\sqrt[3]{R}}{0.827\rho} = \frac{\sqrt[3]{T°}}{\rho} \qquad ...(2.3a)$$

(Original K factor was formulated on °R and ρ at (60°F)).

where R = average boiling point °K. or T° : Rankine
ρ = sp. gravity at 15.6°/15.6°C

For paraffinic base of crude oil K is 12.5 and above, and if it is less than or equal to 10, the base of the crude is denoted as naphthenic. For a fraction, usually average boiling point will coincide with 50% boiling point when Engler slope is less than 3. Watson and Nelson replaced the term average boiling point by molal average boiling point, but retaining the index values for characterisation purpose[2b]. Molal average boiling point is better suited for a wide boiling fraction with a slope more than '3' in Engler distillation.

If the characterisation factor is 12.5, the crude is classified as paraffin base and less than 10, is said to be napthenic base.

Intermediate values are regarded as mixed base crudes. K-value of a mixture is the sum of the components 'K' value multiplied by their weight fractions in mixture.

$$K_{mix} = K_1W_1 + K_2W_2 + K_3W_3 \qquad ... (2.3b)$$

where, W_1, W_2, W_3 are weight fractions of different components.
K_1, K_2, K_3 are respective characterisation factors.

2.1.2 Correlation Index

Like characterisation factor correlation index is also related to boiling point and gravity, however its use is marginal:

$$\text{Correlation Index (C.I)} = \frac{48640}{T_B} + 473.7\,\rho - 456.8 \quad ...(2.4)$$

where T_B = boiling point (molal)°K

ρ = sp. gravity at 15.6°C/15.6°C

Correlation index for paraffins is taken as zero and for aromatics, it is taken as 100. The value is not quantitative hence a relative indication of the groups is only possible.

2.1.3 Distilation Characteristics

Crude Assay Analysis

Distillation characteristics of a crude are assessed by performing a preliminary distillation called 'True Boiling Point' analysis (TBP). This pulse test enlightens the refiners with all possible information regarding the percentage quantum of fractions, base of crude and the possible difficulties beset during treatment operations etc. Information supplied by this distillation forms the basis of design of distillation columns and thus represents the veridity of crude distillation.

TBP Apparatus

Pursuits for development of such kind of stills were left with Peters, Podbielniak stills. A less costly laboratory still with proper adjustments and controls may also be tried, when situation demands. A thorough rectification system constituting 10 to 12 stages forms the operation. Modern stills are instituted to distill at least 2 liters of stock per charge. Larger capacities are preferable as good amounts of higher boiling stocks can be obtained for testing. Heat losses in industrial operations are managed by lagging, but in such small stills hot air circulation through the jacket, surrounding the rectifier section, ensures the minimum thermal losses. Top of rectifier is usually fitted with an efficient air cooled reflux condenser, which also indicates the temperature of the fraction leaving for the condenser system.

The condensate collects into a graduated cylinder, which is in fact connected to a vacuum producing device. The vacuum device must have sufficient displacement capacity to ensure satisfactory operation.

For this reason, a displacement of 0.2 to 0.5 cubic meters of air or vapor is required for stills of capacity 2 to 5 liters. It is also essential that distillation proceeds at a constant rate. Too high a rate results in flooding and lower rates produce imperfect fractionation. Observations have revealed that rate of distillation in the beginning is very high, but the rate of distillation decreases with increasing boiling point of the stock. Hence a careful heating and condensation system is required. A good rate of distillation prescribed for atmospheric operation is 1% distillation in 2 minutes and under vacuum it drops to 1% in 3 to 5 minutes. Further, the system in no way should suffer while removing the condensate during vacuum distillation.

The necessity of vacuum distillation is mainly to prevent the thermal degradation of high boiling stocks. The high boiling—stocks are prone to crack when exposed to a temperature of 370°C or more. Vacuum distillation reduces the boiling point of the fraction, and for every pressure there is a definite boiling point. These boiling points obtained under the exposed pressures or vacuum should be corrected back to the normal pressure (760 mm) operation.

A sketch of the TBP apparatus is furnished in Figure 2.1

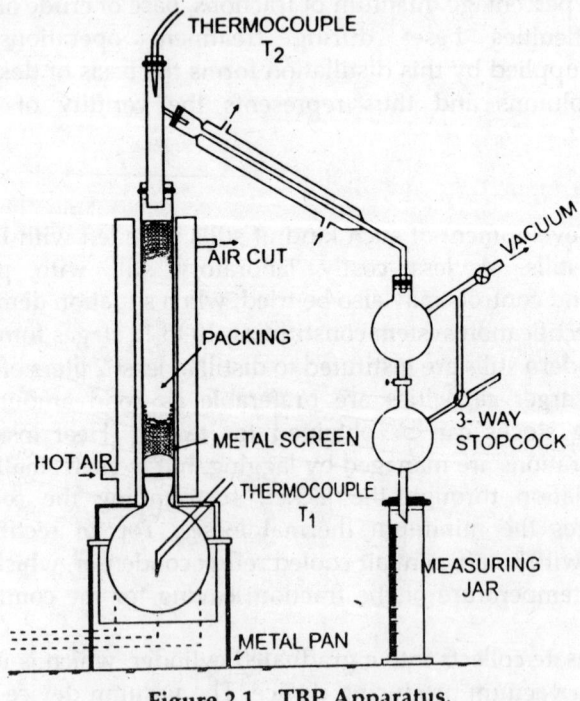

Figure 2.1 TBP Apparatus.

Boiling point corresponding to a particular pressure can be converted to normal boiling point with the help of Figure 2.2a, Figures 2.2b and (c) are useful when very high pressures or low pressures are involved.

The data collected are usually presented in the graphical form as shown in Figure 2.3 and goes by the name TBP curve.

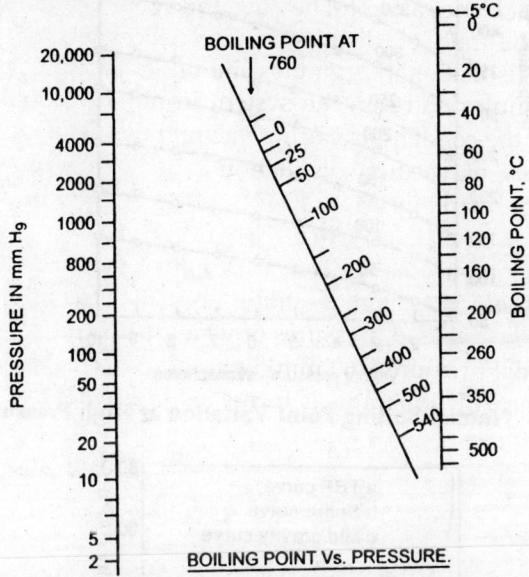

Figure 2.2a   Boiling Point Vs. Pressure
Taken from Petroleum Refinery Engineering W.L. Nelson (c)
Used with the permission of the McGraw Hill Book Comp.

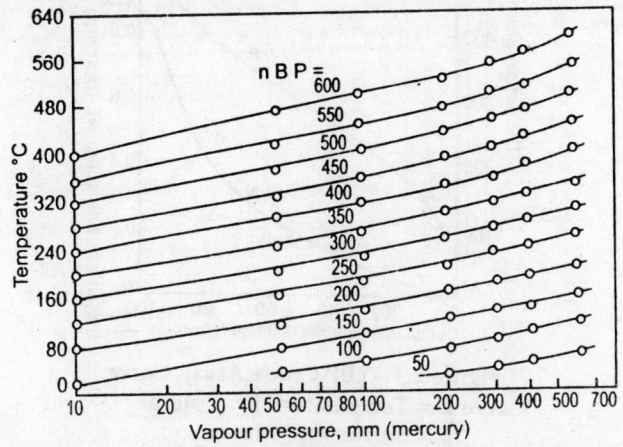

Figure 2.2b   Normal Boiling Point Variation at Low Pressure.

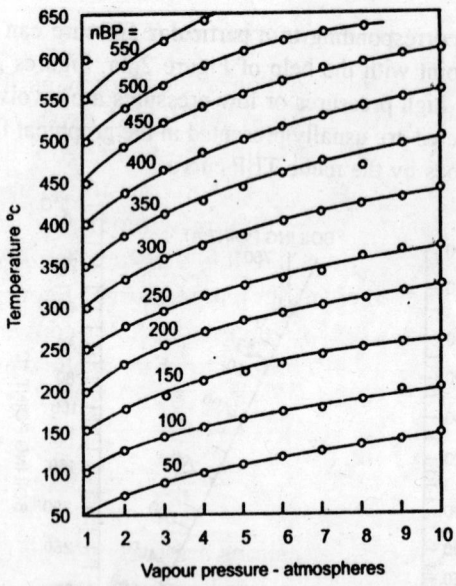

Figure 2.2c **Normal Boiling Point Variation at High Pressures.**

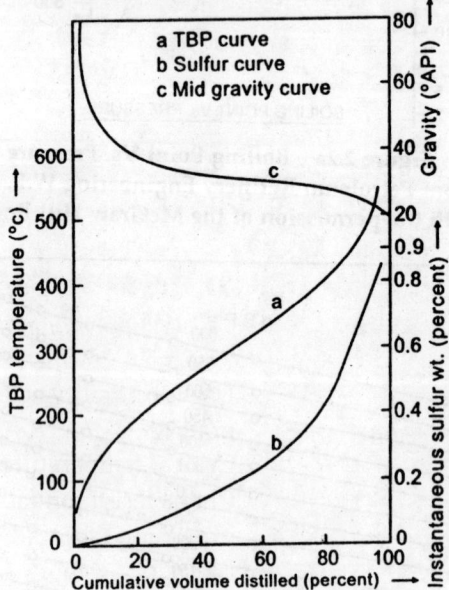

Figure 2.3   TBP Crude Assay Curve
Curve a = Temperature *Vs.* Volume
b = Sulfur *Vs.* Volume
c = Gravity *Vs.* mid percent

2.1.3.1 Equilibrium Flash Vaporisation (EFV)

While in continuous fractionation system such as TBP, always a thorough contact between vapor and liquid exists; in equilibrium flash vaporisation, vapor is kept cohesively with liquid at some temperature and a sudden release of pressure quickly flashes or separates the vapor from the mixture without any rectification. By successive flash evaporations like this the stock can be progressively distilled at different increasing temperatures. Experimentally this type of distillation may not be suggested as inter-conversion from one type of distillation (say TBP to EFV) to other is always available.

2.1.3.2 Engler Distillation: (ASTM-D-86)

It is also known as ASTM distillation and is supposed to be like EFV, a nonfractionating distillation system, distinguishing itself as differential distillation. All fractions are specified by this distillation only. It is a simple distillation carried out with standard ASTM flasks of 100, 200, 500 ml capacity. The data obtained is similar to TBP data and interconvertability is resorted wherever essential.

2.1.3.3 Humpel Distillation

Information about this in modern literature is not abundant, it is considered as a semi-fractionating type of distillation like Saybolt's. Where TBP data is insufficient, this can be used, as it nears the TBP characteristics of distillation.

Different[3] types of boiling points and functional values

Average boiling point	Physical property for which it is distinct
1. Volume average	Use for liquid viscosity, specific gravity
2. Weight average	Critical temperature
3. Molal average	Characterisation factor, thermal expansion of liquids
4. Mean average	Molecular weight, sp. gravity Heat of combustion, sp. heat, etc.
5. Cubic average	For additive properties; viscosities are additive when expressed on cubic average

2.1.4 Average Boiling Point

Because of a fraction boiling regularly or haphazardly over a wide range, it is essential to represent the boiling point of the fraction by some method, which can specifically denote certain physical properties. For narrow boiling cuts (TBP slope less than 2) all the above mentioned boiling points tend to be equal.

Volume average boiling point is based upon the boiling temperature of different cuts of the fraction. Usually cuts are chosen at regular intervals.

Thus for whole crude, it may be represented as:

(Volume average BP)

$$T_B = \frac{t_{10\%} + t_{20\%} + t_{30\%} + t_{40\%} + t_{50\%} + t_{60\%} + t_{70\%} + t_{80\%} + t_{90\%}}{9}$$

$$... (2.5a)$$

If such data is not available then it may be defined as

(Volume average BP) $\quad T_B = \dfrac{t_{30\%} + t_{50\%} + t_{70\%}}{3} \quad$ ... (2.5b)

where all percentages are in volumes.

If the fraction is boiling over a narrow range of temperature then $t_{50\%}$ may be accepted as average boiling point. ($t_{10\%}$ means the temperature indicated when exactly 10% of the stock is collected during distillation.)

Weight average: In this case instead of volume, weight fraction is chosen for evaluating boiling point. Thus

(Weight average BP) $T_B = \dfrac{t_{10\%} + t_{20\%} + \ldots\ldots\ldots t_{90\%}}{9} \quad$ ... (2.6)

where percentage calculations are based on weight.

Molal average: This is based upon boiling temperature at different mole fractions. It is no doubt a perilous job as determination of molecular weight for each cut is not practiced.

(Molal average) $T_B = \dfrac{tx_1 + tx_2 + tx_3}{x_1 + x_2 + x_3} \quad$ ... (2.7)

where x_1, x_2, x_3 are mole fractions.

t_1, t_2, t_3 are corresponding boiling points.

Mean average: This is the temperature at which some physical properties like specific heat, specific gravity etc. of a fraction are found out by taking the mean of temperature levels:

sp. heat at temp. $t' = c'_p$ and $t'' = c''_p$

average sp. heat at $\dfrac{(t' + t'')}{2} = \dfrac{(c'_p + c''_n)}{2}$

Mean average and average boiling point coincide for a blend of two components and these are related to TBP slopes as shown in Figure 2.6a and b.

Cubic average: Some properties like viscosity seem to be additive when cubic average is taken into consideration, rather than mean average, or molal average thus,

$$(\text{cubic average})T_B = [V_a(t_a)^{1/3} + V_b(t_b)^{1/3} + V_c(t_c)^{1/3}]^3 \qquad \dots (2.8)$$

All these boiling points are interconvertable.

Interconvertability of boiling points can be worked out by knowing the slope of distillation curve of a fraction. The method of finding out the slope for ASTM/TBP/EFV is the same.

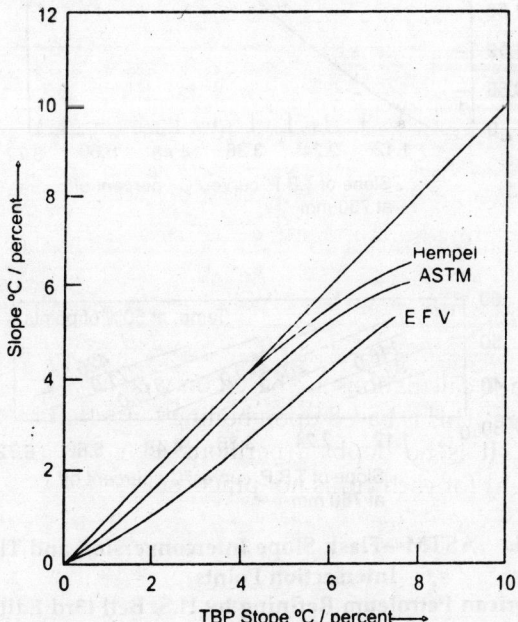

Figure 2.4   Different Slopes and Interconversion.
(By Courtesy Oil and Gas Journal)

TBP slope is given as $\dfrac{t_{70\%} - t_{10\%}}{60}$ i.e. $°t$/percent; where 70% and 10% are volumetric boiling points on vaporization curve. The conversion of TBP slope to ASTM or EFV slope can be done with the help of Figs. 2.4 or 2.5a, b. Fifty percent boiling points of ASTM curves can be calculated by knowing the slope of TBP curve of a crude and is 50% boiling point[4]. The correction factor is incorporated in Table 2.1. Similarly from ASTM or TBP slopes, EFV 50% boiling point can be found out from Figure 2.6a. Slopes of different boiling point curves are also easily interconvertable and Figs. 2.4 and 2.5a, b help in such conversions.

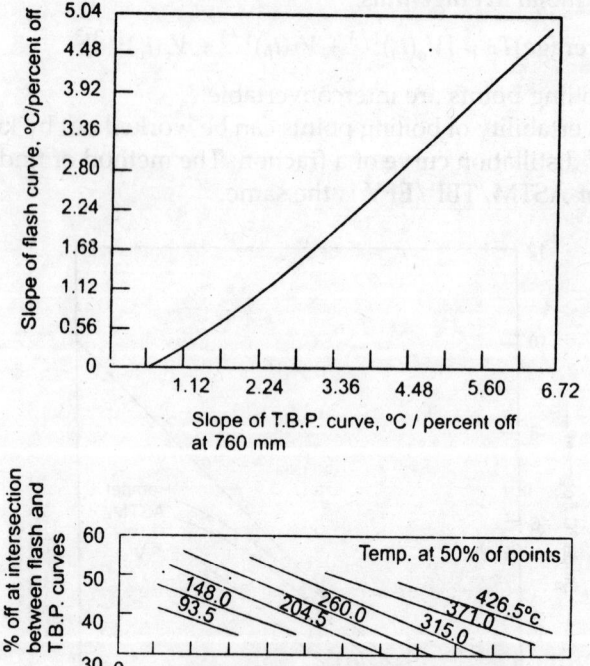

Figure 2.5a   ASTM—Flash Slope Interconversion and Their Intersection Points.
From American Petroleum Refining by H.S. Bell (3rd Edition) with Permission from Brooks/Cole Publishing Co. Owners of Van Nostrand Company.

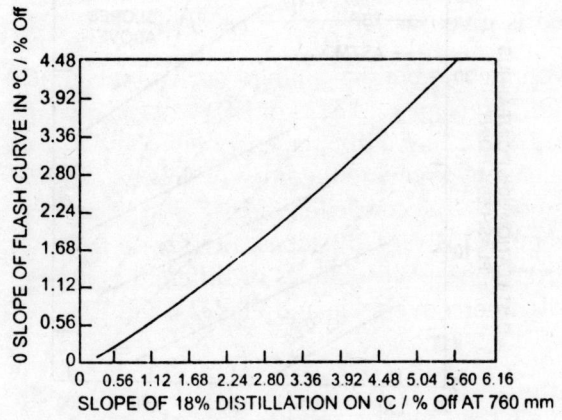

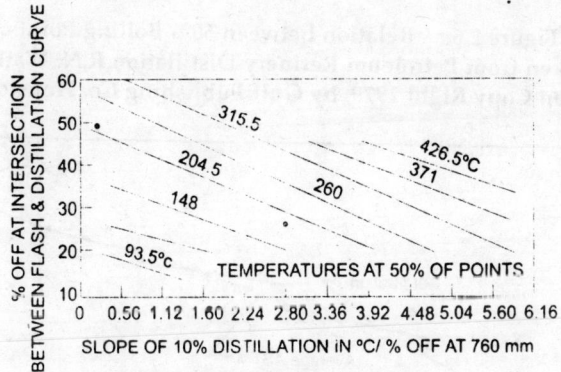

Figure 2.5*b* TBP—Flash Slope.
Interconversions and their Intersection Points
(From American Petroleum Refining by H.S. Bell
3rd Edition. With Permission from Brooks/Cole Publishing Co.
Owners of Van Nostrand Company).

2.1.4.1 Boiling Point—Pressure

Under equilibrium, for every liquid there is a definite vapor pressure for a definite temperature and the vapour pressure increases with increase of temperature and vice-versa; though not linearly. Interestingly when vapour pressure and temperature are plotted on a semi-log paper, a straight line relation is revealed, this relationship although may change from groups to groups like alcohols, ketones, paraffins, aromatics etc. it follows in case of hydrocarbon groups without any exception.

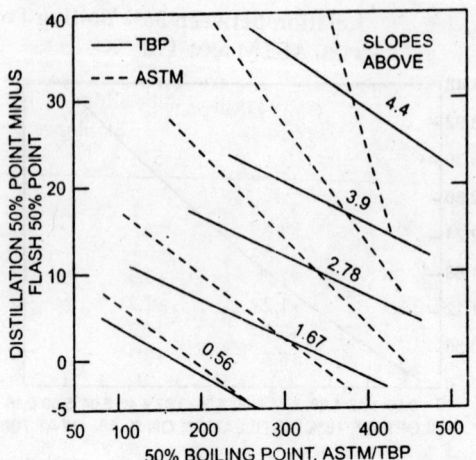

Figure 2.6*a*   Relation between 50% Boiling Points.
(Taken from Petroleum Refinery Distillation R.N. Watkins
2nd Edition Copy Right 1979, by Gulf Publishing Co. Houston, Texas).

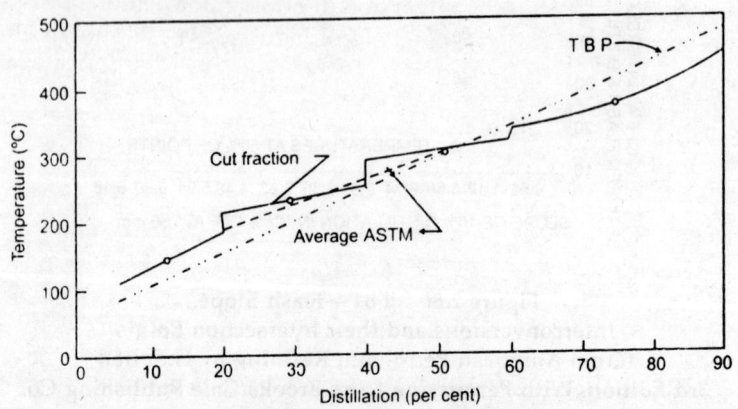

Figure 2.6*b*   ASTM cut fraction curves
T.E. Daubert[47] correlated ASTM 50 and TBP 50% points as TBP
50% = 0.87180 (ASTM 50%).

Clausius-Clapeyron equation can also be used where there is data available on latent heat of vaporization

$$\ln \frac{P}{P_0} = \frac{\lambda}{R}(1/T_0 - 1/T) \qquad \qquad ... (2.9)$$

where, P_0, P are vapour pressure at T_0, T absolute temperatures

**TABLE 2.1 Relation between 50% Boiling Point of
ASTM and TBP**

TBP 50% BP	Correction to be added to TBP 50% BP TBP Slopes				
	1	3	5	7	9
°C					
37.8	−2.8	−9.5	−	−	−
93.3	−1.2	−3.3	−6.7	−	−
119	−0.56	−1.7	−3.3	−13.4	−25.0
204	+0.56	+1.2	−1.2	−9.0	−17.0
260	+1.2	+2.2	+1.3	−4.06	−10.0
315.5	+2.2	+5.0	+3.3	−0.56	−5.6
371	−	7.8	+8.3	+5.6	+0.56
428	−	−	+11.7	11.0	9.0

λ = molal latent heat of vaporization at T

R = gas constant in corresponding units

Relations of Cox, Perry[5], Meyer[6], are worth mentioning in this connection. The boiling points at high and low pressures are also available in literature.[7] Conversion of boiling point at one pressure to another can be done with the help of Figure 2.2*b*, *c* as shown already.

Problem 2.1

Use the following data of a crude; Trinidad blend, to plot crude assay curves. Assess the quality of crude and various fractions.

Trinidad blend

Whole crude	Gravity °API	33.6	(0.8570)
	Sulfur, wt.%	0.23	
	Pour point, °C	+14	
	Viscosity (SUS) at	20°C	47
	at 56°C	38	
	C_4 and lighter, vol. %	0.58	

Light straight run	Range, TBP°C	C_5 – 93	
	Yield, vol.%	3.28	
	Gravity, °API	69.6	(0.7035)
	Sulfur, wt.%	0.01	
	Motor octane, clear	73.2	

Cat. reformer feed	Range, TBP, °C	93–175.0	
	Yield, vol.%	11.71	
	Gravity, °API	46.9	(0.7932)
	Sulfur, wt.%	0.20	
	Arom+naph, vol.%	69.7	

Heater-oil dist.	Range, TBP, °C	175.0–286	
	Yield, vol.%	30.51	
	Gravity, °API	34.8	(0.8508)
	Sulfur, wt.%	0.10	
	Blend pour pt. °C	1	
	Cetane index	40.2	
Furnace-oil dist.	Range, TBP, °C	286–325	
	Yield vol.%	13.69	
	Gravity °API	31 .3	(0.8692)
	Sulfur, wt.%	0.20	
	Blend pour pt.°C	5	
	Cetane index	48.5	
Light FCU feed	Range, TBP, °C	325–365	
	Yield, vol.%	9.05	
	Gravity, °API	29.0	(0.8816)
	Sulfur, wt.%	0.32	
	CA, wt.%	12.5	
	Nitrogen, wt.%	0.024	
Heavy FCU feed	Range, TBP, °C	365–542	
	Yield, vol.%	26.97	
	Gravity, °API	28.1	(0.8865)
	Sulfur, wt.%	0.37	
	CA*wt.%	12.2	
	Nitrogen, wt.%	0.038	
	Equiv. nickel, ppm	0.1	
Reduced crude	Range, TBP, °C	542+	
	Yield, vol.%	4.22	
	Gravity, °API	20.0	(0.9340)
	Sulfur, wt.%	0.89	
	Rams. car, wt.%	4.8	
	V+Ni, ppm	10	

°CA% = Aromatics

Solution

Figure 2.1 shows the TBP characteristic curves of the crude. Close observation reveals the following:

1. Sulfur and sp. gravity increases regularly from low boiling fractions to high boiling fractions. It can be treated as a low

sulfur-crude (0.23% sulfur on whole crude basis); however desulfurisation is recommended for fractions (especially catalytic reforming feed stocks). Gravity shows that it is of mixed base origin, although confirmation is essential.

2. Fifty percent of the crude can be distilled below 300°C, it shows that the crude is dominant in lighter fractions. It is a source of gasoline of quality and quantity and may be utilised even without desulfurisation. Good solvent of end point 180°C can also be obtained.

3. Diesel fuels or heating oils (upto 350°C) form the main products (upto 40%). Diesels are of good quality (cetane no. 40) and pour point is also convenient for regular usage.

4. Vacuum distillation is expected to separate the catalytic cracking feed stock usually known as gas oil. The residue from the vacuum distillation column may be utilised as bitumen rather than for coking operations, as it contains less carbon residue it may not give good quantity of coke.

Problem 2.2
Use the TBP data of the above problem and find out TBP, ASTM, EFV slopes.

Solution
TBP slope is found out by knowing the temperatures of distillation at 70% and 10% (Figure 2.1) off.

$$\text{TBP slope} = \frac{355 - 138}{60} = 3.61 \ °C/\text{per cent}$$

With the help of Figure 2.4 slope of TBP is converted to ASTM i.e. 3.06°C/per cent. Similarly Figure 2.4 gives slope of TBP to slope of EFV i.e. 2.1°C/percent or Figure 2.5a, b are also helpful.

Problem 2.3
Find the percentage intersection of TBP and ASTM & TBP and EFV from the data of problem 2.1.

Solution
Firstly the 50% BP on TBP curve is found out as equal to 302°C. TBP slope is given (from above problem).

From Figure 2.5a the 50% B.P. and slope of the curve will give the percent off intersection points between flash curve and TBP i.e. 39% similarly for ASTM it is obtained from Figure 2.5b.

From the points of intersections, with the corresponding slopes, straight lines can be drawn as shown in Figure 2.6b

Problem 2.4

From the data of problem No. 2.1, compute the vaporisation curves of ASTM and EFV for the whole range. Repeat it, by choosing cuts of 5 to 20%, 20 to 40%, 40 to 60% and 60 to 90% to explain the nature of cut fraction curve of ASTM.

Solution

Use the slopes obtained in problem 2.2. Locate the 50% boiling point with the help of Table. 2.1

Slope of TBP 3.61°C/per cent

50% TBP point 302°C.

Corresponding ASTM correction is +4°C approximately. Add this to 50% TBP point *i.e.* 306°C, which is ASTM, 50% boiling point. At this point draw a straight line with a slope of 3.06°C/percent to get ASTM curve. Similar procedure is adopted for EFV also. EFV slope is 2.1°C/per cent and 50% EFV point is obtained from Figure 2.6a. For a cut fraction 5 to 20%; the slope and 50% boiling point of this fraction is computed assuming that this fraction represents 100% fraction.

So the slope is found out by locating 70% and 10% distillation points of this fraction which correspond to

$$\frac{20 \times 70}{100} = 14\% \text{ of the fraction}$$

$$\frac{20 \times 10}{100} = 2\% \text{ of the fraction}$$

(add initial 5% to make 7%)

Hence the slope is $\dfrac{t_{14\%} - t_{7\%}}{60} = \dfrac{160 - 110}{60} = 0.83°C/percent$

The approximate mid point is $\dfrac{5\% + 20\%}{2} = 12.5\%$

At 0.83 TBP slope and 50% b.p. (150.8°C)

ASTM Slope is (Figure 2.4) = 0.48

Converting TBP 50% point to ASTM 50% point as in the above case (or use Figure 2.7b) ASTM 50% b.p.=151.8°C.

IBP (5% point on TBP curve) = 98.9°C.

Corresponding IBP of ASTM is found out from Figure 2.7a. Similarly FBP of TBP is converted to FBP of ASTM with the help of

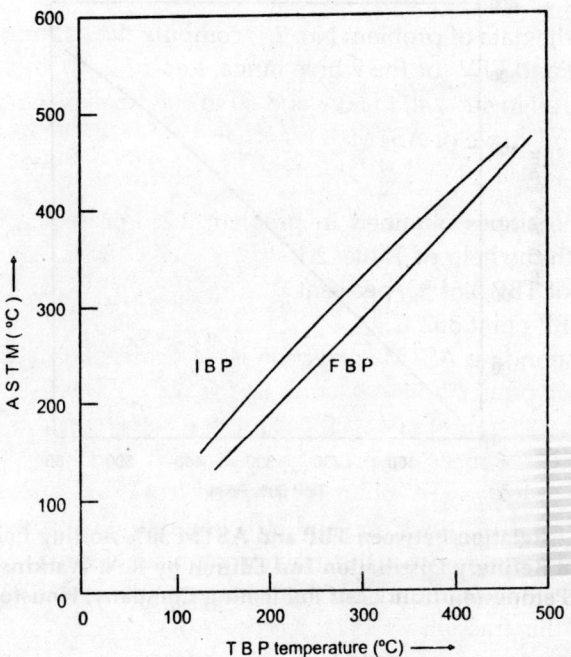

Figure 2.7a Relation between IBP & FBP of ASTM and TBP. From Petroleum Refinery Distillation, 2nd Edition by R.N. Watkins (c) 1979 used with Permission from Gulf Publishing Company, Houston, Texas.

Figure 2.7a. When is not possible to obtain any of these IBPs or FBPs, arithmetic technique can be used to compute.

ASTM(IBP) = ASTM (50%) b.p-ASTM slope × 50%
         = 151.8-0.48 × 50 = 117.8°C.

Like this for all these cuts, an attempt is made to convert into ASTM fractions, as shown below:

    1st cut 5-20%

        ASTM Slope   = 0.48
        ASTM 50% b.p = 151.8°C
        ASTM IBP   = 117.8°C

FBP of 1st fraction (TBP) forms the IBP of 2nd cut

	2nd cut (20-40%)	3rd cut (40-60%)	4th cut (60-90%)
TBP slope	0.44	0.51	1.19
ASTM slope	0.28	0.36	0.72
TBP 50% b.p.	232°C	296°C	375°C

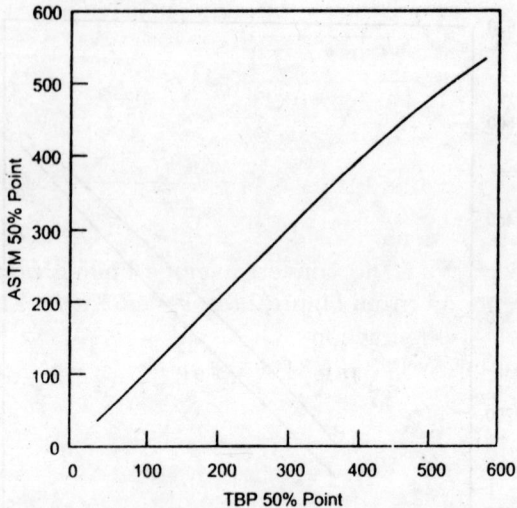

Figure 2.7*b*   Relation between TBP and ASTM 50% Boiling Points (From Petroleum Refinery Distillation 2nd Edition by R.N. Watkins (*c*) 1979 used with Permission from Gulf Publishing Company, Houston, Texas).

ASTM 50% b.p.	225°C	296°C	362°C
TBP (IBP)	195°C	268°C	330°C
TBP (FBP)	268°C	330°C	456°C
ASTM (IBP)	220°C	286°C	330°C
ASTM (FBP)	250°C	312°C	430°C

Such a cut fraction curve is shown in Figure 2.8*b*.

Problem 2.5

From the data of problem 2.1, find the characterisation factor of the crude on the whole basis; also on the basis of choosing three fractions of equal volume.

Solution

Evaluate first. Average boiling point, on the whole crude

basis, *i.e.* $= \dfrac{t_{10\%}........t_{90\%}}{9}$

$$= \dfrac{132 + 193 + 235 + 270 + 305 + 330 + 362 + 407 + 455}{9}$$

$$= 298.8°C.$$

Average ρ : 0.8570. Obtain 'K' by substituting this value in

$$K = \frac{\sqrt[3]{(298.8 + 273)}}{0.8570 \times 0.823} = 11.66 \qquad \ldots \text{I}$$

Average boiling point on the basis of $\dfrac{t_{30\%} + t_{50\%} + t_{70\%}}{3}$ = 374°C slope

of TBP = 3.81°C/percent.

correction to be applied to convert average boiling point to Molal average boiling point (from Figure 2.8 a, b = – 62°C $i.e.$ 312°C.

Average density of the fraction

36°API	= 0.8448	gm/cc
32°API	= 0.8654	-do-
29°API	= 0.8816	-do-
Average	= 0.8673	gm/cc.

$$K = \frac{\sqrt[3]{312 + 273}}{0.827 \times 0.8673} = 11.66 \qquad \ldots \text{II}$$

I and II show the 'K' factor same

Now choose 3 cuts as Cut No. 1 0 to 30%

Cut No. 2 30 to 60%

Cut No. 3 60 to 90%

Evaluate the slope of each cut and average boiling point of each cut. Apply the necessary correction.

	TBP slope °C percent	Av. B.P °C	Correction for converting to MAVBP	ρ
Cut No. 1	1.98	166.6	–31 °C	0.769
Cut No. 2	0.83	271.0	–16°C	0.8602
Cut No. 3	1.20	366.5	–16°C	0.8844

For Cut No. 1 Molal average boiling point = 166.6-31 + 273

408.6°K

$$K = \frac{\sqrt[3]{408.6}}{0.827 \times 0.769} = 11.70$$

Cut No. 2 $\quad K = \dfrac{\sqrt[3]{528}}{0.827 \times .8602} = 11.40$

Cut No. 3 $\quad K = \dfrac{\sqrt[3]{613.5}}{0.827 \times 0.8844} = 11.62$

Average K value 11.57

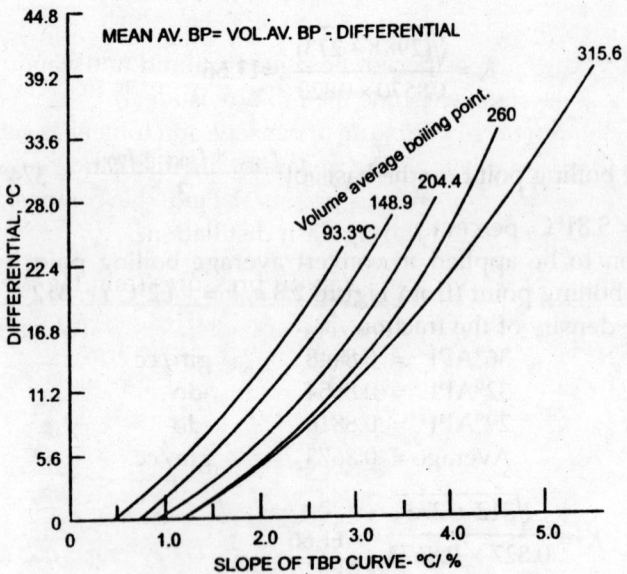

Figure 2.8a, b   Mean Average Boiling Point

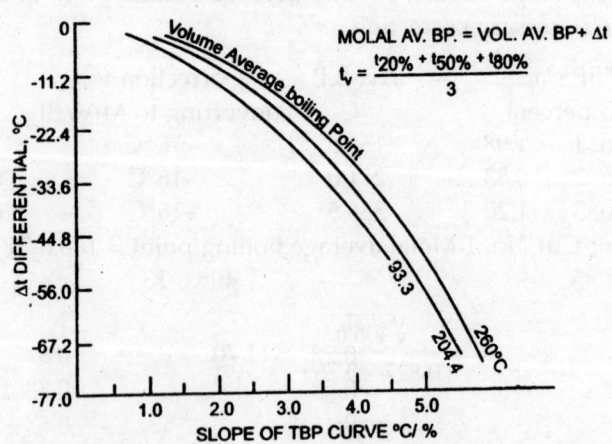

Figure 2.8a, b   are Reprinted from p. 33
Petroleum Refining
James H. *Gary-Glenn. E. Handwork*
Vol. 5 Courtesy of Marcel Dekker Inc. N.Y. 1975.

Problem 2.6

   (a) Find the composition of benzene in liquid and vapour from the given data. (Assume the mixture is ideal).

   (b) An equimolecular mixture of benzene and toluene is subjected to flash distillation. The separator in distillation apparatus operates at atmospheric pressure. Find the composition of vapour and liquid during flash distillation.

Data	Vapour pressures (mm Hg.)	
Temp. °C	Benzene	Toluene
80.2	760	—
85	877	445
90	1,016	405
95	1,168	475
100	1,344	557
105	1,522	645
110.3	1,800	760

Solution

   (a) The composition of liquid that boils at 760 mm Hg. pressure is calculated by the equation

$$x_A p_A + x_B p_B = 760$$
$$x_A = 0.78 \text{ or } x_B = 0.22$$

$$Y_A = \frac{x_A p_A}{p} = \frac{(0.78 \times 877)}{760} = 0.9 \text{ or}$$

$$Y_B = 0.1.$$

Similar calculations at all other temperatures show:

T °C	X_A	Y_A
80.2	1	1
85	0.78	0.9
90	0.58	0.78
95	0.41	0.63
100	0.26	0.45
105	0.13	0.26
110.3	0	0

   (b) Flash vaporisation is done by raising the temperature of liquid and quickly separating the vapour which was in equilibrium with liquid. The slope of vaporisation is connected to fractional vaporisation and given by

$$S = \frac{1-f}{f}$$

Giving values for 'f, S values can be found out.

f	S
0	∞
0.2	–4
0.4	–1.5
0.6	–0.67
0.8	–0.25
1.0	–

The slope is drawn from 'xF' (0.5 = xB) and allowed to cut the BP diagram and x_e and y_e are found out. The flash temperature is noted as t_f. (see Figure 2.8c.)

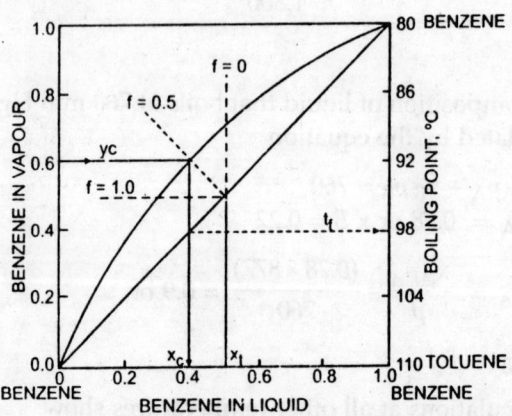

Figure 2.8c : Flash Vaporisation (Composition of Vapour-liquid and Flash Temperature).

T_f	f	Slope	Benzene equilibrium composition	
			Liquid	Vapor
92.2	0		0.5	0.71
93.7	0.2	–4	0.455	0.97
95.0	0.4	–1.5	0.41	0.63
96.5	0.6	–0.67	0.365	0.585
97.7	0.8	–0.25	0.325	0.54
99.0	1.0	0	0.29	0 50

Thus doing successfully flash vaporisation, the enrichment of toluene can be done.

2.2 THERMAL PROPERTIES OF PETROLEUM FRACTIONS

Some of the important thermal properties of petroleum are discussed in this chapter.

Specific Heat

Specific heat of petroleum fraction lies in the range of 0.3 to 0.85 and depends upon temperature and gravity. Lighter fractions have higher values. With increasing density the specific heat decreases. In absence of relevant data, the following correlations can be used. Bureau of standards formula[8]

$$\text{Sp. heat} = \frac{1}{\rho}\ (0.4024 + 0.00081\ t) \qquad \dots (2.10)$$

Where ρ = sp. gravity at 15.6/15.6°C and t = temp. °C.
With API gravity, temperature and characterisation factor (K) (for oils only) the specific heat correlation is given below:

$$\text{Sp. heat}[9] = [0.355 + 1280\ (°\text{API})\ 10^{-6} + (503 + 1.17 \times °\text{API})$$
$$(32 + 1.8t)\ 10^{-6}]\ (0.05K + 0.41) \qquad \dots (2.11)$$

Where K = characterisation factor; t = °C
Further, for pure components, the specific heats are listed over a temperature range of 200 to 6000°K by Duran[10] *et al.*

Watson and Fallon gave the following relation for petroleum vapours[11] :

$$\text{Sp. heat} = (0.045\ K - 0.233) + (0.44 + 0.01777\ K)\ (18t + 32)\ 10^{-3}$$
$$-0.153 \times (1.8t + 32)^2 \times 10^{-6} \qquad \dots (2.12)$$

where K : characterisation factor
       t : °C
In processing calculations, the use of specific heat is not immense; as heat content of the fractions is more versatile and available.[12]

2.2.1 Heat of Combustion

Hydrogen is distinct for its heat of combustion *i.e.* 183.6×10^3 kJ/kg. Obviously higher hydrogen content fractions will be designed to have more heat of combustion. Thus in the order of paraffins to aromatics, the heat of combustion decreases.

Sherman-Kropff relationship showing heat content with °API is worth mentioning.[13]

$$\text{Heating value kJ/kg} = 43,434 + 93.2 \,(\text{API-10}) \qquad \text{... (2.13)}$$

Gases can be classified by their calorific values. Replacement of one by another is possible as shown by Wobbe number.[14]

$$\text{Wobbe No:} \quad \frac{\text{Heat of combustion}}{(\text{sp. gravity})^{0.5}}$$

For different manufactured gases like coal gas, SNG etc. Wobbe No. lies in the range of 21.5 to 38.7

For Natural gas :

Low calorific value	37.1-43.7
High calorific value	43.4-52.4
LPG	72.0-83.5

where

Calorific value is given in K cals/liter and Sp. gravity (Air = 1) in Kg/Cu.M

Gas modulus is another term conveniently used to replace one gas to another (by increasing the pressure at which it burns).

$$\therefore \text{Gas modulus} = \frac{p^{1/2}}{\text{Wobbe No.}}$$

An example may be cited here. A manufactured gas of 25 W. No burning at 1 m. bar can be replaced in a burner with natural gas of W. No. 50.

$$\underset{\text{Mfg}}{\frac{1}{25}} = \underset{\text{NG}}{\frac{P^{1/2}}{50}}$$

The natural gas is thus permitted to burn at a pressure 4 times that of manufactured gas.

2.2.2 Latent Heat of Vaporisation

Latent heat for a fraction varies with the temperature (or pressure) at which vaporisation takes place. This is commonly related to molecular weight, °API and molal average boiling point. Any two of the above parameters are sufficient to give latent heat of vaporization. Latent heat can also be converted from normal boiling point to any other temperature upto critical condition.

The relationship is expressed as:

$$L = \lambda L_B \frac{T}{T_B} \qquad \qquad \text{... (2.14)}$$

L = latent heat at any temperature (T_{absolute})
L_B = latent heat at temperature ($T_{B \text{ absolute}}$)
γ = correction factor depends upon reduced
temperature.

At the critical conditions, the latent heat vanishes. As the critical temperature of high API materials are lower than low API materials, the latent heat curves for these materials cross one another.

Heat of vaporisation for any compound can be estimated at any temperature as shown by the following formula

$$\Delta H_v = = \Delta H_{T_2} \left(\frac{T_1 - T_2}{T_c - T_2} \right)^n$$

T_c is critical temperature and n varies from 0.38 to 0.466. Most of the organic compounds except alcohols and oxides have a value of 0.38

Problem

Propane has a heat of vaporisation of 101.8 cal/g at -42.1°C. Its critical constant is 96.7°C. Find the heat of vaporisation at 80°C.

Solution

$$\Delta H_v = 101.8 \left[\frac{(273 + 96.7) - (273 + 80)}{(273 + 96.7) - (273 + 42.1)} \right]^{0.38}$$

$$= 45.6 \text{ cal/gm}$$

2.2.2.1 Latent Heat of Fusion

The heat of fusion is approximately fifty per cent of latent heat of vaporisation. For waxy distillates and waxes, the average latent heat of fusion may be taken as 167 to 170 kJ. per kg.

2.2.3 Thermal Expansion

Like all liquids, petroleum fractions also suffer in loss of density due to thermal expansion. Coefficient of expansion is very much required to find out the volume of the container, which is exposed to frequent changes of temperature and this is related to API gravity as given below[15]:

Gravity range API	Mean coefficient of expansion (change in volume for 1 °C)
0-14.9	0.00063
15-34.9	0.00072
35-50.9	0.00090
51-63.9	0.00108
64-78.9	0.00126
79-88.9	0.00144
89-93.9	0.00153
94-100	0.00162

2.2.4 Spontaneous Ignition Temperatures

Definition: The temperature at which a material can catch fire and burn continuously without the aid of external firing agencies. Paraffins have less ignition temperatures and this precious property is regarded as most useful for the performance of diesel fuels.

Aromatics have high ignition temperatures. In gasoline engines adiabatic compression of the mixture does not raise the temperature to that extent, hence sparking is required for detonation. Gasoline engines this way differ from Diesel engines. Ignition temperature remarkably decreases with increasing molecular weight of the series as well as with the oxygen content of molecule.

Self ignition Temperatures of some Hydrocarbons[16]

Paraffins	Temp. °C	In presence of Oxygen	Napthenes	Temp. °C	In presence of Oxygen
Pentane	418.3	292	Methylcyclo -hexane	393.3	
Heptane	258.5	284	Methylcyclo -pentane	468.9	
Decane	252.8	–	Alcohol	551	329
Cetane	235.0	–			
2,3 Dimethyl hexane	437.8				

	Aromatics	Temp.°C	In presence of Oxygen
	Benzene	651.7	639
	Toluene	629.4	

m-Xylene	628.9
p-Xylene	690.6
α-Methyl naphthelene	565.6
Decahydronaph-thalene	271.1

Spontaneous ignition temperatures, as determined by Moore,[17] are slightly different from the self ignition temperatures mentioned above. Self ignition temperatures fall with slow oxidation while they increase with the addition of TEL and oxidation inhibitors.

2.2.5 Viscosity and Viscosity Index (VI).

Viscosity is an important property of all lube oils. Lubrication assists in removing the frictional forces between two moving bodies. Viscosity is defined as the force in dynes required to maintain 1 sq. cm plane, with a unit velocity gradient from another similar plane separated by a distance of 1 cm. The relative viscosity is the ratio of the viscosity to that of water at standard temperature. In all refinery operations instead of absolute viscosity kinematic viscosity is used.

$$\text{Kinematic Viscosity} = \left(\frac{\text{Viscosity absolute}}{\text{Sp. gravity}} \right)$$

All the laboratory viscometers like U-tube, Fenske, Redwood, Engler give kinematic viscosity only. However, Saybolt Universal viscometre gives the time in seconds for 60 cc sample efflux, through a standard orifice at the given temperature and is expressed as seconds Saybolt universal (SSU). Engler results are expressed in Engler degrees which gives almost absolute viscosity. Although viscosity is a measurable property of all fractions, it is no way essential for lighter fractions like gasoline, kerosene, naphtha, except for calculating the power consumption in pumping or in atomising purposes.

Lubricants are specified by another derived property namely viscosity index, which indicates the variation characteristics of viscosity with temperature. Dean and Davis had given the following formula[18]:

$$\text{VI.} = \frac{L - U}{L - H} \times 100$$

where L = viscosity of a reference oil of zero viscosity in deg. at 37.8°C

H = viscosity of a reference oil of 100 viscosity index at 37.8°C

$$U = \text{viscosity of test sample at } 37.8°C$$

Viscosity index is explained in the graphical fashion as shown in Figure 2.9. Two standard liquids L and H are chosen. L is having a VI. of zero and H is having a VI. of 100. The starting viscosity for these samples is known by finding the viscosity of test sample U at 98.9°C. Thus all the three samples have the same viscosity (V) at 98.9°C, but exhibit different viscosities at 37.8°C as L, H and U. By substituting the values in the above equation, the VI. can be found out. From this relation, it is apparent, that VI. can be negative or positive or more than 100. L and H can be obtained from standard tables. Commonly paraffinic base oils have high VI. (100 for reference), conversely napthenic base oils are given the value 0 (for reference). The higher the VI. the better is the viscosity temperatures characteristics. Usually the 100 VI. oils are good to start at lower temperatures than zero viscosity index oils, of course with low pour points as shown in Table 2.2a and b.

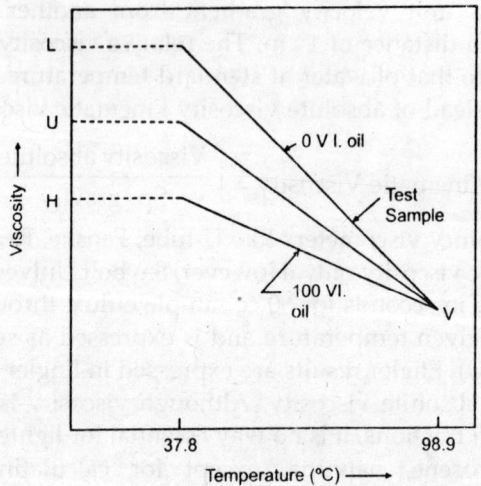

Figure 2.9   Viscosity Index-graphical Explanation.

TABLE 2.2a Lowest Temperature for Easy Starting °C.

	0 VI.	40 VI.	100 VI.
SAE-20	2.78	−16.6	−23.3
SAE-50	+10		−6.67

Even in case of diesel fuels, easy starting is related to viscosity and pour point as shown in Table 2.2b.

TABLE 2.2b   Diesel Engine-Starting Conditions.

Air Temp.°C	Max. viscosity (SUS) at 37.8°C	Max. pour point °C	Cetane number
–28.9	5.6	–34.4	90
–23.3	7.8	–28.9	83
–12.2	13.9	–17.8	69
– 1.1	21.1	–6.7	56
15.6	71.1	10.0	36

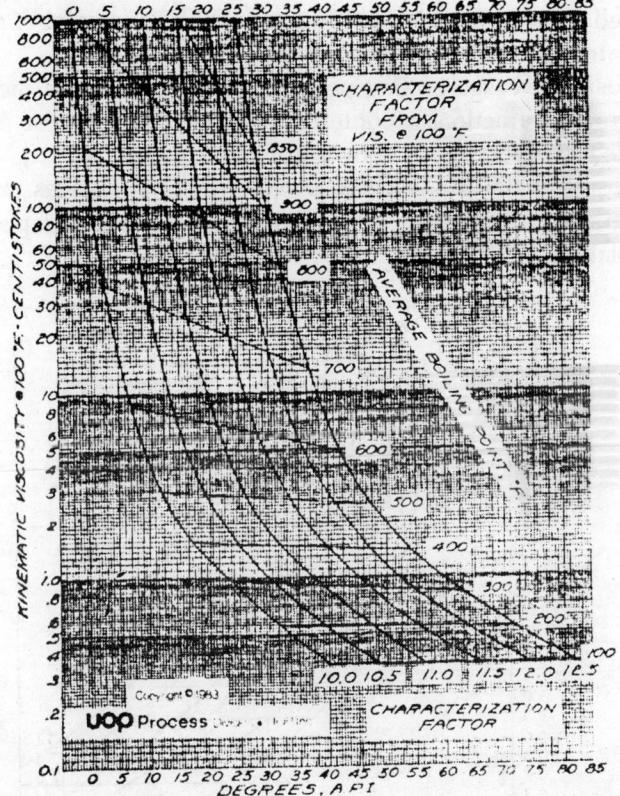

Figure 2.10a   Relation between Characterisation Factor, Viscosity at 37.8°C, Average Boiling Point and Gravity (API)
(Figure 2.10a and b   Taken from American Petroleum Refining by H.S. Bell (3rd Edit.)
With Permission from Brooks/Cole Publishing Co.
Owners of Van Nostrand Company.

126

viscosity variation is given by Walther's equation:

$$\log \log (r + a) = m \log T + b \qquad \ldots (2.15a)$$

$$\begin{aligned}
\text{where} \quad T &= °K \\
r &= \text{centistocks} \\
a &= 0.7 \text{ or } 0.8 \text{ if } r > 1.5 \\
m &= \text{slope} \\
b &= \text{constant}
\end{aligned}$$

This describes, the viscosity variation characteristics of an oil with temperature. On the basis of this equation ASTM charts were prepared. This is essential specially when the pour point of oil is very high (intrapolation to get viscosity at 37.8°C).

Viscosity, average boiling point and characterization factor of all the petroleum fractions are interrelated. Any two of these properties can give the other one, as shown in Figs. 2.10a and b.

Viscosity is not an additive property, hence it does not follow arithmetic rule, however it can be estimated by extrapolation or intrapolation, when represented in ASTM charts.

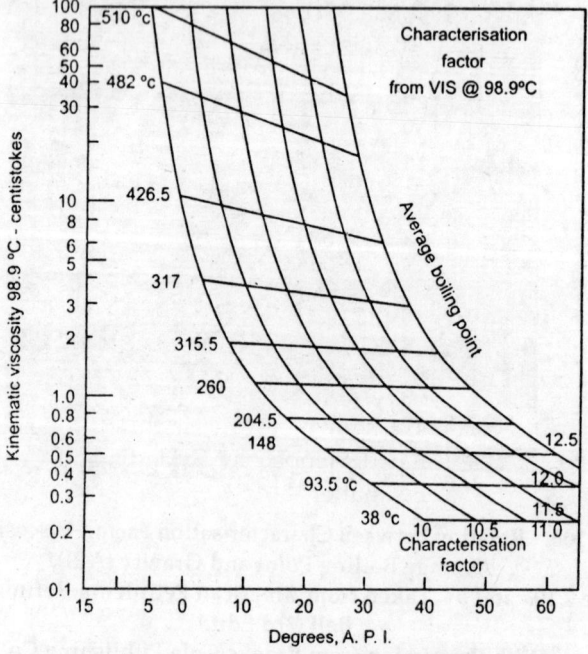

Figure 2.10b   Relation between Characterisation Factor,
Viscosity at 98.9°C, Average Boiling Point and Gravity (API)

Viscosity of hydrocarbon vapors is related to critical properties by the formula[22]

$$M_c = 7.7 \sqrt{M} \frac{P_c^{2/3}}{T_c^{1/6}}$$ ... (2.15b)

M_c = critical viscosity (micropoises)
M is molecular weight
T_c is critical temp. $°K$ and P_c is the critical pressure in atmospheres.

2.2.5.1 Viscosity-gravity Constant

This index of classification connects the viscosity and gravity of the sample by the formula

$$VGC = \frac{10G - 1.0752 \log (V - 38)}{10 - \log (V - 38)}$$ ... (2.16a)

where G is the sp. gr. at 15.6°C, V= SSU at 37.8°C
where the measurement of viscosity is not possible at 37.8°C an alternative formula is used.

$$VGC \text{ for heavy oils} = \frac{G - 0.24 - 0.022 \log (V - 355)}{0.755}$$ ... (2.16b)

where V is at 98.9°C
Naphthenic oils have VGC, 0.9 and above, while paraffinic oils have 0.8. This is due to specific gravity, which is more for naphthenic oils than paraffinic oils. A nomogram based on the above formula is presented[21] in Figure 2.11.

2.2.6 Thermal Conductivity

Much interest was not shown towards this property. Group contribution techniques as developed by Sakiadis and Coates[23] and Bondi[24] are helpful in this matter.

Problem 2.7
Raw pressible waxy distillate (A) and dearomatised pressible waxy distillate (B) of Assam Oil Company are having the following kinematic viscosities at 98.9°C and 65.6°C. (a) Find the viscosity indices of the two samples, (b) Find the composition of blend which can give a viscosity of 16 CS at 78.9°C

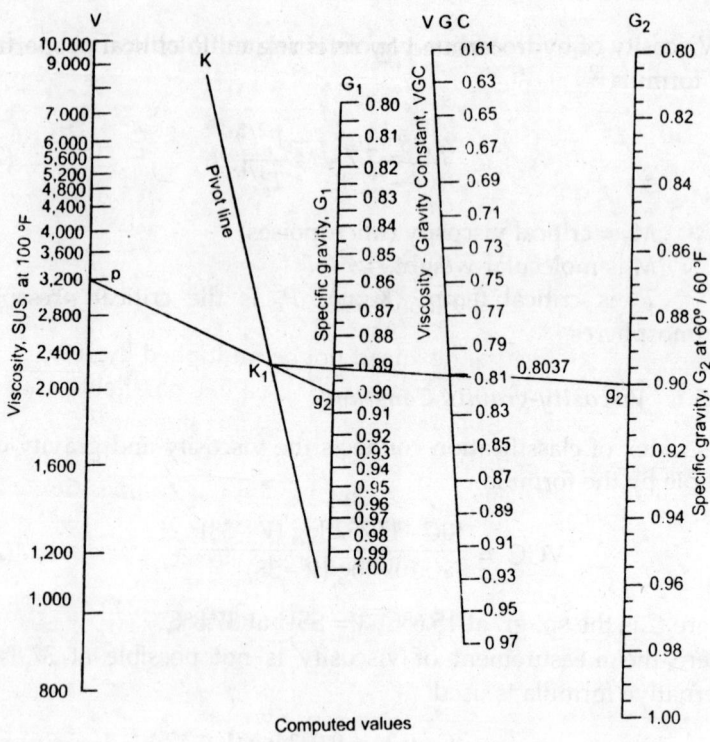

Figure 2.11 Viscosity Gravity Constant (Graphical Method).
Note : G, & G$_z$ are same sp. gravites of the same sample.

(a) Raw pressible waxy distillate K.V. at 65.6°C= 24.5 cS
at 98.9°C= 4.1 cS

(b) Dearomatised pressible waxy distillate K.V.
at 65.6°C= 53.5 cS
at 98.9°C= 7.3 cS

Solution
Refer Figure 2.12

(a) Plot the above readings on a semi-log graph (ASTM chart).

Find the viscosities of the samples at 37.8°C by extension. The viscosities are read as:

For Sample A : 106 cS at 37.8°C.
-do-   B : 276 cS at 37.8°C.

Viscosity index is calculated by referring to IP tables.

For sample

$$A \text{ VI.} = \frac{28.06 - 106}{28.06 - 21.40} = -11.70$$

$$B \text{ VI.} = \frac{91.666 - 276}{91.666 - 55.273} = -5.066$$

Negative viscosity index need not be multiplied by 100.

(b) For this part of the problem, draw a line parallel to x-axis from 16 Cs (K.V), so that the line intersects data lines at $P \& Q$.

Draw a line parallel to y-axis from 78.9°C, such that the line intersects the above at '0'. Now the blend of '0' composition will give the required viscosity at 78.9°C

% B in the blend to give the required viscosity

$$= \frac{PO}{OQ} \times 100 = \frac{1.5}{2.2} \times 100 = 68\%$$

at 78.9°C

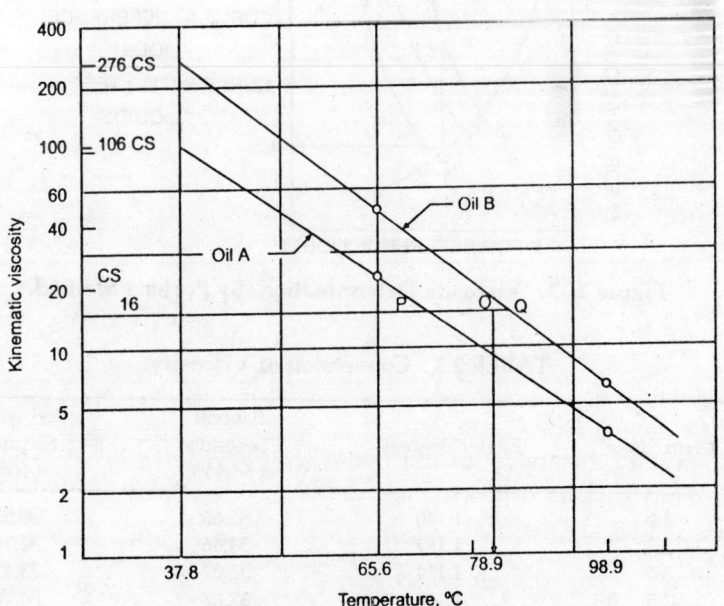

Figure 2.12   Viscosity of Blends.

130

Also viscosity at 37.8°C can be calculated by the correlation given by R.E. Hersh et al[19].

$$KV_{37.8} = 2.808 \left[KV_{99.6} - \frac{0.5}{KV_{99.6}^{1.33}} \right]^{1.5} \quad \text{for paraffinic oils}$$

$$KV_{37.8} = 2.45 \left[KV_{99.6} - \frac{0.213}{KV_{99.6}^{1.08}} \right]^{1.8} \quad \text{for naphthenic oils}$$

Viscosity can be determined by different instruments and different methods and all can be converted to SSU or centi Stokes (Table 2.3).

Porter developed a method to find the viscosity of a component at different temperatures by a plot similar to Duhring's. According to this plot the viscosities of a test liquid at different temperatures are plotted with reference to a temperature viscosity variation curve for a reference liquid. It follows that the viscosities of a test sample and reference liquid will be same at definite temperatures of these two. This is shown in Figure 2.13.

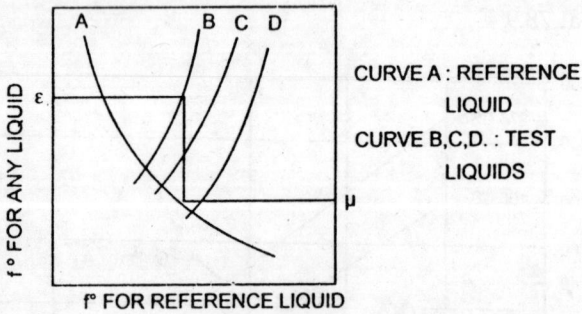

Figure 2.13   Viscosity Determination. By Porter's Method.

TABLE 2.3   Conversion of Viscosity.

Centi Stoke	Engler degrees	Saybolt Seconds at 54.44°C	Red wood Seconds at 60°C
2.0	1.140	32.66	30.95
2.5	1.182	34.46	32.20
3.0	1.124	36.07	33.45
3.5	1.266	37.67	34.70
4.0	1.308	39.17	35.95

4.5	1.350	40.78	37.20
5.0	1.400	42.48	38.45
5.5	1.441	43.98	39.80
6.0	1.481	45.59	41.05
6.5	1.521	47.19	42.40
7.0	1.563	48.79	43.70
7.5	1.605	50.44	45.00
8.0	1.653	52.10	46.35
8.5	1.700	53.80	47.75
9.0	1.746	55.51	49.10
9.5	1.791	57.21	50.55
10.0	1.837	58.91	52.00
12.0	2.020	66.03	58.10
14.0	2.219	73.54	64.55
16.0	2.434	81.25	71.40
18.0	2.644	89.37	78.45
20.0	2.870	97.69	85.75
25.0	3.455	119.10	104.70
30.0	4.070	141.20	124.40
35.0	4.695	163.5	144.2
40.0	5.355	186.0	164.3

To get the viscosity of a test liquid at $t°C$ one has to proceed horizontally till the corresponding curve is reached and then run down vertically to meet the reference curve; horizontal reading on viscosity scale at the point will give the viscosity of test liquid.

Viscosity of pure components can be estimated by Souder's[20] method.

$$\log (\log 10\mu) = \frac{1}{M} \rho - 2.9 \qquad \text{... (2.18)}$$

where i is a parameter given for different liquids

M = molecular weight

ρ = specific gravity at $t°C$

μ = viscosity in CPs. at $t°C$

˙ I in Souder's equation is calculated from the group contributions as given below:

Atom	H	O	C	N	Cl	Br	I
Contribution	+2.7	+29.7	+50.2	+37	+60	+79	+110
Double bond			−15.5				
5 member ring			−24				
6 member ring			−21				

Side groups of 6 member rings

For components of molecular weight less than 17 = −9

For components of molecular weight greater than 17 = −17

ortho, para position   +3
meta position   −1

OH	+57.1	COO	=	+90
NO₂	+80			

Problem 2.8

Estimate the viscosity of phenol at 80°C (ρ =0.9692)

Solution

For 5 H atoms....	5 × 2.7	13.5
6 Carbons atoms	6 × 50.2	301.2
OH		+57.1
Ortho position		+3
(Mol. wt. = 94)		374.8
For one six ring		−21
3 double bonds (−15.5 × 3)		−46.5
one side group mol. wt. greater than 17		= −17.0
	Total	= −84.5

$$l = 374.8 - 84.5 = 290.3$$

$$\log[\log 10\,\mu] = \frac{290.3}{94} \times 0.9692 - 2.9$$

$$= 2.991 - 2.9 = 0.091$$

$$\mu = 1.7 \text{ cps. at } 80°C$$

Thomas method is based upon the Reduced Temperature and is given by the relation[20]

$$\log \frac{8.569\mu}{\rho^{0.5}} = \theta\left[\frac{1}{T_r} - 1\right] \qquad \text{... (2.19)}$$

where T_r is Reduced Temperature

θ = smoothing group coefficient

Smoothing group (atoms)	Contribution
C	−0.426
H	+0.249
O	+0.054
Cl	+0.340
Br	+0.326
I	+0.335
Double bonds	+0.478

C_6H_5	+0.385
S	+0.043
CO	+0.105
CN	+0.381

Problem 2.9

Estimate the viscosity of benzene, given
Specific gravity (ρ) = 0.875 at 20°C, T_c (critical temp.)
= 562°1 K

Solution

For 6 carbon atoms $\quad 6 \times -0.426 = -2.556$
For 3 double bonds $\quad 3 \times 0.478 \ = \ 1.434$

For 6 H atoms $\qquad 6 \times 0.249 \ = \dfrac{1.494}{2.928}$

$= 2.928 - 2.556 = 0.372$
$= 237 \times 0.917 = 0.8715$

$$\log \frac{8.569}{(0.875)^{0.5}} = 0.372 \left[\frac{562.1}{293} - 1 \right]$$

μ = 0.87 cps. at 20°C

2.2.7 Thermo viscosity

This is given by following relations:
Thermo viscosity = 15 + 148 × kinematic viscosity or
Thermo viscosity = 46 SSU − 1,183 where SSU is determined at 15.6°C.

Thermo viscosity is not of much importance, however for fractions having less viscosity, the expression looks to be convincing. Industrially, thermo viscosity is important for illuminating oils. Frequently thermo viscosity is expressed in terms of ring number given as

$$\text{Ring No.} : \frac{\text{Thermo viscosity}}{5} - 10(46 - \text{API}) \qquad \text{... (2.20)}$$

Usual ring number for most of the illuminating oils is in the range of 60 to 130.

2.3 IMPORTANT PRODUCTS—*PROPERTIES AND TEST METHODS*

Distillation of crude yields number of cuts of varying boiling ranges. These fractions or cuts are processed suitable to utility when they are known as products. In the order of increasing boiling points, the main fractions from distillation of crude, are

Gas/light-naptha/gasoline	Solvents/Fuels/motor spirits, etc.
Kerosenes (light distillates)	Jet fuels/illuminating oils/heating oils
Gas oils or middle distillates	Diesel fuels, cat-cracking oils, etc.
Heavy distillates	Lubes/Waxes/fuel oils, etc.
Residuums	Asphalt, Bitumen etc.

2.3.1 Gas

Gas from petroleum is classified under several names like natural gas, associated gas, dissolved gas, casing head gas and off gas from refinery.

2.3.1.1 Natural Gas

Its name indicates that it is readily available in nature, almost as a finished product. It contains mainly varying proportions of methane. It may be accompanied by other dry gas fractions like ethane and propane to a small extent. In addition to these combustibles some inerts like CO_2, N_2, noble gases are also present. The proportion of methane ranges from 85% and goes upto 98%.

2.3.1.2 Associated Gas

This is obtained from oil reservoirs and this exists as a separate gas cap over liquid phase. Though the gas mainly consists of methane and to some extent ethane and propane, the proportions vary depending upon the reservoir conditions. When the gas phase is taken out, it may still contain some liquid hydrocarbons mainly of volatile range like butane and pentane, which when condensed are termed as natural gas liquids (NGL).

2.3.1.3 Dissolved Gas

Gas may be present in the liquid hydrocarbons mainly in the dissolved state depending upon the formation pressure. When the pressure is decreased, this dissolved gas comes out of the oil. Gas production upto 10% crude produced is not uncommon with the oil reservoirs.

It is fair to strip off such dissolved gas before crude is transported to long distances by means of pipelines or tankers. The remaining dissolved gas is first to come out of the distillation column because of higher temperature than the surroundings.

2.3.1.4 Casing Head Gas

Gas that has escaped through oil well Christmas trees is termed as casing head gas. It is also more or less similar to natural gas but contains less % of methane and high percentage of ethane and propane than natural gas. It is a by-product of oil production.

2.3.1.5 Refinery off Gas

In a refinery, gas is formed in cracking and reforming operations due to the thermal degradation of liquid hydrocarbons. During stabilisation of wild gasolines or processed gasolines, the gases are vented. Thus the gas is mainly a mixture of saturates and unsaturates and quantity is also not assessible.This forms a major source of heat energy for refinery, as well as feed stock for petrochemicals. In fact, without any exclusion, all these gases can be utilised for petrochemical industries.

All the gases contain impurities like CO_2, N_2, mercaptans, H_2S, water vapour, suspended impurities etc. First three paraffins are gases at room temperature. Mixture of methane and ethane is called dry gas, while propane-butane mixture is called wet gas. Where petrochemical industries are not instituted, dry gas would find its use mainly in the refinery fuel system. Wet gas is usually liquefied and sold for commercial purposes. Butane is diverted to gasoline streams as a blending component, as butane has more commercial value when blended with gasoline.

2.3.1.6 Liquefied Petroleum Gas (LPG)

The gas that is vented from refinery distillation units, is processed and conveniently stored after liquefaction. For domestic heating purposes, it is supplied in small cylinders (15 kg or 12 kg), while for industries tanker supplies are called in. This gas is known as liquefied petroleum gas as it is stored in vapor liquid mixture. Ease of handling, smokelessness, good and steady heating rates are some key points that made this fuel a popular kitchen aid to the housewife of moderr times. Rising demands of this fuel in domestic and industrial circle: are met by the refinery by installing processing units.

Different types of mixtures resulting from propane-butane blends are sold as LPG. In fact propane itself can be utilised for this purpose and hence the dearth of availability of butane does not deter the manufactures.

Most important property of this fuel mixture is vapour pressure. These fuels are graded according to vapour pressure and temperature at which 90% or 95% is evaporated. LPG is produced in many grades as listed below:

1. Predominantly butanes
2. Butane—propane mixtures, mainly butanes
3. Butane—propane mixture (equal volumes)
4. Less butane—propane mixture
5. Butane—propane mixture (more propane)
6. Predominantly propane

2.3.1.7 Important Tests

Vapour pressure of the sample is foremost important. This is found out by Reid's—Vapor pressure apparatus (See Figure 2.14). The test is described in gasoline testing. Characteristic vapor pressure[25] of these mixtures at 38°C varies from 5 to 13 kg/cm^2. 95% evaporation point at 760 mm Hg shall be kept less than 0°C. In tropical regions, a fair amount of butane may be mixed so that 95% evaporation point shall be around 0°C.

Natural Gas Processing Supplies Association (NGPSA), standards elucidates tests, properties and methods for handling such mixtures. While storing these gases, hydrocarbon dew point is reduced to such a level that condensation by releasing the pressure cannot be experienced. In addition to satisfactory vapor pressure and evaporation point, the gas should be free from corrosive compounds. Hydrogen sulfide, Carbonyl sulfide are such contaminants which are kept below 5 to 6 grains per 10 cubic meters; further it should also be free from moisture. Usual gross calorific value is in the order of 38 to 82 M^2J/M^3. For facilitating leak detection, the gas is mixed with small amounts of odourous mercaptans.

A typical LPG composition of an Indian Refinery is given below:

LPG (Butane-propane mixtures)

Vapour pressure at 65°C = 10 to 26 kg/cm2
Vapour pressure at 38°C ± 0.1 = 6.7 kg/cm^2
95% evaporation temperature = –2°C

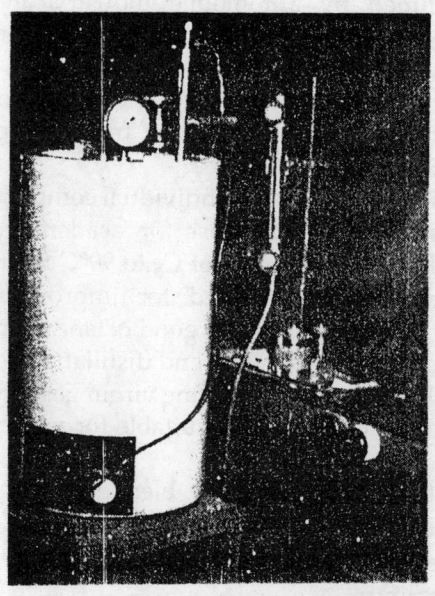

Figure 2.14 : LPG Testing (Reidvapor Pressure).

Sulfur % by weight	— 0.02
H$_2$S	— Nil
Moisture	— Nil
Composition	
Ethane	— traces
Propane	— 24.7%
i-butane	— 36.7%
Butane	— 38.5%
Pentane	— Nil

LPG being a manufactured gas its composition remains same. In the case of natural gas the composition varies from mine to mine, even in the same mine from time to time.

2.3.2 Gasoline

Naphtha or Gasoline (petrol as is known in India) is the next fraction to follow gas. This is a volatile fraction and is known as motor spirit. The boiling point ranges from 37°C to 180°C. Gasoline is a finished product, while raw fraction is termed as naphtha or light boiling fraction. There are different types of gasolines produced by the refineries (approx. 40 types) and almost 90% of the product is exclusively consumed by automobile industry and the rest by aviation industry. Most refineries produce two primary grades, regular and premium, both either leaded or unleaded. The product is so important that the integration of refineries has become a reality to produce as much as 60-70% of gasoline from crude, especially in countries like U.S.A.

Gasolines are never analysed for individual components, yet about, 1500 compounds were isolated just for academic interest. Light straight run gasoline (LSR) consists of C_5 to 90°C fraction. This is the fraction which cannot be reformed for improvement in octane number. This cut as such contributes good octane number (with high susceptibility to lead) and bears light end distillation characteristics of gasoline. LSR is also known by the name virgin gasoline and does not have desired qualities or properties suitable for usage in automobile industry.

Hence straight run gasolines are blended with the processed gasolines obtained by various methods, these are:

1. Reformate gasolines,
2. Coker gasolines,
3. Alkylated gasolines,
4. Polymer gasolines,
5. Cracked gasolines and
6. Hydrocracked gasolines.

2.3.2.1 Tests for Gasoline

The important tests prescribed for gasolines are:

1. ASTM distillation,
2. Reid vapor pressure,
3. Octane number,
4. Gum content,
5. Sulfur content, etc.

ASTM DISTILLATION (15: 1448–P : 18, ASTMD-86)

In this test 100 ml of sample is distilled in a standard flask at a uniform rate of 5 cc per minute. The distillate is condensed in a brass tube condenser, surrounded by a water bath kept at 0°C by ice-water mixture. First drop from the condenser must be available in 5 to 10 minutes after heating started, at which the recorded temperature is mentioned as initial boiling point (IBP) of the sample. The vapor temperature is recorded at each successive 10 cc distillate collected in a measuring cylinder. The test continues in the same way till 95% of fraction is condensed. At this juncture, the heat intensity may be increased to obtain the maximum boiling point also known as end point (EP) (See Figure 2.15).

Figure 2.15   ASTM Distillation.

Fluctuation in temperatures is common when last two or three ccs of sample are distilled. When the bottom of the flask shows dryness, the temperature recorded corresponds to final boiling point. The distillate collected shall not be less than 98% and the difference is accounted as loss; usually ascribed to light ends. This method of distillation is followed for most of the product specifications. When middle distillates like diesel are to be tested, instead of 100 ml. sample, a 200 ml. sample is taken. Similar adjustment of heating rates are followed to obtain IBP in 10 to 15 minutes time. This test is conducted with standard equipment only. The boiling points in this method need not be adjusted to 760 mm pressure; although it is preferred at times of dispute. ASTM distillation specifies the evaporation characteristics of gasoline. Accordingly gasolines fall mainly into 3 types: Type A, Type B and Type C. In all these types 10%, 50% and 90% boiling point along with IBP and FBP are mentioned.

Ease of starting is governed by IBP to 10% range boiling point. Lower IBPs are preferred for cold climates. A drawback of very low IBP is vapour locking of the engine due to high evaporation. This has been to some extent been alleviated by fixing vapour pressure, which is furnished by entire boiling stock. Further, evaporation losses can also be related to vapour pressure. When gasoline tanks are filled up, the vapour that occupied the tank has to be displaced and losses due to such kind of phenomenon are called' breathing losses' which are inevitable in all filling operations. Breathing losses are in addition to evaporation losses due to ambient temperature. (Table 2.4) with all these considerations, to ease starting, winter gasolines and summer gasolines are produced to satisfy the seasonal specifications. Rate of acceleration follows after the starting operation which is best judged by the mid region boiling range (say 30-60%). Rapid warmup is not very essential. It is generally influenced by prevailing weather conditions. Warmup period is defined as the time required to develop fully the engine power. Usual running of engine to 6-8 Km, should be a well considered warm up period and this is closely governed by % distilled at 70°C, with reference[26] to minimum ambient temperature as shown below:

% Distilled at 70°C	3	11	19	28	38	53
Minimum ambient temp, in °C	26.7	15.6	4.4	6.7	–17.8	–29

This explains that 11% should be distilled at 70°C, if the ambient temperature is 15.6°C, for maintaining the required warm up. Evaporation losses and vapour locking are guided by RVP and ambient temperatures, as shown below:

Ambient Temp.°C	15.6	21.1	26.7	32..2	37.8	43.3	48.9
Max. RVP (lbs/in^2)	12.7	11.0	9.4	8.0	7.0	6.0	5.3
(kgs/cm^2)		0.879	0.782	0.66	0.56	0.496	0.423	0.38

ASTM distillation curve closely reveals the following:

If 10% boiling point is low; then

Flash point,	—	low
Ease of starting,	—	excellent
Vapour losses,	—	very high
Vapour locking,	—	high

30-60% boiling point range gives information on

(a) Warm up period (b) Good evaporation

TABLE 2.4 : Relative Equilibrium Volatility Losses of Gasoline of RVP : (0.6 kg/cm²)

Sl. No.	Relative losses at		Altitude (Meters)	Atmospheric pressure (a) N/sq. m²	Temp.°C	Losses at altitude and ambient temp.
	38°C	50°C (Sea level) (b)				
1	2.55	2.65	1,800	81,000	3	2.22
2	3.63	3.77	4,600	57,000	-14.7	3.02
3	6.27	6.52	8,500	33,000	-40	4.64
4	13.79	14.32	14,000	15,000	-50	10.12

(a) Ernest E. Ludwig, Applied Process Design for Chemical and Petrochemical Plants (GPC).
(b) Taken loss at sea level and 30°C as 1%.

If the range of boiling point is high gasoline will not be evaporating normally. If 90% boiling is high, then crank case dilution also increases as the fuel is not volatile at the working conditions of carburettor (Figure 2.16).

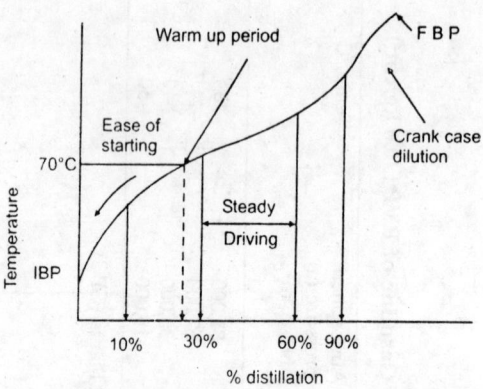

Fig. 2.16 : ASTM Distillation Characteristics.

2.3.2.2   Reid Vapour Pressure (IS: 1448-39)

Vapour pressure of all volatile fractions is measured by the above apparatus at ($38° \pm 0.1$ °C). The apparatus consists of two chambers. The lower chamber is in the form of cylindrical bomb for holding the test sample. Above this, there is air chamber which is a hollow cylindrical space, designed to possess four times the bomb volume. The top portion of the chamber is fitted with a bourdon gauge for pressure indication. For the test, the bomb is first filled with the sample up to the brim and immediately the valve is closed and connected to the air chamber. The apparatus is now immersed in a water bath kept at $38° \pm 0.1$°C. The maximum pressure indicated by the gauge is counted as the Reid vapour pressure of the sample. LPG, gasoline, naphthas, jet fuels are tested like this. It is found that true vapour pressure in most of the cases is higher than the indicated RVP, and the variation for different fractions is different. Maximum difference is indicated in case of crude oils. Relative expression 'True vapour pressure to RVP' is also sometimes indicated and for gasolines it is in the range of 1.03 to 1.45, while for crudes it goes upto 10.

During testing, the following correction is applied for accommodating the changes in water vapour pressure corresponding to temperature variation from air to bath temperature.

TABLE 2.5   U.S. Volatility Trends of Gasoline.

| | Regular | | | | Premium | | | |
	Winter 1956-1966		Summer 1956-1966		Winter 1956-1966		Summer 1956-1966	
Reid vapour pressure	0.773	0.822	0.605	0.633	0.773	0.836	0.619	0.639
10% Evaporation Temp°C	47.3	43.3	53.9	51.7	46.8	43.3	53.3	51.7
50% Evaporation Temp°C	100	93.3	104	97.9	98.4	98.9	103.6	102

Correction : $\dfrac{(P_a - P_t)(t - T)}{T^\circ K} - (P_T - P_t)$

where, P_a—atmos. pressure, P_t—vapour pressure of water at $t^\circ K$, t—air temp. $^\circ K$, T—Temp, at which test is done $^\circ K$, P_T—vapour pressure of water at $T^\circ K$.

Volatility tests are intended for smooth performance of engine under all weather. The importance can be judged by U.S. Volatility.

2.3.2.3   Octane Number

Octane number is defined as percentage volume of i-octane (2, 2, 4-trimethyl pentane) in a mixture of i-octane and n-heptane that gives the same knocking characteristics as the fuel under consideration. Knocking is due to untimely burning of fuel in a spark ignition engine, which results in loss of power and sometimes it is powerful enough to cause damage to engine parts. With the advent of petrol engines of high compression ratios the tendency of knocking has also increased. Knocking is mainly due to two reasons, one assigned to engine and the other to fuel. The fuel part is only dealt here. Being a blend, gasoline responds in different ways, even in the same engine, depending upon the components present. Iso-paraffins and aromatics have high octane numbers while n-paraffins have very low value (heptane 0), unsaturates do have high octane values but not preferred due to gum contribution. Naphthenes usually lie in between i-paraffins and aromatics. Even among n-paraffins also the octane number of the lower carbon series is very high (methane has 104) and that is the reason why natural gasolines or LSR (C_5 -100°C) have high octane ratings. Octane number is influenced by different factors like speed of the engine, ambient weather conditions, altitude, combustion chamber deposits and coolant temperature.

Knock rating is tested by CFR Engine for different purposes and the conditions of test are furnished below:

Method	Research	Motor	Aviation	Aviation supercharge
Designation CFR	F_1	F_2	F_3	F_4
Engine rpm	600 ± 6	900 ± 9	1200 ± 12	1800 ± 45
Air intake°C	75 ± 1	38 ± 1.4	75 ± 2	75 ± 2
Coolant temp.°C	100 ± 1.5	100 ± 1.5	140 ± 5	140 ± 2

Motor method (P. 26, ASTM D357-58) gives the octane rating of high way driving (high speeds), while research method (P. 27, D 908-58) gives for city driving (low speed) conditions F_3 & F_4 methods are for aviation gasolines.

Some additives like tetraethyl or methyl lead are extensively blended into gasolines to boost the octane number.

Octane number estimation for leaded gasolines was furnished by McCoy[27] as

Octane number = A (% aromatics in fuel) + B

where A & B are constants.

The constants A, B satisfy the following equation

$$100 A + B = 120.3.$$

When   A = 0.526   —   B is 67.5
                       0.464   72.3
                       0.44   75.0

After analysing vast data, the above relation was formulate·l. The octane number variation with aromatics resulted in a straight line and A and B are the Slope and intercept of such a line. Ambient conditions— Octanes[28] number.

Any engine during service has to encounter different weather conditions. For smooth running it is better to adjust the octane rating of the fuel under such ciːcumstances.

Altitude: With increasing altitude a fuel of less octane value is satisfactory. For every 100 meters increase in altitude drop of one unit RON in fuel quality is allowed.

Humidity

The performance of the engine is again responsive to humidity conditions. Engine working on high humidity conditions requires less octane fuel. For an increase of 10% humidity at 25°C a drop of 0.5 RON in fuel quality is permitted.

Engine Speed

High speed engines required low octane value fuel. For an increase of 300 RPM in an engine speed, a decrease of one unit RON of fuel quality is permissible.

Air Temperature

Cold countries require less RON fuel than hot countries. With a rise of 12°C in air temperature, an increase of one RON in fuel is required to compensate the effect of air intake temperature.

Coolant Temperature

For every 6°C increase in coolant temperature, the engine demands an increase of one RON in quality of fuel. By this one can judge the importance of coolant.

Spark Advance

This is purely a mechanical adjustment requiring 1.5 RON increase for 1° advancement of sparking.

Combustion Chamber Deposits

Running of an engine over sufficiently long periods, invariably results in deposits. It is essential to increase the octane number of fuel by one unit for every 5000-10000 km of run.

Alkylated gasolines and reformed (catalytic) gasolines have high octane numbers. While polymer gasolines contribute to medium extent (60-80) RON, gasolines produced by cracking and hydrocracking techniques yield about 60 RON. The light ends of any gasoline produced by thermal decomposition of higher stocks give good value RON, still they are not preferred because of the large quantities of unsaturates. Now Anti Knock Index has been used widely, it is defined as the average of RON and MON. AKI for premium is 88 and Regular MS is 84.

Ethers increase the antiknock characteristics of gasoline 10% volume of MTBE can contribute an increase of 2.4 units of RON and 1.6 MON for reformate while an increase of 1.3 in RON and 0.6 in MON are reported for cracked gasoline. Similarly for 10% MTBE RON increase of 4.3 and MON increase of 3.4 for reformate and 2.3 in RON and 1.4 for MON for cracked gasoline are observed. Even upto 15%

MTBE can be added to gasoline as its volatility properties are similar to $C_5 - C_6$ isomerates B.P. $> 55.2°C$ and $\rho = 0.745$. Practically there is no difference in consumption of fuel. The production of MTBE depends upon the availability of Butene from FCC units. Hence a blend of TAME and MTBE would be advisable. 3-methyl butene can also take part in the reaction to produce TAME.

2.3.2.4 Oxidation-stability (Gum Content)

Motor gasolines are often stored for six months or even longer before use. So it is essential that they should not undergo any deleterious change under storage conditions. Further gasolines should remain free from deposits during discharge from the fuel tank of a vehicle to the cylinder of its engine, otherwise these harmful deposits are built up in the tank, fuel lines and in valves and cause extensive choking. Gasoline manufactured by cracking processes contain unsaturated components which may oxidise during storage and form undesirable oxidation products. An unstable gasoline undergoes oxidation and polymerisation under favourable ambient conditions, forming gum, a resinous material. This in early stages of formation may remain in solution and gasoline but due to further chemical changes may be precipitated.

Gum formation appears to be the result of a chain reaction initiated by radicals such as peroxides and catalysed by the presence of metals, particularly copper, which might have been picked up during refining and handling operations.

This deterioration of gasoline can be eliminated for a substantial period of time by addition of small quantities of antioxidants known as inhibitors. Inhibitors retard the oxidation of the olefin constituents of the fuel. Some such inhibitors are normally phenols or amine compounds, and small quantities between 0.001 and 0.02 percent by weight are usually sufficient to ensure good stability.

Gum is of two kinds, one is known as existent-gum (preformed) while the other is called potential gum (ultimate). These two gums are estimated by two different methods. Existent-gum, as already available in solution, is to be separated by evaporation of the volatile gasoline. This is estimated by ASTM D381; IP 131 test. In this test, a sample of 100 ml. is evaporated with the aid of a heated jet of air, while the sample is maintained at 160-166°C After the evaporation of the sample the residue is weighed. Usually gum content should not exceed 6-10 milligrams per 100 ml. of gasoline.

Potential or ultimate gum content of the gasoline is evolved by oxidising the sample with oxygen under restricted conditions, thus assessing possible gum formation during storage.

Induction period for gum formation is estimated by Bomb method (ASTM D525; IP 40: P. 28). This employs an atmosphere of oxygen at initial pressure of (7 kg/cm^2) over the test sample of 100 ml. in a specially designed bomb. The bomb is fitted with a pressure gauge for indication of pressure, another connection is provided for feeding oxygen. The bomb and its contents are kept in a water bath and heated continuously at 100°C till the test is over. Due to slow oxidation gum formation begins and the pressure of oxygen in the bomb slowly falls. The time required for fall in pressure by 0.15 kg/cm^2 within fifteen minutes is taken as break point. The difference in time from placement of the bomb to break point is the required induction period. More induction period means more the stability of gasoline, hence less chance of gum formation. A minimum figure of 240 minutes usually ensures satisfactory limit for storage. Induction period in hours is equivalent to months that the gasoline can be stored under commercial terms. Further, extent of inhibitor (additives) to be added is guided mainly by induction period.

2.3.2.5 Sulfur

In addition to corrosiveness and pollution, sulfur compounds are extremely harmful to the susceptibility of gasolines to tetra ethyl lead (TEL).[29a] More TEL is required if sulfur components are present in gasolines. The effect of overcoming sulfur compounds is done by adding more[30, 31] lead. The effect of sulfur on lead can be predicted by the relation:

$$L = \frac{a_0 - a}{a_0} \times 100$$

where, a_0 = TEL present in gasoline with sulfur compounds,

   a = TEL present in gasoline, free of sulfur compounds,

   L = % TEL excess required to boost the octane number of sour gasoline to the same extent as sweet gasoline.

Sulfur is also termed as noncorrosive sulfur, provided the sulfur exists in alkyl disulfide form only. Most of the sulfur in crude is concentrated in heavier fractions, lighter fractions require relatively less cumbersome treatment techniques. Total sulfur is estimated by bomb method (1 P. 6/60, P. 33) or lamp method (P. 34) and mercaptan sulfur as silvermercaptide (P. 36 : IP 104/53).

2.3.3 Additives for Gasoline

While modern gasoline engines have been designed for efficient use of gasoline, still, gasoline engine operation is not entirely trouble free. Complete combustion of gasoline in engine is rare. Motor oils used for lubrication cause deposits in the combustion chamber. These deposits (presumably of high carbon ring compounds) enter the crank case of the engine to form lubricating oil sludges along with unburnt gasoline. Other source of trouble is effluent through positive crank case ventilation which causes deposits in the delivery end of the carburettor. Thus a clean combustion is remote; hence additives are preferred to reduce the deposits as well as pollution belches. The effect of sulfur in different forms on lead requirement is presented in Figures. 2.17 *a, b, c, d.*

Different types of additives are blended into gasoline, in fact in all petroleum fractions, to give uninterrupted and smooth service. Additives for different functions are classified as follows:

1. Anti-icings and detergents
2. Inhibitors for (a) corrosion, (b) oxidation
3. Combustion aids
4. Anti-knocks
5. Colours and dyes.

2.3.3.1 *Anti-icings and Detergents*

During cold and wet weather, spark ignition engine has a tendency to stall till the combustion chamber is fairly warmed up.

Formation of ice crystals begins when the ambient temperature is −5 to 10°C, at a relative humidity of 65%. This phenomenon is observed even when gasoline is free of moisture, so, obviously moisture is contributed by air. During vaporisation of fuel in carburettor, there is evidently a good temperature drop. These two facts result in uninvited formation of ice crystals. For prevention of these formations, additives like glycols, mono ethers methanol, propanol are added.

It is true, where moving parts are present a timely lubrication is demanded. Carburettor detergents take care of smooth function of carburettor. For free piston movement; rapid erosion of deposits in combustion chamber, lubricants and some scavengers are essential. The deposits are mainly due to decomposition of lube oils and carbon formation by frictional heat and combustion heat.

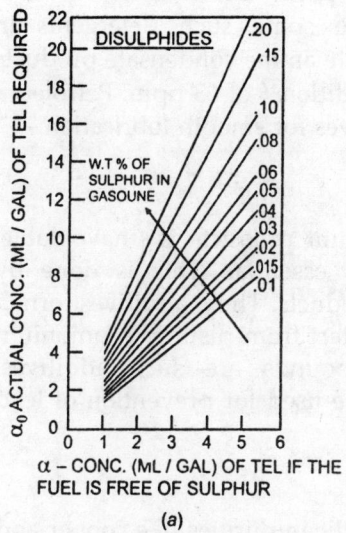

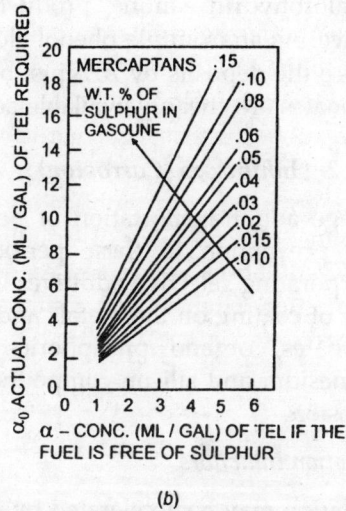

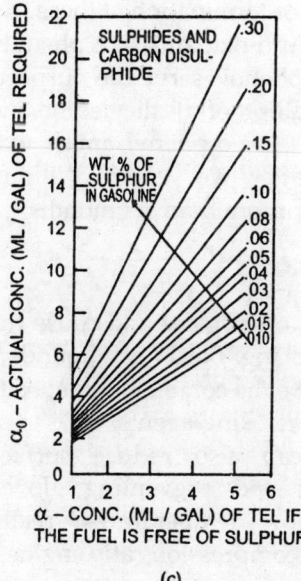

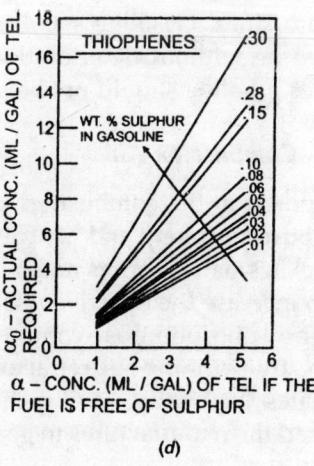

Figure 2.17 *a, b, c, d*   **Effect of Sulfur Compounds on Lead
Requirement**
(a) Disulphides (b) Mercaptans (c) Sulfides & CS_2 (d) Thiophenes

Maleic anhydride, polyamines, poly isobutylenes, phenol epichlorohydrin—amine products are some such detergents in service. As an example phenol-aldehyde amine condensate products reduce the deposits by 73% just by addition[32] of 63 ppm. Petroleum sulfonates are cheaply available additives for smooth lubrication.

2.3.3.2 Inhibitors (Corrosion)

Storage and transportation of petroleum products is unavoidable. Rust prevention in these periods is essential. This is done by incorporating selective additives in products. These additives form a layer of coating on the metals and protect from rusting. Ammonium sulfonates, organo phosphoric compounds are such additives. Magnesium and silicon compounds are used for prevention of lead corrosion.

Oxidation Inhibitors

Oxidation may be accelerated by metallic impurities like copper and its alloys unsaturates always have an inclination to react with atmospheric oxygen leading to gum formation. This is due to free radical mechanism. Hence the inactivation or termination of these free radicals ensures inhibition to oxidation. Anti-oxidants like phenols, cresols, phenylene diamines, alkyl amino phenols serve the purpose. Copper corrosion can be prevented by addition of disalicyledine and aminopropanes. Gasoline stabilised with n-oxydiphenyl amine was found to be without deterioration, even after two years[33]. Without additives gasoline should not be stored for more than six months.

2.3.3.3 Combustion Aids

The deposits in the combustion chambers often glow and cause the fuel to burn untimely and contribute knocking. Boron compounds[34] (organic) in small amounts not only decrease the combustion deposits but also increase the effectiveness of anti-knocking agents.

Organo phosphorous compounds[35] can also reduce surface ignition by forming a refractory layer and preventing glows. Phosphates of lead, having higher glow temperatures, really impressed the manufactures to go for high compression ratio engines.

2.3.3.4 Anti-knocking Agents

Tetraethyl lead is the most important of all the additives of gasoline. Though working mechanism is not convincingly reported, it is assumed to decompose into free radicals of organic nature and lead. The ions of lead try to stop the propagation of flame, thus establishing an instantaneous combustion, which favourably decreases the knocking effect, but pollution due to oxides of nitrogen increases. Further, lead compounds accumulate in engine parts. This has to be met by suitable scavengers like dibromomethane additive. With stringent pollution laws, refineries do produce unleaded gasolines of premium and regular grades. Tetramethyl lead is also employed in place of tetra ethyl lead. Modern tendency is to add a single suitable additive that can take care of all these additives. One such additive gaining importance is methyl cyclopentadienyl manganese tricarbonyl.

Further, with increasing consciousness of environment protection agencies, in every country, toxic levels of lead and hydrocarbons are carefully monitored. Smog causing hydrocarbons are as much unwanted as lead, hence gasolines blended with methyl tertiary butyl ether[46] (MTBE) have been introduced. By adding 10% MTBE to gasoline an improvement of 5-6 ON and reduction in benzene and CO emissions in exhausts are obtained. Further, the front ends increase and back ends decrease in volatility (i.e. reduction FBP) temperatures have made the gasolines to have the RVP less by 0.1 to 0.2 kg/cm^2, thus eliminating storing difficulties etc. to some extent. Oxygenates like alcohols and ethers are added to gasoline.

2.3.3.5 Dyes

Though dyes do not have special contributions towards gasoline properties, still they are incorporated for obvious reasons, namely to identify, to assess if there is any deterioration in gasoline during storage and to impart pleasing hues especially when cracked gasolines are blended. Pink colour is used for automobile gasoline. While light blue colour is added for aviation gasoline.

2.3.3.6 Aviation Gasoline

Gasoline used in aircrafts has more stringent specifications than motor spirit, although, it contains all the additives meant for motor gasoline. Sulfur content and paraffin content are more injurious to this fuel, as the operation under—60°C is severely hampered due to crystallisation of paraffin compounds. Sulfur is presumably kept below 0.05%, to eliminate the possible pollution and corrosion

TABLE 2.6 Specifications: Motor Gasoline, 83 Research.

Sl. No.	Characteristics	Method	Unit	Requirement
1.	Colour	Visual	—	Orange
2.	Copper strip corrosion for 3 hours @ 50°C	P : 15		Not worse than No. 1
3.	Density @ 15°C	P : 16	g/ml	To be reported
4.	*Distillation:*			
	Initial Boiling Point		°C	To be reported
	Recovery upto 70°C, Min		% vol.	10
	Recovery upto 125°C, Min		% vol.	50
	Recovery upto 180°C, Min		% vol.	90
	Final Boiling Point, Max		°C	215
	Residue, Max		% vol.	2
5.	Octane Number (Research Method), Min	P : 27	—	83
6.	Oxidation Stability, Min	P : 28	Mints	360
7.	Residue on Evaporation, Min	P : 29 (Air Jet Solvent Washed)	mg/100 ml	4.0
8.	Sulphur, Total, Max	P : 34	% wt	0.25
9.	Lead Content (as Pb), Max	P : 37	gm/1	0.56
10.	Reid Vapour Pressure @ 38°C, Max	P : 39	kg/cm²	0.70

TABLE 2.6(1) Summary of Specification Requirements for Aviation Gasolines.

Properties		Av gas specifications					
		DERD 2485 Issue 8 December 1978			MIL-G-5572E Amendment 3 20th March 1975		
		Grade 100LL	Grade 100	Grade 115	Grade 80/87	Grade 100/130	Grade 115/145
1	2	3	4	5	6	7	8
Colour (see Notes 1 and 2)		Blue	Green	Purple	Red	Green	Purple
Knocking Ratings							
Lean Mixture — Motor Method	min	99.5 ON	99.5 ON	—	—	—	—
or Aviation Method (see Note 3)	min	—	—	115 PN	80 ON	100 ON	115 PN
Rich Mixture	min	130 PN	130 PN	145 PN	87 ON	130 PN	145 PN
Tetraethyl Lead (TEL), ml/USG	max	(2.00)	(3.00)	(4.60)	0.50	3.00	4.60
or as Lead (Pb). g Pb/1	max	0.56	0.85	1.28	0.14	0.85	1.29
Aromatics, % vol	min	—	—	5.0	—	5.0	5.0
Net Heat of Combustion, Btu/1b	min	(18700)	(18700)	(18900)	18700	18700	18900
or Amine Gravity Product	min	7500	7500	9000	7500	7500	9800
Specific Gravity, 80°/60°F or density 15°C (kg/1) or Gravity API degrees			Report			Report	
Distillation							
Initial Boiling Point, °C (°F)	min		Report			—	
Fuel Evaporated, % vol at 75°C (167°F)	max		10			10	
			40			40	

(Contd.)

TABLE 2.6 : (Contd.)

105°C (221°C)	min		50
135°C (275°F)	min		90
End Point, °C(°F)	max	170 (338)	170 (338)
Sum of 10% plus 50%			
Fuel Evaporated Temperatures, °C (°F)	min	135 (307)	135 (307)
Residue, % vol	max	1.5	1.5
Loss, % vol	max	1.5	1.5
Freezing Point, °C (°F)	max	−60 (−76)	−60 (−76)
Reid Vapour Pressure, 1b/in^2 (kPa)	min	5.5 (38)	5.5
	max	7.0 (45.5)	7.0
Copper Corrosion, Classification	max	1	1
Sulphur, Total % wt	max	0.05	0.05
Existent Gum, mg/100 ml	max	3.0	3.0
Oxidation Stability/Ageing			
Gum 16 hours			
Potential Gum, mg/100 ml	max	6.0	6.0
Precipitate, mg/100 ml	max	2.0	2.0
Water Reaction			
Interface Rating	max	2	2
Separation Rating	max	2	–
Volume Change, ml	max	2	2
Antioxidants (See Note 6). mg/l	max	24	24

(Contd.)

TABLE 2.6(1) : (Contd.)

ASTM D910-76			AFORJOS Check List Issue 8' December 1987				Test methods	
Grade 80	Grade 100	Grade 100LL	Grade 80	Grade 100LL	Grade 100	Grade 115	IP	ASTM
9	10	11	12	13	14	15	16	17
Red	Green	Blue	Red	Blue	Green	Purple	17	D2392
–	–	–	–	–	–	–	236	D2700
80 ON	100 ON	100 ON	80 ON	100 ON	100 ON	115 PN	(236)	(D2700)
87 ON	130 PN	130 PN	87 ON	130 PN	130 PN	145 PN	119	D909
0.50	4.00	2.00	0.50	2.00	3.00	4.60	248, 228	D2547, D2599
0.14	1.12	0.56	0.14	0.56	0.85	1.28	or 270	or D3341
–	–	–	–	–	5.0	5.0	156	D1319
18720	18720	18720	18700	18700	18700	18900	12 or 193	D240 or D1405
–	–	–	7500	7500	7500	9800	2 and 160	D611 and D1298
							160	
				Report	Report		–	
				–		123		D237
				Report	Report			D96
	10			10				
	40			40				
	50			50				
	90			90				
	170 (338)			170 (338)				
	138 (307)			138 (307)				
	1.5 (see Note 4)			1.5 (see Note 4)				
	1.5			1.5				

155

(Contd.)

TABLE 2.6 (1) (Contd).

9	10	11	12	13	14	15	16	17
	−58 (−72)			−60 (−76)		16		D2386
	—			5.5		69 or		D323 or
	7.0			7.0		171		D2551
	1			1		154		D130
	0.05			0.05		107		D1266 or D2622
	10.0 (see			3.0		131		D381
	4.0 Note 5)			6.0		138		D873
	—			2.0				
	—			2		289		D1094
	2			2				
	24 (see Note 7)			24	—			—

Notes:

(1) Refer to specifications for full details of dye requirements

(2) DERD 2485 species visual test with lp 17 (Lovibond) as reference method only

(3) Aviation Ratings converted from Motor Method results.

(4) ASTM D910 and 'Check List' require that distillation residue is not acid by ASTM D1093.

(5) By agreement this is acceptable alternative to 5 hour ageing gum test required in ASTM D910.

(6) Refer to specifications for full details of permitted antioxidants.

(7) Applies only to fuel required to meet 16 hour ageing gum test. Max antioxidant content is 12 mg/l for fuel tested by 5 hour ageing gum method (see Note 5).

problems with the advent of jet engines, the demand for aviation
gasoline is decreasing.

Quantities of Additives

Additive	Quantity
1. Anti-knocks, TEL	0-3 ml/5 liters
2. Anti oxidants	1-8 kgs/1000 bbl
3. Metal deactivators (salicyledene)	0.5-2 kgs/1000 bbl
4. Corrosion-preventers	10-15 ppm
5. Preignition preventers/ Combustion aids	0.01-0.02%
6. Anti icing agents	1/2-1%
7. Dyes and colours	traces

Though the specifications often change due to engine designe
Clean Air Act, somebasic dimensions remain same. Civilian needs
demand one variety of gasoline while defence requires stringent
formulations as shown in Table 2.6 (1).

Cracked products in aviation gasolines are not permitted due to
their chemical behaviour of giving inimical deposits; i-paraffins and
naphthenes are preferred and form bulk of constituents; n-paraffins
and aromatics follow. Aromatics (Max. 12%) are also not very much
desirable (their presence is greeted very much in motor gasoline), as
they do not possess high anti-knock characteristics under lean
mixtures, but their performance is excellent during take off (rich
mixtures). Iso octane and other alkylated products are in immense
demand because they suit in close boiling ranges of aviation gasoline
(35-170°C) with high octane member.

All the tests needed for motor gasoline are well applied to this also.
As a special case the sum of 10% and 50% ASTM distillation
temperatures must not be less than 135°C. This is required to control
excessive volatility. Final boiling point is kept in the vicinity of 170°C.
Reid vapour pressure is less than motor gasoline and is kept around
0.4 to 0.49 kg/cm².

2.3.4 Aviation Turbine Fuels (ATF), Jet Fuels

Modern jet engines use fuel similar to kerosene. It is a most flexible
fuel in its boiling range (upto 300°C). All properties which are
desirable for kerosene are considered to be sufficient. Pour point of
this fuel is of extreme importance and should not be higher than

−30°C as international flights always visit lands of caprice climates. High smoke point is essential for clean combustion. Smoke is contributed by aromatics and is made of high carbon to hydrogen ratio molecules called carboids which can be distinguished from coke. Naphthenes have good smoke point and illuminating characteristics, hence preferred in these fuels. To elevate the smoke point, aromatics should be either converted to naphthenes or physically removed. It is thus required to keep aromatics in limits of tolerance upto 20%. Unsaturates estimated by Bromine number are kept below 5% to limit gum formation. Sulfur is fairly tolerated, unlike gasoline. Different aviation grades like JP—1, JP—2, JP—3, JP—4, JP—5 are used in U.S.A. while in India two grades K 43 and K 50 are produced (See Table 2.7).

Smoke point is an important property of jet fuel. Sometimes instead of expressing the quality of jet fuel by smoke point alone, it is expressed in terms of smoke volatility index defined as:

$$\text{Smoke volatility Index} = \text{Smoke point} + 0.42 \times (\% \text{ distilled at } 204°C)$$

The advantages gained by this expression are still not clear. The tests prescribed for this fuel are :

(1) Volatility (ASTM)
(2) Pour point
(3) Smoke point (combustion quality)
(4) Water retention and separating properties

Volatility had already been discussed; pour point and smoke point are discussed in the kerosene and lube oil tests.

Jet fuels being denser and viscous than gasoline, they tend to retain fine particles and droplets of water for a considerable amount of time. Although filtration system can remove fine particles, it is no use for water droplets; water droplets can be removed by either heating or adding demulsifying agents. Maximum permissible limits of impurities are for fine particles 1 mg/litre and moisture content 30 ppm. Free water in jet fuels can be detected by field kits like Esso-hydro kit, Mobel Moisture Indicator. The standard water reaction test (IP 289) is used as in the case of aviation gasoline. The clarity of fuel is measured by photo electric device.

Noticeable phenomenon during high altitude flights and at high speeds is that fuel gets charged. To disperse these static charges antistats are added.

TABLE 2.7 Specifications of Aviation Turbine Fuels.

	Aviation Turbine Fuel	
	Grade K-43	Grade K-50
Acidity in grains	Nil	Nil
Aromatic % vol. max	20	20
Bromine number, max	5	5
Colour Saybolt, min	8	8
Copper strip corrosion 2 hrs. at 100°C. Not less than No. 1 Distillation		
(a) I.B.P.	Not limited but to be reported	
(b) Recovery upto 20% by volume	-do-	
(c) Recovery upto 20%	200°C minimum	
(d) Recovery upto 50% by vol.	Not limited but to be reported	
(e) F.B.P. max	300°C	300°C
(f) Residue, % age vol. max.	2	2
(g) loss percent by vol. max.	1.5	1.5
Flash point min.	38°C	38°C
Cold test temp. max.	–40°C	–50°C
Residue on evap. max. mg/100 ml	3	3
Gum accelerated (16 hrs.) mg/100 ml. max	6	6
Cal. value Cal/gm. (min)	10167	10167
Density at 15°C	0.775 to 0.825	0.775 to 0.825
Sulfur by vol.	0.005	0.005
Sulfur total, percent by wt.	0.02	0.2
K.V at –17.8°C (CS)	6	6
Water reaction Change in vol. of ether layer max.	1 ml.	1 ml.

The paraffin rich fuels can allow micro organisms to survive, for this reason, when these fuels are allowed to be stored over great period, biocides are added. Other additives are also used as in the case of aviation gasolines. Additives and their functions are presented below:

Additives	*(Purpose) Quantity*
Poly sulphones	Antistats 1 PPM
Organic boron compounds	Biocides (20 PPM element boron)
Phosphoric compounds	Rapid combustion aids in ram jets (200 PPM)
Glycol, polyhydric alcohols	Anti icing agents upto 1%

TABLE 2.7.1 Summary of Specification Requirements for Jet-A-1, Kerosene Type, Aviation Turbine Fuels.

Properties	Jet A-1 specifications				Test methods		
	DERD 2494 Issued 8 October 1978		IATA Guidance Material (Amended November 1978)	ASTM D1655-78	AFORJOS 'Check List' Issue 8 and Addendum 8/1 of December 1978	IP	ASTM
	Kerosine Type		Kerosine Type	Jet A-1	Jet A-1		
1	2	3	4	5	6	7	8
Composition							
Total Acidity, mg KOH/g	max	0.015(0.012)	–	–	0.015(0.012)	(273)	D3243
Total Acidity, mg KOH/g	max	–	0.10	0.10	–	–	D974 or D3242
Aromatics, % vol	max	22	20 (see Note 1)	20 (see Note 1)	20 (see note 1)	156	D1319
Olefins, % vol	max	5			5	156	D1319
Sulphur, total % wt	max	0.30	0.30	0.30	0.33	107	D1266
Sulphur, Mercaptan, % wt	max	0.001	0.003	0.003	0.001	104	D1323 or D3227
or Doctor Test	max	Negative	Negative	Negative	Negative	30	D484
Volatility							
Distillation							
Initial Boiling Point, °C		Report	–	–	Report		
10% vol at °C(°F) Fuel Recovered	max	205	204(400)	204(400)	204(400)	123	D86
20% vol at °C		Report	–	–	Report		
50% vol at °C		Report	Report	Report	Report		

(Contd.)

TABLE 2.7.1 : (Contd.)

161

1	2	3	4	5	6	7	8
90% vol at °C		Report	Report	Report	Report		
End Point °C (°F)	max	300	288(550)	300(572)	288(550)		
Residue, % vol	max	1.5	1.5	1.5	1.5		
Loss, % vol	max	1.5	1.5	1.5	1.5		
Flash Point, °C (°F) (see Note 2)	min	38	38(100)	37.8(100)	38(100) or 40 (104)	170 or 303	D3243 or D56
Density at 15°C kg/l	min / max	0.775 / 0.830	— / —	— / —	0.775 / 0.830	160	D1298
Specific Gravity at 60°/60°F	min / max		0.775 / 0.239	0.775 / 0.640	0.775 / 0.830	160	D1298
Gravity API	max / min	51 / 37	51 / 37	51 / 37	— / —		D287
Fluidity							
Freezing point, °C (°F)	max	−50 (−58)	−50 (−58)	−50 (58)	−50 (58)	16	D2386
Viscosity at −20°C (−4°F) cSt	max	8.0	8.0	8.0	8.0	71	D445
Combustion Calorific Value, net BTU/1b	min	18400	18400	18400	18400	12 or 193	D240 D2382 or D1405
or Specific Energy net J/g	min	42800	(42800)	(42800)	42800	160 & 2 (193)	D611&D1298 (D1405)
or Aniline Gravity Product	min	5250	(5250)	(5250)	5250	(57)	D1322
Smoke Point, min	min	25(26)	25	25	25		D1740
or Luminometer Number	min	45	45	45	45	(57)	D1322
Smoke Point, min PLUS Naphthalene	min	19(20)	20 (see Note 1)	20 (see Note 1)	20 (see Note 1)		D1840
or { Content, % vol	max	3	3	3	3		—

TABLE 2.7.1 : (Contd.)

1	2	3	4	5	6	7	8
Corrosion							
Copper Corrosion, Classification	max	1	1	1	1	154	D130
Silver Corrosion, Classification	max	1	–	–	1	227	–
Stability							
Thermal Stability by:							
CFR Coker							
(Preheater Temp. 300°F, Filter Temp. 400°F)						197	D1660
(Fuel Flow 6 lb/hour, Duration 300 min)							
Change in Pressure Drop in 5 hours, inches Hg	max	–		3	3	–	
Preheater Deposit, Classification	max	–		<3	<3	–	
or by Alcor JFTOT						323	D3241
(Max Heater Tube Temp 260°C)							
(Fuel System Pressure 3.45 MN/m²)							
(Fuel Flow 3 ml/min, Duration 150 min)							
Filter Pressure Differential, mm Hg	max		25	25	25	25	
Tube Rating (visual),							

(Contd.)

TABLE 2.7.1 : (Contd.)

1	2	3	4	5	6	7	8
Classification	max	<3 'no Peacock' or 'Abnormal' colour deposits	<3	<3	<3 'no Peacock' or 'Abnormal' colour deposits		
Contaminants							
Copper Content, µg/kg (see Note 3)							
Existent Gum, mg/ 100 ml	max	150	–	–	150	225	—
Water Reaction	max	7	7	7	7	131· 289	D381 D1094
Interface Rating	max	1b	1b	1b	1b		
separation Rating	max	2	2	2	2		
Water Separometer							
Index Modified WSIM						—	D2550
Fuel with static Dis-sipator Additive	min	70	—	—	—	—	
Fuel without Static Dissipator Additive	min	85	—	—	85		
Additives							
Anti-oxidant, mg/1							
Hydrotreated Fuels	min max	17 Mandatory 24	Optional —	Optional 24	17 Mandatory 24	17 Mandatory 24	
Non-Hydrotreated Fuels	max	24 Optional	24 Optional	24 Optional	24 Optional		
Metal Deactivator, mg/1	max	5.7 Optional	5.7 Optional	5.7 Optional	5.7 Optional		

(Contd.)

TABLE 2.7.1 : (Contd.)

1	2	3	4	5	6	7	8
Corrosion Inhibitor (see Note 4)					By Agreement		
Anti Static Additive (ASA-3), mg/1	max	By Agreement	By Agreement	By Agreement	By Agreement		
Conductivity-at time/ temperature of delivery to aircraft if ASA-3		1.0	(1.0)	(1.0)	1.0		
present cu (pS/m)	min	50	50	50	50	274	D2624
	max	450	300	300	300		

Notes

(1) Temporary relaxations allow supply of fuel with aromatics content up to 22% max and smoke point (BSTM) down to 10 mm min.

(2) Results by ASTM D56 (Tag method) can be at least 1-2°C above these obtained by ASTM D3243/P303 (Setallash method) and IP 170 (Abel method).

DERD 2404, ASTM D1655 and 'Check List' specify IP 170 as 'referee method'.

(3) Only for fuels treated by copper sweetening process.

(4) Details of approved Corrosion inhibitors and concentrations can be obtained by reference to the specifications concerned.

TABLE 2.7.2 Summary of Specification Requirements for Jet B/JP4 Wide Cut Type, Aviation Turbine Fuels.

Properties	Jet B/JP4 specifications						Test method	
	DERD 2486 Issued 9 December 1978		IATA Guidance Material Amended November 1978	ASTM D1655-78	MIL-T 5624 K and Amendment 1 Nov 1976	AFORJOS 'Check List' Issue 8 and Addendum 8/1 of December 1978	IP	ASTM
	Wide Cut Type		Wide Cut or JP-4 Type	Jet B	Grade JP-4	Jet B		
1	2	3	4	5	6	7	8	9
Composition								
Total Acidity, mg KOH/g	max	0.015(0.012)	0.10	–	0.015	0.015(0.012)	(273)	D3243
Aromatics, % vol	max	25	20 (see Note 1)	20 (see Note 1)	25	20 (see Note 1)	156	D1319
Olefins, % vol	max	5	–	–	5	5	156	D1319
Sulphur, total % wt	max	0.30	0.30	0.30	0.40	0.30	107	D1266, D1552
Sulphur, Mercaptan, % wt	max	0.001	0.003	0.003	0.001	0.001	104	D1323
or Doctor Test	Negative		Negative	Negative	Negative	Negative	30	D484
Volatility								
Distillation								
Initial Boiling Point, °C		Report	–	–	Report	Report	123	D86
Fuel Recovered 10% vol at° C		Report	–	–	Report	Report		(see Note 2 ref gas chroma-

(Contd.)

TABLE 2.7.2 : (Contd.)

1	2	3	4	5	6	7	8	9
20% vol at °C (°F)	max	143	143(290)	144(290)	145(293)	143(290)		tography method
50% vol at °C (°F)	max	190	188(370)	190(370)	190(374)	188(370)		D2887
90% vol at °C (°F)	max	245	243(470)	243(470)	245(473)	243(470)		
% recovered at 400°F (204.4°C)	max	-	-	-	Report	Report		
End Point, °C		270	-	-	270 (518)	Report		
Residue, % vol	max	1.5	1.5	1.5	1.5	1.5	1.5	
Loss, % vol	max	1.5	1.5	1.5	1.5	1.5	1.5	
Specific Gravity at 60°/60°F	min	0.751	0.751	0.751	0.751	0.751	160	D1298
(or Density at 15°C kg/1)	max	0.802	0.802	0.802	0.802	0.802	160	D1298
Gravity, °API	max	-	57	57	57	57	-	D287
	min	-	45	45	45	45		
Reid Vapour Pressure, 1b/in²	min	2(14)			2(14)	2(14)	69 or	D323
	max	3(21)	3(21)	3(21)	3(21)	3(21)	171	D2551
Fluidity								
Freezing Point, °C (°F)	max	58(−72)	−50(−58)	−50(−58)	−58(−72)	−58(−72)	16	D2386
Combustion								
Calorific Value, net, BTU/1b	min	(18400)	18400	18400	18400	18400	193	D1405 or D2382
or Specific Energy, net J/g	min	42800	(42800)	(42800)	42800	42800	12	D240, D611 & D1298
or Aniline Gravity Product	min	5250	(5250)	(5250)	5250	5250	2 & 160	
Hydrogen Content, % wt	min	138 Optional (see Note 3)			13.6 (see Note 4)		338	D3701 or D3343
Luminometer Number	min		45	45		45		D1018 or D1740
or Smoke Point, mm	min	20(21)	25	25	20	25	(57)	D1322
or { Smoke Point, mm	min		20	20		20		D1322
{ Plus Naphthalene Content, max	max		3	3		3		D1840
or % vol								

(Contd.)

TABLE 2.7.2 : (Contd.)

1	2	3	4	5	6	7	8	9
Corrosion								
Copper Corrosion, Classification	max	1	1	1	1	1	154	D130
Stability								
Thermal Stability by:							197	D1660
CFR Coker (Preheater Temp 300°F, Filter, Temp 400°F, Fuel Flow 6 lb/hour, Duration 300 min)								
Change in Pressure Drop in 5 hours, inches Hg	max	—	3	3	—	—		
Preheate: Deposit, Classification	max	—	<3	<3	—	—		
or by Alcor JFTOT (Max Heater Tube Temp 260°C)							323	D3241
(Fuel System Pressure 3.45 MN/m^2)								
(Fuel Flow 3 ml/min, Duration 150 min)								
Filter Pressure Differential mm Hg	max	25	25	25	25	25		
Tube Rating (visual) classification	max	<3, no 'peacock' or 'Abnormal' colour deposits	<3	<3	<3	<3, no 'Peacock' or 'Abnormal' colour deposits		

(Contd.)

1	2	3	4	5	6	7	8	9
Contaminants								
Copper content μg/kg (see Note 5)	max	150	—	—	—	150	225	D481
Existent Gum. mg/100 ml	max	7	7	7	7	7	131	D2276
Particulate matter, mg/litre	max	—	—	—	1.0	—	216	(see Note 6)
Filtration Time, mins	max	—	—	—	15	—	—	
Water Reaction								
Interface Rating	max	1b	2b	1b	1b	1b	289	D1094
Separation Rating	max	2	2	2	1	1		
Water Separometer Index modified (WSIM)								
Fuel with Static Dissipator Additive	min	70	—	—	70	70	—	D2550
Fuel without Static Dissipator Additive	min	85	—	—	70	85		
Additives								
Fuel System Icing Inhibitor, % vol	min	—	—	—	0.10 Mandatory	—	—	(see Note 7)
	max	—	—	—	0.15	—	—	
Anti-oxidant, mg/l Hydrotreated Fuels	min	17 Mandatory	Optional	Optional	17.2 Mandatory	17 Mandatory	17 Mandatory	
	max	24	—	24	24	24	24	
Non Hydrotreated Fuels	max	24 Optional	Optional	24 Optional	24 Optional	24 Optional	24 Optional	
Metal Deactivator, mg/l	max	5.7 Optional	Optional	5.7 Optional	5.7 Optional	5.7 Optional	5.7 Optional	
Corrosion Inhibitor, mg/l (see Note 8)	max	By Agreement	By Agreement	By Agreement	Mandatory	By Agreement	By Agreement	

(Contd.)

Table 2.7.2 : (Contd.)

1	2	3	4	5	6	7	8	9
Anti-Static Additive (ASA-3), mg/l	max	1.0	(1.0) By Agreement	(1.0) By Agreement	(1.0) By Agreement	1.0 By Agreement		
Conductivity — at time/temperature of delivery to aircraft if ASA-3 present cu (pS/m)	min max	50 450	50 300	50 300	50 300	50 300	274	D2624

Notes:

(1) Temporary relaxations allow supply of fuel with aromatics content up to 22% max.

(2) ASTM D2887 may be used as an alternative in MIL-T-5624K only, with following limits:
20% vol at 130°C max 90% vol at 250°C
50% vol at 185°C max. End Pt. 320°C max.

(3) By NMR method (ASTM D3701) only.

(4) Smoke point 20 mm by ASTM D1322 is alternative to Hydrogen content. If latter is calculated by ASTM D3343 method refer to Footnote 6 of MIL-T-5624K for distillation temperature details by ASTM D2887.

(5) Only for fuels treated by 'copper sweetening' process.

(6) Method described in specification.

(7) Method described in specification.

(8) Details of approved Corrosion inhibitors and concentrations can be obtained by reference to the specification concerned.

For jet fuels adequate information has not been provided by smoke point alone; a better method is by prescribing luminometer number (ASTM D-1740). Satisfactory luminometer numbers for some grades are given as

For $JP_1 = 51$
               $JP_4 = 60$
               $JP_7 = 75$

Further, the smoke point of fuel can indicate the percentage hydrogen content of fuel[27] as shown by the relation

$$H_2\% = 12.99 + 0.04 \times \text{(smoke point in mm)}$$

2.3.5 Naphthas

These fractions are highly volatile and fall in the boiling range of motor spirits.These are mostly used as solvents in paints, perfumery and other industries. Solvent grades are produced by distilling wide cut naphthas into small boiling range cuts. Naphthas are not suitable for combustion, because of the rapid flame propagation, resulting in explosions. Cuts boiling below 80°C do not have any aromatics, hence their solvent power is also less. Such fractions are sent for cracking operation. 80-120°C fraction is reformed to improve octane number and this goes as a blend into straight run gasolines.

General requirements of solvent naphthas are
1. Almost water white in color
2. Non-corrosive nature
3. High stability
4. High solvent power (Kauri Butanol test ASTM D1133)
5. Low boiling point
6. Low acidity

Suspended impurities are removed by filtration. Solvent power is determined by Kauri Butanol number. The volume of solvent that causes a standard Kauri Butanol gum in butyl alcohol to become a 10 point type illegible when viewed through this solution. Another method of estimation of solvent power is by Aniline point. Harvey[29a] Mills relationship for KBN is given by

$$KBN = 99.6 - 0.806\,G - 0.09912\,A + 0.0755\,(340 - B \times 0.56) \quad \text{(For}$$
less than 50)

where G = Sp. Gravity
       A = Aniline point °C
       B = Mid boiling temperature °C

2.3.6 Kerosenes

Kerosene is the general name applied to the group of refined petroleum fractions employed as fuel and illuminant. All these fractions have approximate boiling range 150–250°C. These are uniform close cut distillates, low in viscosity, with a good degree of refinement to be fairly stable, light in colour and free from smoky, ill-smelling substances. It is highly desirable that sulfur be kept as low as possible. Kerosene is used as illuminating oil in domestic needs both in wick burners and mantle burners. Though obsolete these days, it was used in railway signalling purposes. Such kerosenes were special blends of relatively high boiling fractions with oils of marine animals.

Tests and properties of kerosene:

1. Flash point and Fire point
2. Smoke point (Burning quality)
3. Volatility
4. Sulfur content
5. Aniline point

Above listed properties are cardinally judged for kerosenes of all grades. Specifications for superior kerosene are given in Table 2.8.

TABLE 2.8 Specification Kerosene (Superior).

Sl. No.	Characteristics	Method	Unit	Requirements
1.	Acidity, Inorganic	P : 2	mg KOH/gm	nil
2.	Burning Quality:	P : 5	mg/Kg of	20
	Char Value, Max		oil consumed	
	Bloom on Glass Chimney			Not darker than grey
3.	Colour (Saybolt), Min	P : 14		21
4.	Copper Strip Corrosion			Not worse
	for 3 hours. @ 50°C	P : 15		than No. 1
5.	Distillation:			
	Recovered @ 200°C, Min		% vol.	20
	Final Boiling Point, Max		°C	300
6.	Flash Point (Abel), Min	P : 20	°C	35
7.	Smoke Point, Min	P : 31	mm	20*
8.	Sulphur, Total, Max	P : 34	% wt	0.25**

* For supplies to Defence, the smoke point of product shall be 21 mm minimum.
** For supplies to Defence, the total sulphur, per cent weight, shall be 0.20 maximum.

2.3.6.1 Flash Point

This is defined as the minimum temperature at which the vapours from oil sample will give a momentary flash on application of a standard flame under specific test conditions. Abel (P: 21; IP 170/59) Pensky-Martens (P : 21; IP 34/58) Cleveland are the test apparatus frequently used for the purpose. Significance of the flash point is that it can predict the possible fire hazards during transportation, storage and handling. Early tendencies were to incorporate more volatiles in kerosenes, thus inviting unusual fire hazards. This necessitated a strict vigilance by way of fixing flash point. The flash point of marketable kerosene should be above 42°C.

Pensky-Martens Apparatus (P : 21; IP 34/58) and Test:

The apparatus consists of a cylindrical cup made of brass of approximate dimensions 50.8 ± 1.27 mm × 55.88 ± 1.27 mm (height inside) with a thickness at the bottom 2.41 ± 0.64 mm. Inside the cup at 2/3 rds height from the bottom there is a sudden and slight tapering up to the top of the cup. This looks like a ring and guides as a filling level for the sample. The top position acts as a vapour-air chamber.

This cup is provided with a lid, and the lid is actually made of two metal discs, one sliding over the other. The lid also includes (a) stirring device (b) two flame-holders, one test flame and other pilot flame (c) provision for thermometer (d) spring handle. By turning the spring handle, it is possible to slide one lid over the other whereby the exactly cut chords align with each other, exposing air-vapour mix of the cup to flame. In fact, there are three chord openings, central one is meant for flame introduction, while the other two act as air introducers. When the handle is turned test flame is also simultaneously lowered into the central chord opening to explode air-vapour mixture. The test flame is issued from an opening of 0.69 mm dia. The whole cup is heated by air-bath, which is primarily heated by electric power.

The test sample is filled up to the mark, and the temperature of the oil is slowly and uniformly raised at a rate of 3°C per minute. Test and pilot flames are lighted. Occasional stirring is done. The spring handle is rotated at every 1°C raise in thermometer; till a bluish light flash is noticed. Heating should be maintained at the prescribed rate.

Abel flash meter is used for highly volatile samples, whose flash point shall be less than 40°C. Pensky-Martens closed cup can be used for medium flash liquids, when a high flash point liquid is to be tested opencup flash meter is convenient.

Flash point is essentially dependent up on the light end characteristics of a fuel. Approximate relationship between flash point and IBP —10% boiling range is given by Nelson[29b] as:

$$\text{Flash point} = 0.64 \, T°C - 61.95$$

where, $T°C$ is the temperature at 10% distillation point.

Flash point is regarded as vapour property hence the ambient pressure directly influences the flash point. Thus one is reminded to note the ambient pressure before such a test is conducted. Correction for pressure may be accorded by applying the following relation:

Flash point $T°C$ corrected to 760 mm. Hg

$$= \text{Flash point } t°C + \frac{760 - P}{30} \; ; \text{ where } t°C \text{ is at P mm. Hg pressure}$$

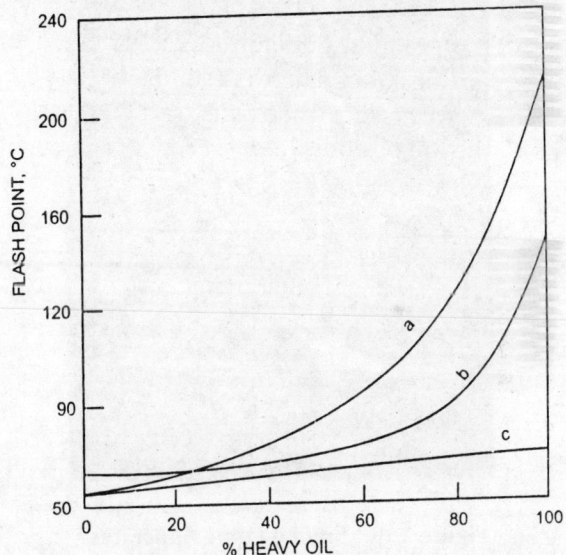

Figure 2.18 Flash Point of Blends, a is the curve of a blend of 53 and 212, b is the curve of a blend of 62 and 142, c is curve of a blend of 53 and 68.

where P is the pressure in millimeters at the time of observation. Flash point of blends does not follow any linearity: however the influence of high volatiles (range of 0-10% distillation) is predominant. Instead of laborious theoretical correlations it is always easy to determine the flash point of mixtures experimentally as shown in Figure 2.18.

2.3.6.2 Fire Point

This closely follows the flash point. The test is carried out in open cup rather than a closed one. Cleveland apparatus (ASTM D 92) offers the advantage of open flash point and fire point. Fire point temperature is noted when the oil vapours can burn continuously for 5 seconds when tested in flash point apparatus and it occurs after the flash point, by 3-4°C. For domestic needs a very high flash point above 50°C is also not desirable. The easy ignition is credited to volatiles only and decrease in volatiles enhances the flash point causing difficulty in ignition.

Figure 2.19   Smoke Point Apparatus.

2.3.6.3 Smoke Point

Smoke point is an indication of clean burning quality of kerosene. Illumination depends upon the flame dimension although it is not related to flame height. Many paraffins may be gifted with better flame heights but illumination may be poor. Smoke point is defined as the maximum height of flame in millimeters at which the given oil will burn without giving smoke. Thus smoke point test (P: 31, IP 57) enables the burning quality to be measured by adjusting the wick height to give proper non-smokey flame (Figure 2.19). Illumination is supposed to be characteristic of the components of fuel usually not measured. Different flame heights are obtained due to the presence of different components such as paraffins, naphthenes and aromatics. Aromatics contribute smoke, hence removal of aromatics increases the smoke point. Naphthenes with side chains are inevitably retained to give good illumination. In India, marketable kerosene should possess a smoke point 18 mm. Smoke point may be predicted approximately by knowing the group composition of oil, given as:

$$\text{Smoke point} = 0.48\ P + 0.32\ N + 0.20\ A.$$

where P, N, A are % of paraffins, naphthenes and aromatics in the sample.

A dogma of high smoke point meant a better quality fuel is not always acceptable as it cannot specify any illumination characteristics. In recent literature stress is given to this luminosity number (ASTM D 1740) and prescribed as a test for kerosene. This indicates luminosity of the flame and accompanying radiation problems in combustion chambers. Smoke point apparatus with modifications to include a photo electric cell can measure the flame radiations. Luminosity number is expressed just like octane number. The constituents taken here are tetralin (O luminosity number) and iso octane (100 luminosity number). The blend of these two should match with the luminosity of fuel.

2.3.6.4 Burning Quality Test (P : 5; ASTM D182)

This indicates the ability of kerosene to burn steadily and cleanly over a long period. Kerosene is tested by burning in a standard wick lamp for a period of 24 hours, in a room free from air currents. The average rate of burning, change in shape of flame, the density of chimney deposits are the prime factors by which the quality of kerosene is assessed. A good quality kerosene (after burning for 16 hours) should present the chimney as clean as possible (only slightly clouded) and the wick should not have any appreciable hard incrustations with as large flames as in the beginning.[36]

The continuity of the initial illumination over long periods in a lamp rests on the normal flow of kerosene to the wick end. Quantity of oil available at the tip of wick depends on the height of wick over liquid surface, texture of wick and properties like viscosity and surface tension of oil. Even though surface tension of different oils (kerosenes) may not vary much, viscosity varies. Stabilisation of flame is governed by viscosity. A good oil of 650 c.c. must burn at least for 120 hours.

176

2.3.6.5 Volatility

The nature of distillation range estimated by ASTM D86 is of significance with regard to burning characteristics. Ten percent boiling point reveals the flash point which in turn, indicates the ease of ignition; specially when kerosene is used in pump stoves, this plays a very important role in continuous support of flame. Mid boiling range contributes towards viscosity. Often flare up in oil heaters is observed if mid-boiling range is not properly constituted. Sixty percent distillation at 200°C ensures steady performance of oil in heaters. Minimum of 15% distillation at this temperature is enforced to guarantee adequate volatility[41a].

2.3.6.6 Sulfur Content

High sulfur content is inimical due to its combustion products. When large amounts of such fuel is burnt, accumulation of these oxides results, offering wayward problems of corrosion and pollution. Total sulfur can be estimated by bomb method (ASTM D 1266; IP 107). Maximum permissible amount of sulfur is 0.13%; in all kerosenes.

2.3.6.7 Aniline Point

(IP 2/56; IS (1448) P : 3) This test indicates qualitatively the amount of aromatics present in kerosene. This method is fully discussed in diesel fuels.

2.3.7 Diesel Fuels

Diesel oils are the fractions in the boiling range of 250-320°C; and fall under gas oil fractions. These are basically divided into two classes as high speed diesels and low speed diesels.

Classification of diesel oils is done according to speed and loads of the engine as given below:

Low speed	Below 300 RPM	For heavy loads at constant speeds.
Medium speed	300-1000 RPM	Fairly heavy loads moderately constant speeds.
High speed	above 1000 RPM	Load and speed vary.

Volatility test for these fuels is of trifle value; however it is advisable to conduct ASTM distillation test to infer the boiling range.

High percentage of light ends (upto 10% distillation) are traits of low flash points. Even though flash point has no direct influence upon performance of fuel, still its minimum value should be maintained to prevent possible fire hazards in handling and storage.

TABLE 2.9 Specification: High Speed Diesel Oil.

Sl. No.	Characteristic	Method	Unit	Requirement
1.	Acidity, Inorganic	P : 2	mg KOH/mg	Nil
2.	Acidity, Total, Max	P : 2	mg KOH/mg	0.50
3.	Ash, Max	P : 4	% wt	0.01
4.	Carbon Residue (Ramsbottom), Max	P : 8	% wt	0.20
5.	Pour Point, Max	P : 10	°C	6
6.	Copper Strip Corrosion for 3 hrs @ 100°C	P : 15		Not worse than No.1
7.	Density @ 15°C	P : 16	g/ml	To be reported
8.	Diesel Index	P : 17		45 min
9.	Distillation: Final Boiling Point Recovery @ 366°C, Min	P : 18	°C	To be reported 85/90
10.	Flash Point, (Abel), Min	P : 20	°C	33/38*
11.	Kinematic Viscosity @ 37.8°C	P : 25	cS	2.5–7.5
12.	Sediment by Extraction, Max	P : 30	% wt	0.05
13.	Sulphur, Total, Max	P : 33	% wt	1.1**/1.0
14.	Water Content, Max	P : 40	% vol.	0.05
15.	Flame Height, Max	Indian Customs	mm	18

* Subject to agreement between the purchaser and supplier, a lower or higher maximum Flash Point may be accepted.
** Subject to agreement between the purchaser and supplier a lower or higher maximum Sulphur Content may be accepted.

If 10%-50% boiling temperature is high, its warm up period will considerably increase. A high boiling point gives much smoke formation and crank case dilution. For all high speed oils 50% point is kept around 300°C; too low 50% point gives a low viscosity and possibly low heating value too. Table 2.9 gives specifications of high speed diesels.

The general tests recommended for diesel fuels are (in the order of performance).

(1) Pour point
(2) Aniline point—Diesel Index (Cetane number)
(3) Flash point
(4) Calorific value
(5) Viscosity

2.3.7.1 Pour Point

The criteria of pour point fixation depends upon two factors namely climatic conditions and storage (handling). Fuel, at minimum ambient temperature must be free flowing. In India pour point is fixed at 5°C; however in the Himalayan belt, where the climate persists at sub-zero level this may not be satisfactory; hence low pour point oils are essential. It is also observed that at close approach of pour point (within 2 to 3°C), the viscosity increases very much, the result of which is high pumping cost.

2.3.7.2 Aniline Point

Diesel fuels are mainly composed of paraffins although there is no bar for aromatics. Aromatics of this boiling range present in the fuel cause abnormal ignition delay. For this reason an estimation of aromatics is essential. One such method is aniline point which can predict the suitability of oil. Aniline point is defined as the minimum temperature at which equal volumes of anhydrous aniline and oil mix together. Diesel index (IP 21/53; IS P : 17) a derivative of Aniline point is defined as:

$$[0.018 \text{ A.P°C} + 0.32] \text{ API.}$$

Aniline being an aromatic compound freely mixes with aromatics; so a low aniline point indicates, low diesel index (because of high percentage of aromatics). Aniline point can also predict the amount of carbon present in the molecule (aromatics), as given by the formula[37]

$$\% \ C_A = 1039.4 \ n_D^{20} - 470.4 \ d_{20} - 0.567 \ \text{AP(°C)} - 1104.42$$

where n_D^{20}. refractive index at 20°C

d_{20} density at 20°C

Diesel index is a measure of ignition quality of fuel. Diesel engine works on the principle of compression ignition. During compression adiabatically, the air temperature reaches around 600°C, when the fuel in finely atomised form is fed in, it instantaneously explodes. The temperature at which this explosion takes place without the aid of fire is called self ignition temperature. Self ignition temperature is low for paraffins while it is high for aromatics (benzene 651 °C) (See Chapter 2.2). Thus a fuel rich in aromatics burns later causing ignition delay and it gives rise to what is known as diesel knock. For this reason all diesel fuels are processed to have a diesel index in the range of 45 to 55. A high diesel index is also not desirable, as a fuel rich in aromatics gives rise to better calorific value than paraffin rich fuel for equal weights. When the fuel is injected into the chamber of hot air, first the

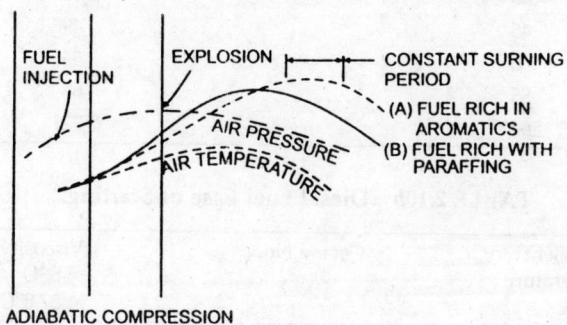

Figure 2.20   Diesel Oil Combustion Performance in Engine.

fuel drops acquire the temperature of the surroundings. This takes some time. The time delay can be decreased by atomising the fuel into fine drops and also by varying the composition of fuel. Viscosity also has its own role in heat transmission. Once the droplets acquire the temperature of self ignition, the droplets explode. A continuous flame front may move if the aromatic rich fuel is fed in—obviously a cause for diesel knocking. Figure 2.20 shows the diesel fuel performance in an engine. During adiabatic compression, the temperature rises and at that instant the fuel is injected. The time taken by the fuel, from injection to explosion is known as delay period. Then the fuel burns at a constant rate. The air pressure and temperature increase after explosion. The two curves A & B marked in the figure show the characteristic performance of two fuels—'B' is for a fuel rich in paraffins and 'A' is for aromatic rich fuel.

Peaks a and b obviously should coincide with crank position, otherwise, knocking results. High speed engines require a fuel of viscosity, 1.4 to 2.5 CS (at 30°C), where as low speed engines may work with even high viscous fuels (5.8-26.4 CS).

Cetane number: (IS P 19; ASTM D613-58T)

Cetane number is a corollary of Diesel index.

It is defined as percentage volume of n-cetane in a mixture of n-cetane and α-methyl naphthalene, which gives the same ignition delay as the fuel under consideration, when tested in a CFR engine.

TABLE 2.10a   Cetane Number Vs. Diesel Index.

Cetane No.	Diesel index
30	26
35	34
40	42
45	49
50	56
55	64
60	72

TABLE 2.10b   Diesel Fuel Ease of Starting.

Ambient Temperature °C	Cetane No.	Viscosity SSU at 37.8°C
-34.4	90	—
-28.9	83	46
-23.3	76	57
-12.2	63	65
-6.7	56	—
-1.1	49	—
+4.4	43	100
+10.0	36	—

TABLE 2.10c   Diesel Fuels—Properties.

	High speed	Low speed	Medium speed
Viscosity at 37.8°C			
SSU Minimum	32		32
Maximum	50	250	70
Sulfur %	1.5	2.0	1.5
Conradson Carbon Residue %	0.2	3.0	0.5
Ash %	0.02	0.04	0.02
Water and Sediment % vol.	0.05	0.6	0.1
Flash point °C	—	62.3	62.3
Cetane No.	50	30	40
Diesel Index	45	20	30

Some people prefer heptamethyl nonane (Cetane No. 15) in place of α-methyl naphthalene.

A relation between diesel index and cetane number is presented in Table 2.10a.

2.3.7.3 Flash Point

It has no real significance on performance of fuel. Knowledge of this is required for safe handling and storing. In India flash point of diesel is kept around 50°C-55°C.

2.3.7.4 Calorific Value

Though high calorific value of the oil is desirable, it cannot be at the expense of diesel quality; hence a fuel having[29b] cal. value 41.83 kJ per gm is sufficient. Apparently paraffins have high calorific value due to higher hydrogen content, but aromatics mainly contribute by bond dissociations. The disparity in calorific value for these materials in this region is not overwhelming.

2.3.7.5 Viscosity

Ease of starting depends upon viscosity and ignition quality. At lower starting temperatures high cetane fuels are required (See Table 2.10b). Diesels used for different speeds must possess the properties as shown in Table 2.10c.

Thermal stability of diesel fuels is important, because the remaining fuel from the combustion chamber is returned back to the reservoir. The test is performed by exposing 50 ml sample kept at 149°C for 90 minutes with an additive. When the sample is filtered through a filter paper, the excessive stain on filter paper is an indication of bad oxidation stability of the fuel. Usually additives like 1-n butoxy-l-ethane, dioctyl-diphenyl amine (upto 20 ppm) are used to improve the stability of fuel.

2.3.7.6 Diesel Additives[38, 39]

Just like gasoline, diesel fuels are also blended with additives to give satisfactory performance. Ignition accelerators are used for reducing the ignition delay. These are generally oxygen donating substances (oxidising type) like nitrates and nitrites of C_7-C_8 acids (amyl nitrate is an example); aldehydes, ketones, ethers and peroxides. These have the tendency to oxidise the fuel. Inhibitors to stop gum formation are the same type as used in the case of gasoline (phenol amines under the trade name Topanol).

Pour point depressors are generally chlorinated paraffins, and these are used to improve the cold-start performance of the engine. Anti-smoke additives like ethyl lubrizols cause less smoke in the exhausts. Alkaline agents and ashless dispersant additives have been used to improve quality of rail diesels. These alkaline agents provide improved piston ring and cylinder line protection particularly with high sulfur fuels.

Antistatic additives are optional because of the ubiquitous demands of diesel; time of storing is not much, hence these may or may not be mixed. Usually these are the chromic salts of alkylated salicylic acid, calcium aerosols, vinyl methacrylate polymers. To prevent microbiological growth dioxaborinanes (boron compounds) N-benzylidenemethyl amine are used.

2.3.7.7 Fuels of Future[48, 49]

Demand for fuels of transportation is increasing in an unexpected rate specially in developing countries the effect of which is increasing air pollution. Different regions will have their own legal and fiscal methods to reduce pollution under a time frame. USA has the most severe gasoline regulations since 1966 under California Air Resources Board (CARB). The legislations relating to fuel quality in all countries is centered around the content of lead, sulfur, benzene and sometimes additives. Clean Air Act calls for reformulated gasoline table 2.6.2

TABLE 2.11 : Euro Specifications of Gasoline and Diesel.

Specifications of	Gasoline		Diesel	
Year	2000	2010	2000	2010
Sulphur ppm	150	50	350	50

Specifications for Diesel	USA in 1993	Europe CEN# 1996	India 1966
Sulfur ppm	500	500	10,000
Cetane no. (D.I.)		46	42
Aromatics % vol			
Density kg/m^3	<876	820-860	
Viscosity CS@40°C	1.9-4.1	2-4.5	2-7.5
Carbon residue			0.2%
Acidity (Inorganic)			0.5 mg KOH
Flash point			33/38°C

Comite European de Normalisation

shows the specifications of gasolines in three different regions. India also had modified the specifications and initiated a classic example of lead free gasoline (0.013 g/l) throughout the country by 2000 followed by reduction of benzene content to 5%. Sulfur content of unleaded gasoline will be brought down to 0.1%; compared to USA where the sulfur and benzene levels are 0.05% and 0.8% respectively. Table 2.6.2 shows the Indian specifications 1995 and 2000 of Motor Gasoline similarly the pollution control is enforced with setting of emission standards for petrol and diesel vehicles.

Indian refineries utilize maximum amount of FCC or cracker lights which are flush with olefins, this gives high amount of unsaturation to gasoline. Though alkylates give high octane, our industry depends upon the aromatics from other sources. Thus the formulation of gasoline may have to be changed in view of benzene regulations. Reformulated gasoline specifications generated by Clean Air Act prompted ARCO and other refiners to build new processing facilities to meet the higher requirements in the toxic control strategy. The first important one being benzene reduction. Cyclohexane is the key operating variable in reforming processes. This cyclo paraffin dehydrogenates to give benzene. Loa Angles Refining LAR has recently installed two dehexanisers[50] towers upstream reforming unit.

India still is supplying diesel with 1% sulphur and is likely to be reduced to 0.25% by 2000. The sulphur content in diesel to come down 0.25%. Diesel hydro desulfurisers (DHDS) units will be set up in 9 m refineries at a cost of Rs. 6000 crores. Mathura Hydro Cracker is completed by 1999 is envisaged for meeting the requirements of diesel.

2.3.8 Lube Oils

The principal source of lubricating oil is the fraction that is left after lighter components, namely gasoline, kerosene, diesel oil during crude distillation. Generally lubes have a boiling point above 350°C and these are obtained as the main products from vacuum distillation units. Residuums, after precipitation of asphaltenes are known as bright stocks and form a good source for lube oils.

2.3.8.1 Composition of Lube Oils

Lubricating oils are composed of paraffins, naphthenes, aromatics and unsaturated bodies. The chief molecular structure of lubes seems to be naphthene rings or naphthenes and aromatic rings arranged in groups of as many as six with paraffinic side chains. The more such chains the more paraffinic the oil is. Naphthenic crudes give more rings with less side chains. Normal paraffins (long chains) in this range are usually waxy in nature and have to be separated for maintaining free flow of oil. Thus a crude oil rich with paraffinic chains gives high viscosity index, less viscosity gravity constant and high stability lube oils. Depending upon the services intended for, lube oils are classified. Nelson[29c] grouped the oils into seven categories as shown below:

(1) Machine and Engine Oils (Neutral oils)
(2) Compounded Oils
(3) Turbine Oils
(4) Cold test Oils
(5) Transformer Oils
(6) Colour Oils (White Oils)
(7) Corrosive Oils

2.3.8.2 *Machine Oils and Engine Oils*

These are all high viscosity index oils as they have to serve under varying temperatures. Medium viscosity oils are suitable for most of the industrial applications. Low viscosity index oils are suitable for compression ignition engines, Aircraft engines require a very high viscosity index oils. These oils are further subdivided into straight, premium, heavy duty, etc. depending upon viscosity index and serviceable temperature. These oils also include spindle oils of (HVI). These are produced from HVI, neutral or bright stocks. Cylinder oils are also HVI grade oils, but at[40] present our country does not produce them. The availability of bright stock and HVI neutral is at present satisfactory.

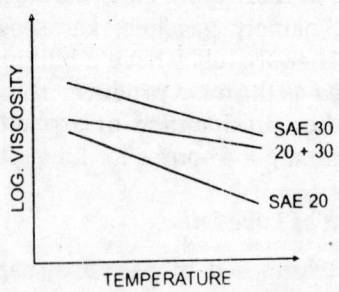

Figure 2.21   Viscosity Variation of Multigrade Oils with Temperature.

Marine grade oils are required to meet critical necessities of performance. Thus, these stocks are produced in our country as blends from indegenous and imported oils. Aviation oils are also special grade oils; such oils are not produced in India at present. The oils which are used in different services are shown below:

HVI oils for planes, high speed operations
MVI oils for petrol, diesel and stationery vehicles
LVI oils for light equipment.

In fact no single oil is suitable for all types of applications. This has called for blends of multigrade oils. All multigrade oils contain viscosity improvers. As an example, a multi grade blend of SAE 30 and 10W, can have the starting viscosity corresponding to that of SAE.10W, while in operation it can show the characteristics of SAE 30. The trend of viscosity variation of such blend is shown in Figure 2.21.

2.3.8.3 *Compounded Oils*

These are lube oils mixed with animal or vegetable oils. These oils exhibit more wettability characteristics which makes them suitable for steam engines, compressors and quenching and tempering operations. Emulsibility is considered to be more important than viscosity index of these oils.

2.3.8.4 *Turbine Oils*

High speed machinery, such as steam turbines, electric motors should have emulsion-free oils. Extreme stability is required for such oils.

2.3.8.5 *Cold Test Oils*

These oils are mainly used in refrigeration and hydraulic systems, specially designed for arctic climates. Pour point is the major important thing compared to all other properties of oil.

2.3.8.6 *Colour Oils*

These are industrial oils used in textiles, food and paper industries. Medicinal oils must be very thoroughly freed from colours and must have high stability. Depending upon the application, the concentrations of sulfur, aromatics and resin bodies are fixed.

SAE Classification for crank case oils[42]
(Viscosity in c s.)
Viscosity range

SAE No.	at 18°C		at 99°C
	Min –	Max	Min – Max
5 W		1,300	
10 W	1,300 –	2,600	
20 W	2,600 –	10,500	
20	–	–	5.7 < 9.6
30	–	–	9.6 < 12.9
40	–	–	12.9 < 16.8
50	–	–	16.8 < 22.7

2.3.8.7 Corrosive Oils

Some oils are blended with corrosive ingredients. These are used in cutting, shaping of metals. Heavy duty oils and extreme pressure lubes are of this category.

Above classification by service is no doubt suitable for almost all applications. Physical properties are to be included for a technical classification. Accordingly SAE classification and Bureau of Ships classification have come into service, which set forth the physical parameters for application.

As lube oils are made for a wide variety of applications it is not possible to make a single product to meet all the requirements. As a consequence lubes are formulated for a particular application, hence large formulations are indubitable.

Whatever may be the special formulation, all lubes are to be submitted to tests, as given below: (Special tests are not mentioned here) :

(a) Flash point   (d) Oxidation stability
(b) Pour point   (e) Carbon residue
(c) Viscosity and viscosity index

2.3.8.8 Carbon Residue

The propensity of cracking is indicated by carbon residue of the oil. Heavy oils being delicate to high temperatures, have a tendency to

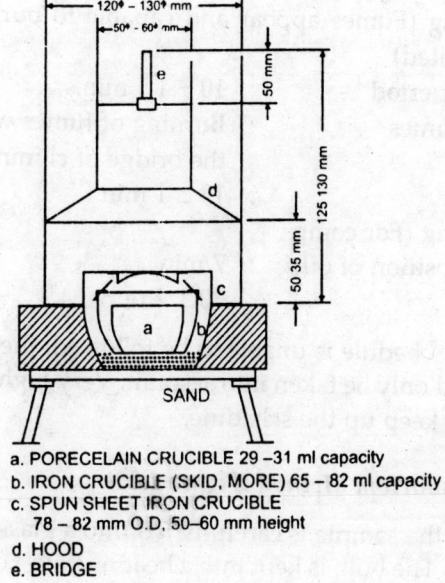

a. PORECELAIN CRUCIBLE 29 –31 ml capacity
b. IRON CRUCIBLE (SKID, MORE) 65 – 82 ml capacity
c. SPUN SHEET IRON CRUCIBLE
   78 – 82 mm O.D, 50–60 mm height
d. HOOD
e. BRIDGE

Figure 2.22   Conradson Carbon Residue Apparatus.

crack, with deposition of carbon. Lube oils are often subjected to such hazardous temperatures. The amount of carbon formed during cracking does ultimately provide an idea of usability of oil at high temperatures. This test is conducted by two methods namely Conradson method (IP13/66) and Ramsbottom method (IP 14/65; IS: 1448 P : 8).

2.3.8.9   Conradson Method

A sample of 10 gms oil is taken into a silica-crucible and heated out of contact with air. The set-up provides the necessary precautions, such that the oil is thermally decomposed out of contact with air. The decomposition is brought by means of high heating rates. As the heating continues fumes appear at the chimney top. The fumes are burned and heating rate is adjusted such that the burning sustains, but the flame is never allowed to cross the bridge of the chimney (See Figure 2.22). After the burning of the fumes further heating is continued. Afterwards, the set is cooled and the silica crucible is weighed to get the weight of carbon deposited. Expressed in percentage gives conradson carbon residue.

Prescribed heating rates:

Initial heating (Fumes appear and capable to burn continuously, when ignited)

Pre-ignition period	:	10 ± 1.5 min.
Burning of fumes	:	Burning of fumes without crossing the bridge of chimney.
	:	13 ± 1 min
Strong heating (For complete decomposition of oil)	:	7 min
Total time	:	30 ± 2 min.

If the above schedule is unable to be followed, the burning period of fumes should only be taken into account. Very high rates of heating are required to keep up the schedule.

2.3.8.10 Ramsbottom Method (Figure 2.23)

In this method the sample is carefully fed into a glass bulb which has a capillary end. The bulb is kept into a heating bath kept at 550°C. The sample is allowed to decompose for 20 minutes. After heating is over the bulb is cooled and weighed to find the carbon formed. Amount of material to be taken is inversely proportional to coking tendency of the oil.

Figure 2.23 Ramsbottom Apparatus.

Expected Carbon residue	Size of sample for experimentation
2%	4.0 gms
2–4%	2.0 gms
4%	1.0 gms

Best quality of oils give less carbon residue under serviceable conditions. Conradson carbon residue of good oils should not exceed 0.2 to 1.0%. Paraffin oils give less carbon residue compared to naphthenic oils.

2.3.8.11 Oxidation Stability

This test reveals the deterioration characteristics of oil on oxidation. Oils are generally in contact with air and hot surfaces, thus oxidation is inescapable. Higher paraffins are easily oxidised compared to naphthenes and aromatics. Major amount of lube oil is paraffin, which is more prone to oxidation. In the laboratory the test is conducted (IP–48/62) by passing air at a rate of 15 litres per hour through a sample of 40 ml kept at 200°C for 6 hrs. Second oxidation period for another 6 hrs. at the identical conditions for the same sample is carried out. After oxidation periods are over, the sample is cooled and its viscosity and carbon residue are found out and compared with the original sample. Papers on the chemistry of oxidation of fuels and its inhibition have been authored by Bettoney et al.[44] and Lod Wick.[45]

All oils of high boiling points are susceptible to this natural oxidation which proceeds by free radical mechanism. So, inhibitors prevent the free radical chain propagation and increases the resistance of oil to oxidation. All oils in general are blended with such additives.

Additives and their functions in lube oils are presented in Table 2.12. Additives supplied by different companies also accompany in Table 2.13.

2.3.9 Transformer Oils

These oils are used in electrical industry mainly for insulating, cooling purposes; additionally these oils protect the equipment from moisture compared to vegetable or coal distillate oils, petroleum oils are found to be more suitable because of high viscosity, thermal stability & hydrophobic nature. Petroleum oils used in electrical industry fall into two categories[41a]; Oils in transformers & circuit breakers etc.

TABLE 2.12   Additives for Lube Oils.

Purpose	Additive	Quantity	Ref.
Pour point depressors (depress the pour point of lubes and other fractions for crude Lauching)	Poly alkyl metha[a] crylates[b] Maleic-fumaric esters[c] Curene[c]	(0-1%) gives 5°C depression (0.1%) gives 10°C 0.2-0.5% for 9°C	a. J.K. Rostogi, 2 LAWASP I.I.T. Bombay b & c: M.C. Dwivedi & R. Vipradas, 2 LWASP d. N.K. Purohit 1st LWASP Nov. 22-23; 1978 I.I.T. Bombay.
	Chloro paraffins Wax-naphthalene phenol condensates	0.1%–5°C	e. C.C. Colyer, 7 WPC, 1967, Vol. IV, p. 294
Oxidation inhibitors (prevents oxidation of unsaturates)	Phenols; chloro acetonitrites Amines	0.4-2%	f. H.M. Noel Pet. Ref. Manual Reinhold, p. 147
Corrosion inhibitors (to prevent chemical attack on metal surfaces)	Higher organic acids alkyl ammonium acid phosphates (Benzyl phosphates)		
Anti-bearing corrosives	Hydrazides, zinc dialkyl dithiophosphate sulfurised terpens	0.5%	
Antiwear improvers (for protection of surfaces of high friction by forming protective films)	Fatty acids, dibenzyl disulfide, Tricresyl phosphates[c]		
Detergents (Clean surface exposures and eliminate deposits) keep oil insoluble material in suspension.	Organometallic compounds, Soaps of Ca, Mg, Ba, etc. Poly amines or sulfonates	2 to 10%	

(Contd.)

TABLE 2.12 : (Contd.)

Purpose	Additive	Quantity	Ref.
Dispersants (to keep sludges in suspension and not allowing them to deposit)	Organometallic compounds of Ca, Sr.	f	
Anti foams	Silicones	0.001%	
Rust preventers (prevents rusting even in presence of moisture)	Soaps of fatty acids; Metal sulfonates		
Viscosity index improvers (improve viscosity temperature relations)	Poly olefins,[d] Styrene polymers Acrylate polymers		
Oiliness (to improve thin film characteristics)	Vegetable and animal oils		
Extreme pressure agents (when heavy loads, and shock loads are encountered)	Esters of fatty acids, sulfonic acids; sulfurised molybdenum esters Penta erithritol oleates and Bisphosphoramides, dichloro, dibenzyl disulfide lead naphthenates.		
Water repellants (Settle water from oil phases)	Long chain fatty acids, silicones; boron compounds or hydrophobic polymers	200-2000 PPM	
Emulsifiers (When emulsions are required).	Sulfonates, naphthenic acids, mono alkyl amino borates, Lithium soaps.		

TABLE 2.13 Additives Supplied by Different Companies.

Gasoline anti-oxidants:	Chemical	Quantity	
Ethyl Corporation 701	2,6 di, tert, butyl phenol	1-4 Kgs/1000 Barrel	
Dupont AO-5	n-n butyl-p-amino phenol	1-5	-do-
Lubrizol 803	2,6 di-t-butyl p-cresol	2.5 to 10	-do-
Amoco 533	2,6 di-t-butyl p-cresol	2.5 to 15	-do-
UOP No. 85	dioctyl-p-phenylene diamine		
Metal deactivators			
Dupont D M D	N,N' disalicylidene 1, 2 diamine pro- pane	0.1 to 1.5	-do-
Lubrizol	-do-	0.4 to 1.2	-do-
Ethyl Corporation M D A	-do-	0.2	-do-
Corrosion inhibitor			
Dupont AFA-1	alkyl amino		
Petrolite Corporation KONTROL-157	alkyl phosphate Not known	0.5 to 10 –	-do-
Ignition Controls			
Amoco 541	Phosphorus Compounds	0.2-0.4 T	
Ethyl Corporation KC-1	Bis-thiono phosphate	0.2 T	
ICC-4	Tery methyl phosphate	0.2 T	
Monsanto CDP	Cresyl diphenyl phosphate	0.2 T	

(T: theoretical requirement to convert lead in gasoline to $Pb_3(PO_4)_2$

Detergents:			
Amoco 572	...	2.5 to 12.5 Kgs/1000B	
Lubrizol 580		1.75	-do-
Anticings Dupont	DMF Chemical	Vol. % .02 to 0.1 Quantity	
Multifunctionals			
Dupont DMA-4	Amine salts of alkyl phosphates	2.5 to 10 Kgs./1000 B	
DMA-5	DMA-4 plus cresyl diphenyl phosphate	5 to 45	-do-
Biocides			
Phillips PFA 55 MB Sohio	Biobar JF	200-1000 PPM 135-270 PPM	

belong to one class; other class of oils are generally used for impregnating cables & papers used in electrical industry. Oils used in transformers are low viscous, thermally stable but having high boiling points, in close ranges. Second type oils are highly viscous and act as insulators only. The deterioration of these oils in service is closely connected with the oxidation characteristics of oils. Oxidation products are mainly acids which cause sludge formations.

Important tests for transformer oils

(1) Acid value (ASTM D 974)
(2) Dielectric strength
(3) Flash point
(4) Pour point
(5) Sludge value
(6) Copper corrosion test

2.3.9.1 Dielectric Strength

High dielectric strength is required for oils employed for insulation purpose. Even though high electric strength does not give any indication of degree of refinement of oil or oil contaminants, it ensures the freedom of oil from water and other adventitious matter. Electric strength or break down test is performed by increasing the voltage at a rate of 2 KV/sec in between electrodes of 12.5 to 13.0 mm dia. Which are placed at a distance of 2.5 ± 0.1 mm. The cell capacity is usually between 300 to 500 ml. The voltage is continuously raised until break down takes place. A satisfactory oil test (IP 20/73) should show a break down strength of around 120 KV/cm.

2.3.9.2 Permittivity (dielectric constant)

It is the ratio of capacity of a capacitor in which oil is dielectric; to the capacity of air when this acts as dielectric. The range of permittivity should lie between 2.1 to 2.3.

2.3.9.3 Sludge Value

Oils contain some products which give rise to sludge formation. Sludge value is found out by precipitating insolubles with heptane. This test is performed before oxidation of the oil and after oxidation of the oil. High sludge formations indicate the negative quality of oil, which also indicates the high percentage of oxidative products.

2.3.9.4 *Copper Corrosion Test*

The use of copper in electrical industry is immense, hence the attitude of oil towards copper should not be calamitous. Satisfactoriness of oil can be verified by heating a copper strip in oil for 3 hours at 100°C and observing the copper strip. Discoloration of the strip is direct indication of corrosiveness of the oil as given below:

Appearance of Sample (Copper strip)

Light orange, looks like a fresh one	—	(1)
Dark orange to red or multi coloured	—	(2)
Multi coloured with greenish tinge	—	(3)
Dark grey to black colour	—	(4)

International Electro Technical Commission (IEC) test involves the blowing of air through oil at 150°C for 45 hours in presence of copper strip. After oxidation, the oil is tested for acidity and sludge.

Indian specifications of transformer oil are given below:

1.	Sludge value (Heptane by wt)	1.2 max
2.	Acidity mg KOH/gm	0.4 K(OH)/gm
	after oxidation for 164 hrs	(fresh)
	at 100°C	2.5 max
3.	Electric strength one min.	40 K.V.
4.	Saponification value mg KOH per gm in oil	1.0 max
5.	Acidity gm KOH/gms of oil	0.03
6.	Viscosity at 27°C CS	27 max
7.	Pour point °C	−10 max
8.	Flash point °C (Pensky Martin Apparatus)	140 minimum
9.	Copper strip corrosion (at 100°C for 3 hrs)	Not more than 1
10.	Resistivity at 90°C	13×10^{12} min
	ohms cms at 27°C	500×10^{12}
11.	Specific Gravity	0.89 max

2.3.10 Bitumen

Bitumen is the residual product obtained from crude distillation unit. It is essentially solid at room temperature and has got very high viscosity. Asphalt is usually a mixture of bitumen in oil, containing much mineral matter. Bitumen natural deposits are also available;

Figure 2.24 Bitumen Softening Point Apparatus.

famous Trinidad pitch lake is such a deposit. Bitumen obtained from the distillation column is poor in qualities. This has necessitated air blowing of bitumen to obtain suitable grades. Most of the bitumen is used in highway constructions, waterproofing and coatings works.

Bitumen is specified by the following two tests

(a) Softening Point.
(b) Penetration index.

These two tests do not describe the properties of bitumen completely. Other tests like viscosity, volatility, durability, ductility are necessary for obvious reasons. Outlines of these tests and significance are mentioned below:

2.3.10.1   Softening Point (ASTMD 1398 IP. 198 Ball and Ring test)

Bitumen being amorphous does not melt sharply. The softening point is found out by the Ball and Ring Test. Bitumen is first melted and casted into discs of two numbers in standard rings. A steel ball weighing (3.5 ± 0.05) gms (dia 9.53 mm) is kept on each casted disc of bitumen. The whole stand carrying these two discs with balls is immersed in a water or glycol bath. As the heating proceeds, softening occurs. The temperature at which the sample detaches from

TABLE 2.14   Bitumen IS Specifications.

Grade	Sp. Gr. 27°C	Flash point °C min	Softening point °C	Penetration	Ductility cm min	Solubility in CS$_2$ % max
75/15	1.0-1.05	200	65-80	10-20	2.5	99
65/25	"	"	55-70	20-30	10	99
75/30	"	"	70-80	25-30	3	99
85/25	"	"	80-90	20-30	3	99
85/40	"	"	"	35-45	3	99
90/15	1.0-1.06	"	85-100	10-20	2	99
105/20	"	"	95-115	15-25	2	99
115/15	1.02-1.06	"	110-120	8-20	2	99
135/10	1.02-1.07	"	130-140	7-12	1	99
155/6	1.02-1.07	"	150-160	2-12	0	-

Grade 75/15 means softening point 75°C and penetration index is 15.

the die and falls; indicates its softening point. The two samples from such rings are supposed to fall within a difference of 5 seconds. Higher the softening point, the better is consistency of bitumen. (see Figure 2.24).

2.3.10.2 *Penetration Index*

The hardness or penetration quality of bitumen is assessed by this test (IP 49/72).

A standard needle is allowed to penetrate under a load of 100 gms, through a sample kept at 25°C for 5 sec. Distance travelled under those conditions indicates the consistency of sample. The distance (easily) travelled would naturally depend upon its load, and softening point. More penetration index naturally impairs the surface applications, where hardness is sole criteria.

2.3.10.3 *Ductility (D 113)*

Bitumen must possess good ductility, when it is used in surface application. The ductility is measured by a ductilometer. Ductility is a measure of the capacity of bitumen to elongate or stretch. This test is carried out by pulling a test piece of bitumen of standard dimensions at a uniform rate (5 cm/min) keeping sample at 25°C. Higher capacity for elongation indicates that the sample is having a high ductility.

Other tests like viscosity, boiling point, ash content are desirable. Suitable solvents are added to solubulise the bitumen and then applied over the surface, when the solvent evaporates a fine layer of bitumen appears. These type of bitumens are called cutback bitumens. These cutback bitumens, after full curing period must possess the same properties as the bitumen from which they are made.

Indian specifications of bitumen are given in Table 2.14.

2.4.1 Residual Fuel Oil

A general classification for the heavier oils, known as No. 5 and No. 6 fuel oils. These are the remaining oils after the distillate fuel oils and lighter hydrocarbons are removed from the distillation towers. It conforms to ASTM Specifications D 396 and D 975 and Federal Specification VV-F-815C. No. 5, a residual fuel oil of medium viscosity, is also known as Navy Special and is defined in Military Specification MIL-F-859E It is used for steam-powered vessels and in power plants. No. 6 fuel oil includes Bunker C fuel oil and is sued for the production of electric power and gasification. Usually it can be a mixture of various residuals from refinery operations.

2.4.2 Bunker fuel quality testing: Usually the following tests are conducted:

Marine fuel deliveries are measured by volume but paid for by mass. Ensuring the fuel quality at the time of delivery and calculating the density is an integral part of good bunkering practices.Measure density and water content to calculate the mass of fuel delivered Ensure the fuel is within the required specification under ISO 8217, Viscosity, Density, Water, CCAI, Compatibility, Pour Point, Salt, Asphaltenes, Ash, Carbon Residue Density/viscosity, water content Flashpoint ,pour point Sulfur Content, Hydrogen Sulfide Microbes Contamination Heavy oils can grow microbes because of wax content. Total Sediment Potential (TSP) , Further corrosive contaminants like chlorides/ nitrates/sulfides are tested for.

2.5. 1 Nelson's Complexity Factor

The index was developed by Wilbur L Nelson in 1960 to originally quantify the relative costs of the components that constitute the refinery. In the Refining Industry, a common index termed as "EDC" - Equivalent Distillation Capacity is defined to calculate the benchmark of manpower/ investment requirement. Perhaps with the modern technology in hand the significance of this may not be lustrous Calculation of EDC is a two-step process. The first step is the multiplication of the capacity of each unit in the refinery with the Nelson's complexity factor and the second is the sum of these products to arrive at the EDC for the refinery in total.

Nelson Complexity Index : Nelson Complexity Index is a measure of secondary conversion capacity in comparison to the primary distillation capacity of any refinery. It is a guide line not only for the investment but also the value addition potential of a refinery or potential for secondary conversions. Nelson assigned a factor of one to the primary distillation unit. All other units are rated in terms of their costs relative to the atmospheric distillation unit.

Nelson's Complexity Index

Unit	Older Reports	1998 Reports
Distillation Capacity	1.0	1.0
Vacuum Distillation	2.0	2.0

Thermal Processes	5.0	2.75
(Categories 1 and 2 - 2.75)		
(Categories 3 to 5 - 6.00)		
Coking	6.0	
Catalytic Cracking	6.0	6.0
Catalytic Reforming	5.0	5.0
Catalytic Hydrocracking	6.0	6.0
Catalytic Hydrorefining	3.0	3.0
Alkylation / Polymerization	10.0	10.0
Aromatics / Isomerisation	15.0	15.0
Lubes	60.0	60.0
Asphalt	1.5	1.5
Hydrogen (Mcfd)	1.0	1.0
Oxygenates	10.0	10.0
(MTBE / TAME)		
Thermal Operations		
Thermal Cracking	3.0	
Visbreaking	2.5	
Fluid Coking	6.0	
Delayed Coking	6.0	
Others	6.0	

Perhaps these indices are only guiding in selecting certain operations. For example it is shown alkylation is more complex than Hydro treatment or hydrocracking. Any one knows the difficulty in handling hydrogen catalysts for hydro processes Similarly fluid coking and delayed coking and other operations like de-oiling or de-waxing are put in the same category which are considerably different in processing. The Nelson Complexity Index provides the nature of refinery units and the cost, without considering floating or operating costs. Fixed cost and environmental fees , local taxes and costs for extra ordinary safety measures will be in addition to these costs.

REFERENCES

1. [15, 29]a, b, c, Nelson, W.L Pet. Ref. Engineering McGraw Hill, p. 80, 189, 153, 131, 60.
2. (a) Watson, Nelson, Murphey, IEC, 27, 1935, p. 1460 & UOP Booklet No. 196
   (b) Watson, Nelson, IEC 25, 1933, p. 880.
3. Maxwell, J.B., Data Book on Hydrocarbons Van Nostrand Book Co., p. 15-25.
4. Van Winkle, M., HCP, April, 1964 p. 139.
5. [9], Perry, J.H., Chemical Engineers Handbook, p. 634.
6. Meyer, P., Trans. Inst. chem. Engrs. 12, 1934, p. 96.

200

7. Rao, B.K.B. & Takhalate, H.R., Chemical Industry Developments (India), March, 1974.
8. Bell, H.S.. American Petroleum Refining Van Nostrand, p. 47.
10. Duran, J.L., Thinh, T.P, Ramalho, R.S. & Kalia-guine, S., HCP Aug., 1976, p. 153.
11. Nat. Petroleum News, June 7, 1944, p. 372.
12. Hadden, ST., HCP July, 1966, p. 138.
13. Sherman & Kroff., J. Am. Chem. Soc. 30, No. 2, 1908, p. 1630.
14. Pomfrett, E. Ken., Gas Service Technology Earnest Benn Ltd., 1978, p. 9.
16. Griswold, J., Fuels, Combustion and Furnaces McGraw Hill, p. 192.
17. Dunstan, Science of Petroleum, Vol. 4, Oxford, p. 2974.
18. Guthrie, V.B., Petroleum Handbook McGraw Hill, 9-11.
19. Hersh, R.E., Fischer, E.K., & Fenske, M.R., IEC, 1935, 27 & 12.
20. Bretznajder, S. Physical Properties of Fluids Pergamon, p. 215.
21. Rao, B.K.B., Oil & Gas J. Dec. 13, p. 69.
22. Carl, L. Yaws., Physical Properties, McGraw Hill, p. 212.
23. Sakiadis & Coates, AICHE 1, 1955, p. 275.
24. Bondi, A., AICHE 8, 1962, p. 610.
25. John R. Hughes. The storage and Handling of Petroleum Liquids 2nd Ed. 1970 Charles Griffin & Co. Ltd., London, p. 10.
26. 7th World Petroleum Congress (1967), Vol. 7, p. 7.
27. MSS Information Corporation, New York, 1976, p. 119 & 124.
28. James H. Gary, Pet. Refining Technology & Economics, Marcel Dekker, p. 10-11.
30. Arunachalam & Rao, B.K.B., J. Chemicals & Petrochemicals, Oct. 1974.
31. Mapstone, C.E. Pet. Refiner, Feb. 1952, p. 132.
32. Raney, M.W., Fuel Additives for I.C. Engines, NDC, New Jersey (1978), p. 38 & 190.
33. Sokolov, V. Petroleum; Mir Pub. Moscow, p. 276.
34. Larson, C.M., Additives for Fuels and Lubricants, Pet. Engg., March 1955, PC-44.
35. Perry Ploss., HCP, Feb. 1973, p. 63.
36. [41]a, b. Allinson, J.P., Criteria for Quality of Petroleum Products. Applied Science Pub. (London), p. 130, 230 Chapter 14.
37. Brooks, T.G., Textbook on Petroleum Hydrocarbons, Vol. I, p. 467.
38. McCoy, R.D., Editor. Current Research in Petroleum Fuels I.
39. Sheflan, L. & Jacobs, M.B. The Handbook of Solvents D. Van Nostrand Corp. 1953, p. 31.
40. Majumdar, A., Proce. of Symposium on Lubricants Additives, Waxes & Speciality Products, IIT Bombay, Nov. 22-23, 1978.
42. SAE Handbook, 1977, p. 13.05.
43. Deen, H.E., Kaestner, A.M, & Stendhi, C.M., Part II. 7th WPC, p. 47.
44. Bottoney, W.E., Burt, J.G. & Schule, H.J., Proce. 6th WPC, 1963, Vol. VI, p. 111.
45. Lodwick, J.R., J. Inst. Petrol, 1964, 50 (491); p. 29.
46. William J. Piel. Energy Progress, Dec. 1988, A.I. ChE. N. York, p. 203.
47. T.E. Daubert; HCP, Sept. 1994, 75.
48. A.K. Jain, Hydrocarbon Technology, 15 May 1996.
49. William H. Keesom, Fred M. Hibbs, Martin A.R. Thornhill and Stuart Simson UOP 15th May, Hydrocarbon Technology, GOI, INDIA.
50. M. Johnston, K. Kumaura, R. Hinkle, and S. Melville HCP June 1999, 57.

PROBLEMS

1. Draw the ASTM distillation plots of marketed gasoline and kerosene. Convert these plots into EFV lines at atmospheric pressure and see whether there is any continuity, if not find the overlap.

2. (a) From the given data, draw TBP graph and assess the qualities of different fractions.

   (b) Convert TBP curve to EFV curve for the above crude at 150 kg/cm^2.

   Java. Indonesian crude
   API: 28.8
   P. Pt: 41 °C
   Sulfur wt: 0.11%
   Vis. cs.
        69°C: 56.1
        45°C: 128.1

Salt Kg/1000 bbl:	70
Wax wt%	23.8
C.R.	8.1%

   Light naphtha
C$_5$-100°C:	2.6% by vol.
Sp. grv.	0.701
Paraffins%	62
Naphthenes%	30
Aromatics%	8
RON	70

   Naphtha
100-155°C	3.0% by vol.
sp. gr.	0.755

Paraffins%	46
Naphthenes%	43
Aromatics%	11

   Kerosene
155-295°C:	13.4% by vol.
Sp. gr.	0.813
Aromatics vol.%	15.5
Sulfur wt.	0.04%
Smoke point	19

Heavy distillate

295–350°C	7.3% by vol.
Sp. gr.	0.834
P. Pt.	15.6°C
Diesel index	72
Catane no.	66
Sulfur wt.%	0.08
Viscosity (CS) at	(38°C) – 553

Residual oil

More than	350°C: 72.9% by vol.
Sp. gr.	0.915
P. Pt	58°C
Sulfur wt.%	0.12
Vanadium Nickel Iron PPM	1.6/0.6/24
Viscosity at 80°C	280 (cst)

3. (a) Estimate the gross calorific value of a kerosene fraction having an API gravity of 48 from eq. 2.13.

(b) Estimate the heat energy required in raising the temperature from 35°C to 82°C. The fraction has a cubic arrange boiling point 204.5°C (use eq. 2.11 and Figure 2.6a).

(c) If this fraction were to vaporise completely, find the specific heat of fraction in vapour phase (use eq. 2.12).

4. A natural gas having a calorific value of 43.4 MJ/M^2 and density of 0.632 (gms/lit) is allowed to burn at 5 milli bars; a replacement with another gas of 25 Wobbe number is sought. Find the pressure at which the gas should be combusted?

5. 3000 kilo liters of gasoline of 74 API has to be stored in a tank of suitable capacity under varying temperatures of a maximum fluctuation 22°C. Assuming an average temperature of 25°C design a suitable storage tank. Find the breathing losses.

6. (a) 150 Neutral oil, is having a viscosity of

$$\frac{134\,SSU \,/\, at\, 38°C}{(28.8\, cs/)} \quad and\ 42.4\ SSU\ at\ 99°C.$$

Find the viscosity index (consult 1P tables for getting standard value of L & H).

(b) From its VGC value classify the nature of oil (API gravity 30).

7. Estimate the viscosities of toluene and ethylbenzene, by Souder's method at 25°C.

8. Find the vapour pressure of a chlorinated methane at 120°C using Claussius—Clapeyron equation.

B.P 76°.7C.

Latent heat of vaporisation at boiling point: 7140 Cals per gm
mole. (Ans: 1440 mm Hg)

9. Vapour pressure of a hydrocarbon is represented by the
following

eqn. $\log P(atm) = \dfrac{-314.49}{T°K} + 8.05$

Calculate the latent heat at one atmospheric pressure and boiling
point 165°K (neglect liquid volume).

(Ans. 1200 Cals/g. mole)

10. The thermodynamic properties of a paraffin gave the following
values at 32°C and 3 Kg/cm^2.

Density of liquid = 0.5635 gm/cc.

Density of saturated vapour = 0.00756 gm/cc.

Latent heat of vaporisation = 85.2 gm cals/gm.

The vapour pressure as a function of temperature was found to
be represented by the equation

$$\log P(atm) = a + \frac{b}{T} + 1.75 \log T + CT$$

where   T = °K

   a = 1.756

   b = –1337.8

   c = 0.004

with the aid of Clapeyron equation test the consistency of the
above relation.

11. Explain how ATF differs from aviation gasoline. What are the
additives required for a good ATF?

12. (a) What are the different additives used in gasoline and diesel
oils?

(b) Explain the effects of paraffin rich fuel in spark ignition
engine.

3

Fractionation of Petroleum

3.1   DEHYDRATION AND DESALTING OF CRUDES

All crudes contain moisture and salts to varying degrees. Water is likely to occur in emulsion form when the crudes are naphthenic or sulfurous. No harm may be expected to the distillation column due to the presence of moisture, as there is always steam in distillation. However, crude has to be dehydrated to remove the salts. Water being good solvent for these salts, the removal is very much effective in the form of brine. Of all the existing salts, chlorides of calcium and magnesium distinguish themselves in playing an invincible role in overhead corrosion. These salts in presence of steam at 150-200°C easily hydrolise generating hydrochloric vapours. These vapours cause corrosion to the equipment. Any crude that contains[1] more than 5 kgs of total salts expressed in terms of sodium chloride per thousand barrels may be regarded as salty crude.

Long standing of crudes may permit the separation of aqua phase along with salts and other suspended impurities (brought during mining operations); in other words dehydration removes all salts. In general, dehydration of crude is practised in two stages; first at the site of mine and later in the refinery.

The following general methods are versatile for dehydration of crudes.

1. Chemical treatment
2. Gravity settling
3. Centrifugal separation

4. Electric desalter

At the oil field, salt is removed by settling or by adding chemicals or by combination of these two. Crudes possessing emulsifying characteristics are not responsive to the settling method; these demand demulsifying agents to increase the coalescence of water drops. Soda ash (0.5 to 5%)[2, 3], sodium hydroxide; salts of fatty acids (0.05 to 4%), petroleum sulfonates are such chemicals; which hasten the agglomeration of water droplets. A good amount of water should be available in crude for such treatment; lack of water demands fresh additions to the extent of 20% even. After adding the chemicals and water to the desired extent the crude mixture is allowed to stand at 75-80°C, at a pressure of 15 Kgs/cm^2 in huge tall tanks. Pressure ensures the retention of volatiles in the crude. Demulsifying chemicals, if necessary are added in very small amounts i.e. few hundred ppm. The storage capacity of such tanks is stupendous and reaches up to 3000 Kilolitres. Good separation into hydrocarbon and aqueous phases results when the crude is allowed to stand for 48 hours (See Fig. 3.1a).

Coalescence is also aided by passing the mixture through towers packed with gravel. The settling techniques are not effective and time consuming. Continuous operations are not possible, with the result large amount of space and equipment are to be isolated for this purpose. Similarly centrifugal separation is also not economical due, to the huge energy requirements and less quantity handled; all these have given the way to electric desalting.

3.1.1 Electric Desalting[4]

Simultaneous desalting and dehydration is achieved in this unit with a spectacular removal of more than 90% salt in just less than half an hour. The principle in this separation is very simple; under a charged electric field the polar molecules orient. A potential of 20,000 to 30,000 volts is applied between electrodes through which crude is passed. Water present in the form of emulsion also coalesces and agglomerates into a stream entrapping all the salts in this process. Brine collects at the bottom of the desalter, while crude floats above and forms a separate stream.

Compactness, efficiency and ease of operation of these units are indubitable and induced every refinery to adopt. Temperature during electric desalting is maintained around 90°C and a pressure of 5 to 6

Kg/cm^2 is also superimposed. Power consumption is also very small, frequently of the order 0.01 Kwhr per barrel.

Electric desalting technique is purposefully employed in treatment of distillate stocks, although the aim is different. Acid or caustic treated stocks are liable to form emulsions. These emulsions are efficiently broken by electrostatic field. Catalyst poisoning salts can also be favourably removed. Thus it can form an alternative method for pretreatment of catalytic cracker feed stocks.

Figure 3.1*b* shows electric desalting technique and Table 3.1 furnishes the information regarding various crude treatment techniques.

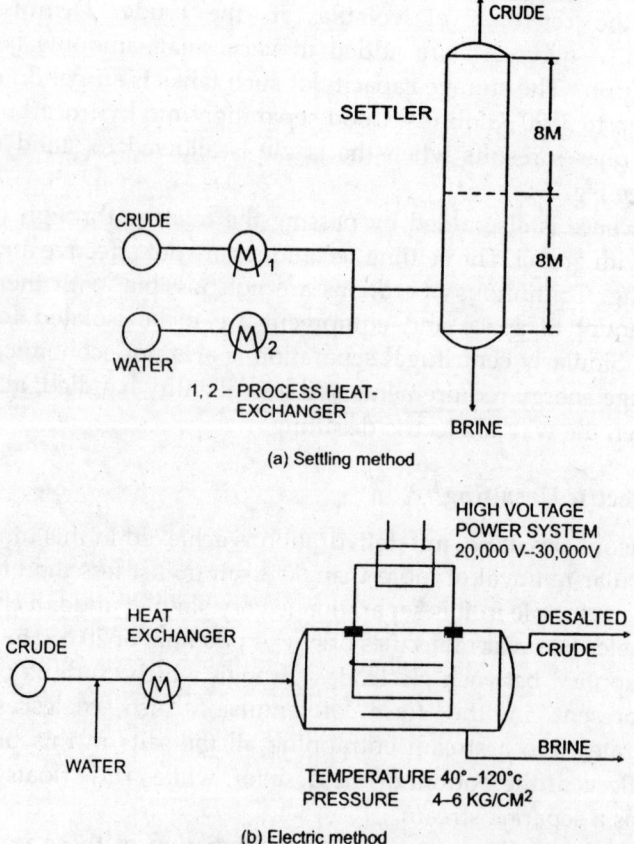

(a) Settling method

(b) Electric method

Figure 3.1*a* Desalting by Settling Method. *(b)* Desalting by Electric Method.

TABLE 3.1   Process Conditions in Treatment of Crudes[5].

	Temperature°C	Type of treatment
Chemical separation	60-90	0.05 to 4% sol. of soaps in water +0.5-5% soda ash
Electrical	60-90°C	20,000 Volts
Gravity	80-90°C	Water added upto 4%
Centrifugal	90°C	20% water added

3.1.2   Pumping of Waxy Crudes

An important factor that stands in the way of pumping crude is its pour point. Pour point is directly related to the wax content of oil. Indian crudes are waxy and hence have relatively high pour points. Middle East crudes which are credited with high sulfur exhibit astoundingly low pour points. Crudes from Assam are characteristically famous for high pour points; (wax content 16%, pour point about 30°C). Although it is not essential that all crudes from the region should display the same qualities, in fact often differences exist in the crudes obtained from the same field (different wells). Pumping such crudes, is extremely difficult. Often the temperature of the environment falls much below the pour point. It is seen that viscosity near the pour point abruptly shoots up, adding to the difficulties in pumping. Such crudes can be transported only after conditioning. At the oil field of Nahorkatiya such conditioning plant exists. This plant is only one of its kind in the world.

The principle behind the conditioning is based upon the behaviour the wax itself. Crude oils containing high wax often respond to specific type of heating and cooling. At a definite rate of heating and cooling, the response to retain the mobile and pumpable characteristics of crude are illustrative and taken advantage of. These characteristics once attained shall last till further drastic conditions of heating are administered. The conditioning of crude starts with cycles of heating and cooling. Firstly, heating is done up to 95°C, followed by cooling[6, 7] in two stages. First cooling goes down to 65°C followed closely by another cooling cycle to 18°C at a rate of 0.5°C per minute. This way, the planned operation successfully modifies the crystal structure of wax to permit the flow. Two such plants are planned by Oil India Limited at Duliajan and Moran.

Other treatments that can be given to waxy crudes are:
(a) Diluent addition   (b) Chemical additives

3.1.2.1 Diluent Addition

Solvents although they can be employed for this purpose, are not used due to the very nature of contamination of the products. Best diluent can be water, but required in huge amounts, needing large pumps and pipe lines. Economic feasibility may be ascertained before such thing is attempted.

Chemical additives: Cheap chemical additives which can depress the pour points, if available economically, then the treatment shall be capturing and becomes an irreplaceable one in industry. This can obviously remove the ills of conditioning plants where huge amounts of energy and time are lost. These additives[8] are also known as flow improvers. These additives act by changing the structure of wax and retard the crystal growth. Obviously modification of wax structure reduces viscosity. Even small quantities of additives. (say 300-800 ppm) can markedly depress the pour point by 10 to 16°C. Usually these additives will be incorporated after preliminary desalting operations at the mine. After stabilisation which lasts for a period of 40 to 60 hrs, the crude will be ready for transportation to refinery. The stabilised crude will be in a position to retain the influence of additives at least for a month's time during which refining may be done to completion. The additive being hydrocarbon in nature, heating crude to distillation temperature can easily decompose.

A note on Indian crudes at this juncture is worthy. Crudes of India being waxy, naturally drain a substantial amount of foreign exchange in getting the flow improvers. This has become imperative on the part of Government to go for native additives. Accordingly indegenous additives like 'Flowcell'[9] developed by Excel Industries and SWAT 104, 105 & 106 developed by Regional Research Laboratory, Jorhat, play a vital role in transportation of Eastern and Bombay High Crudes in the near future. These additives, it seems remarkably lower the pour points by 10 to 15°C at a marvelously low concentrations of 300 ppm. Also simultaneous reduction in viscosity down to 7 to 12 cps helps in easing the pumping characteristics.

3.2 HEATING OF CRUDE: PIPE STILL HEATERS

Refinery processing is basically a fact rested upon effective distillation and condensation. Before the introduction of pipe still heaters (1918) into refinery operations, the direct heating stills (convective type) were used for all heating operations. Non-uniform heating and excessive heating could not be abated satisfactorily in these stills; with these inescapable difficulties, the refinery could not be magnified or utilised for high capacities at short notice, thus, one has to be content with small refinery sizes. The introduction of pipe still heaters, and other chemical engineering appliances into refinery, has tremendously increased the activities in trade. Modern refineries are thus compact in size and accurate in control, with a degree of flexibility in operations for quality and quantity production; hence at short notice the production pattern can be changed without much change in economy. The versatile use of pipe stills from topping of crude to cracking techniques is exciting.

Pipe still heaters are different from old still heaters in one way; the radiant section is separated from convection section. Prudent planning of radiant and convection sections are primary considerations for a pipe still for efficacious operations. The recent developments in pipe still heaters have brought the heat transfer in radiant section as high as 70%. The stack losses and heater losses are brought down to 12% and 5% respectively followed by a close response to control.

Pipe heaters can be conveniently categorised into three types:
(1) Box/Rectangular, (2) Cylindrical and (3) Radiant Wall.

All these furnaces have got separate radiation section and convection section. The most universal classification is based on direction of tubes as well as shape of furnace and mode of application of heat.

3.2.1 About Radiation

In most of the furnaces, the direction of tube is horizontal in all box type heaters and vertical in cylindrical stills. Radiant walls also use horizontal tubes; however tubes can be placed vertically also. The radiant section design is based upon Stefan-Boltzman equation.

Stefan Boltzman Law gives energy radiated per second:

$H = e\sigma AT^4$

Q = radiant energy = Ht J

e = emissivity (0-1)

σ = Stefan-Boltzmann constant
    = 5.67×10^{-8} J/(s-m²-K⁴)

A = surface area of object

T = Kelvin temperature

210

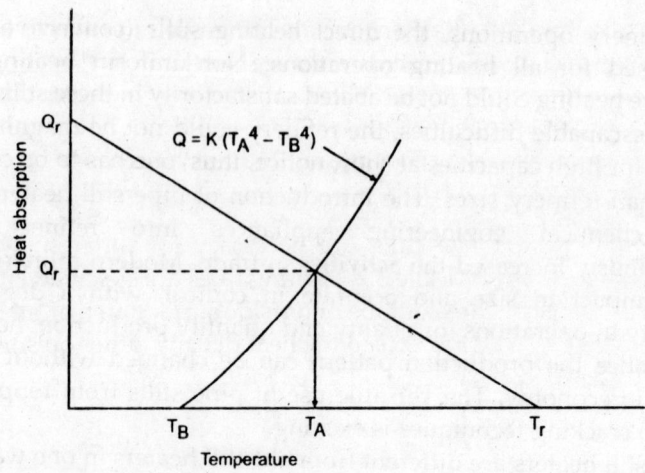

Q_F = Rate of heat release
Q_r = Rate of heat absorption
T_A = Temperature of combustion space
T_B = Temperature of receiving body
T_r = Psuedo flame temerature

Figure 3.2 Locus of Heat Absorption (Stefan Boltzman Equation).

$$Q_r = 4.88 \times 10^{-8} \left[T_G^4 - T_B^4 \right] \qquad \text{... (3.1)}$$

Where T_G and T_B are temperatures of furnace and tube in °K. Q_r radiant heat transfer (K. cals) per unit area (Sq. meter) and time (hr).

The equation is graphically represented in Fig. 3.2. The locus of Q_r is shown to be the point of intersection of tube skin temperature and temperature of combustion space, which incidentally relates the pseudo flame temperature (T_r).

This equation enables one to judge the amount of heat transfer as proportional to the difference of fourth power of temperature of the heat generating body and the receiving body. This differs very much from convection and conduction, where heat transfer is proportional to gradient only i.e. the temperature difference of the operating levels. However, here the properties of material receiving the radiation are also important. The properties are not like conductivity or heat transfer coefficient, as intended in other heat transfers, but the visible colour of the material only. As black body absorbs all the radiation that falls on it, its emissivity is expressed as unity and other bodies (coloured) graded accordingly. Except mono atomic gases, all other gases and vapors aid this mechanism. The angle of vision or in other words, the surface which can directly see the flame is also very much important.

3.2.2 Considerations for Designing Radiant[10] Section

(1) Heat Duty

Different heat transfer rates are necessary for different types of operations as shown in Table 3.2. Heating rates are specified either on the basis of projected area of tube or outside surface area of tube. Care should be exercised not to exceed the heat density either on surface of the tube or in furnace.

TABLE 3.2   Heat Duty for Various Operations.

	Projected area basis	Velocity M/Sec.
	kJ/hr. Sq. meter	
Topping operations	25-50 × 10^4	2.5
Vacuum distillations	25-30 × 10^4	2-2
Mild cracking	14-20 × 10^4	2-3
Severe cracking and coking	25-35 × 10^4	2-3
Treatment operations	20-25 × 10^4	1-2
Cracking of gases for olefins	13-16 × 10^4	8-10

(2) Air Fuel Ratio

Fuel in radiant section does another duty above the normal heating duty, i.e. it assists radiant heat transfer owing to the presence of tri or multi molecular gases. Consequently a fuel rich in hydrogen gives better radiant heat transfer due to the formation of water vapour after combustion. The ratio of water vapor to carbon dioxide usually depends upon the air fuel ratio.

Excess of air fuel ratio no doubt decreases the attainable maximum temperature, but increases the partial pressure of carbon dioxide and water vapor, hence increases heat transfer.

(3) Tube Spacing

Generally tubes are placed in one or two rows in radiant section. More than two rows are not opted due to the following reasons:
(a) The cost of structure support of such rows is prohibitive,
(b) Maintenance cost shoots up,
(c) It is seen that third row absorbs only 8% of heat transfer while first and second rows absorb more than 90%. Thus the amount of

heat transfer is not economically balanced with the overhead charges.

To clarify further, number of rows bring about blanket-effect and all the tubes cannot see the flame; hence a single row is even regarded as sufficient. The spacing of two tubes is measured by center to center distance. Experimental studies showed that when the tubes were spaced at larger distances, the heat transfer was more. In general, in the furnace, the tubes are spaced at a distance of 2 to 3.5 times outside diameter (OD) of the tubes. It may be seen that 5 tubes of 5 cm OD spaced at 2 OD distance will absorb the same amount of heat as 6 tubes spaced at 1.5 OD. So an optimum tube spacing can reward the designer.

(4) Over Burdened Tubes

Some of the tubes (blackened ones in Fig. 3.3) receive heat from radiation as well as from convection, as the gases pass over this bank of tubes. This way the tubes receive more heat than they can bear or is required. Such over heating exacerbates coke formation and for this reason, these tubes are utilised for superheating purposes.

(5) Cross Over Temperature

The net amount of heat transfer in radiant section is dependent upon the available temperature of the gases. Modern furnaces extract 60-70% of the net heat available, in radiant section alone. So the temperature of the gases leaving the radiant section has to be calculated from heat balance.

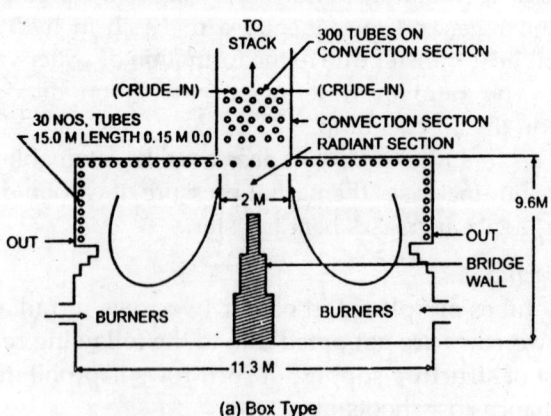

(a) Box Type

Figure 3.3 Types of Furnaces.

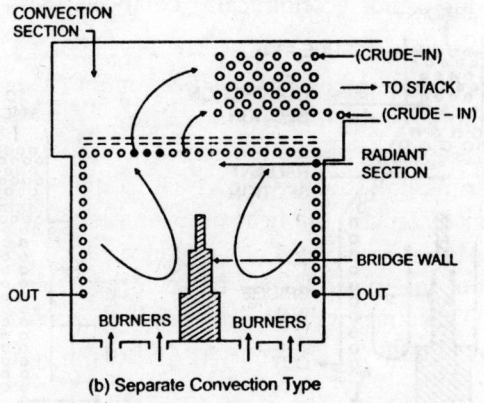

(b) Separate Convection Type

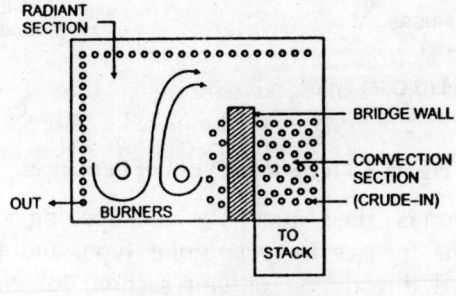

(a) Down Convection Type

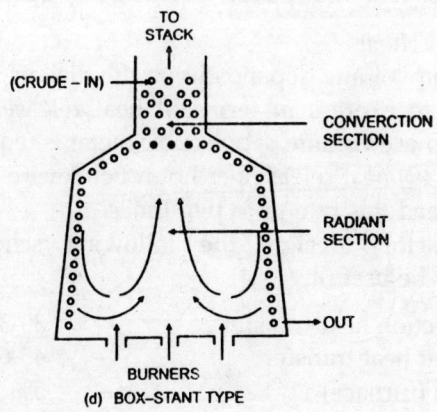

(d) BOX-STANT TYPE

Figure 3.3 (*Contd.*) : Types of Furnances.

As was mentioned earlier, the convection zone is situated away from the radiant zone, the passage leading from radiant section to

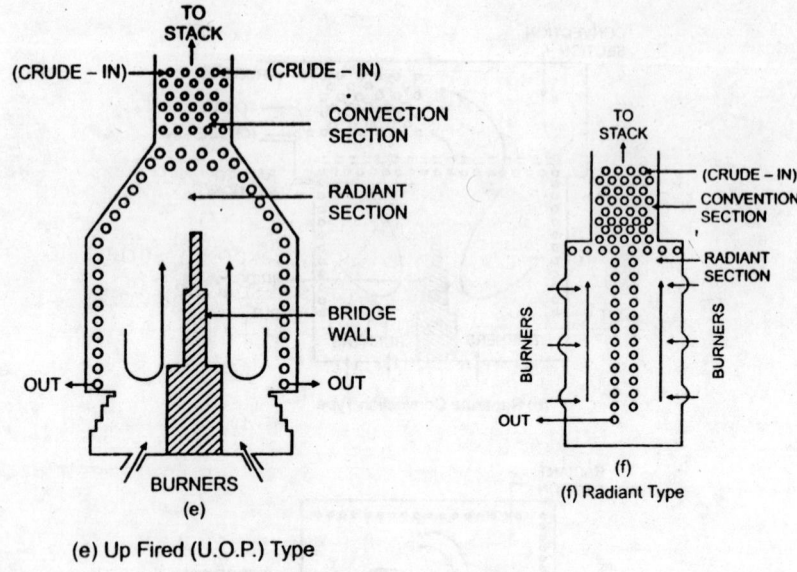

Figure 3.3 (*Contd.*) : Types of Furnances.

convection section is called 'duct'. The duct opening is usually 1/3 of the width of the furnace in rectangular type, and the convection section is placed directly on radiant section; for circular type or separate convection type or down convection type it is better to calculate the duct area on the basis of flue gases generated.

(6) *Combustion Volume*

Radiant section volume depends upon the fuel-air ratio. However, it is customary to express in terms of heat release. For moderate heating rates, furnace volume of one cubic meter is required for 28,000 Kcals (11.7×10^4 kJ) heat release per hour. For severe operations heat density may exceed this rate upto two times.

For a satisfactory design, the following schedule of heat distribution may be usefull:

Convection heat transfer	—	30-35%
Radiant heat transfer	—	45-60%
Losses (furnace)	—	5%
Stack losses	—	12%

Design of a furnace is based upon Hottle, Wilson method and radiant heat absorption is given as[11]:

$$R = \frac{1}{1 + \dfrac{G\sqrt{Q/\alpha A_c}}{S}} \times 100 \qquad \ldots (3.2)$$

where $\quad R$ = % heat absorbed in radiant section

G = Air: Fuel ratio (Wt. basis)

α = Factor to convert actual exposed surface to cold surface 0.986 for two rows at spacing 2.0D

if Q is in kJ/hr, S = 14200 (Area in M^2)

if Q in Btu hr^1, S = 4200 (Area in Ft^2)

if Q in Kcal hr^1, S = 6930 (Area in M^2)

A_c =Area of wall having tubes in front of it.

The trend of α is shown in figure 3.4, depending upon the number of rows of tubes and pitch distance between tubes, the factor varies,[12] for one row of tubes at pitch distance 2D, α is equal to 0.88 and for two rows for same pitch it is 0.986. When more than one row is present, α factor can be estimated to total number of rows of each individual row to total rows. When two rows are there, the first row (nearest to wall) will have an α = 0.69 while the next row will have α equal to 0.3.

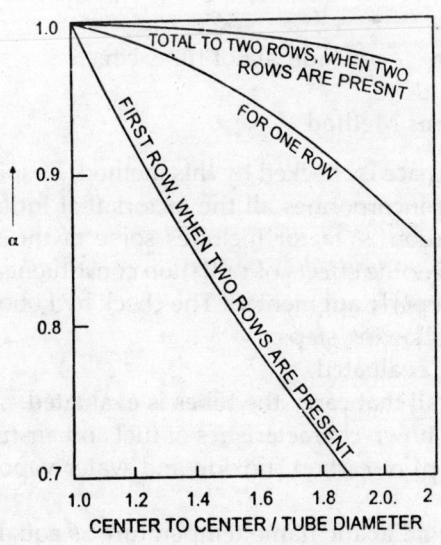

Figure 3.4 α-Variation with Tube Spacing and Number of Rows.

When area of wall (A_c) with mounted tubing in front of it is calculated, it is obviously immaterial how many rows are present as they do not change the significance of arrangement.

Lobo and Evans[13] method gives a good procedure for checking the furnace performance which is already in working condition but, it is seldom used for design (method) calculations.

The remaining heat from the gases going out of radiant section is significantly extracted in convection zone.

Convection section resembles any other heat exchanging system, and the design data of such equipment[14] may furnish valuable information. Unlike radiation coefficient, convection coefficient depends upon many factors and the coefficient derived by conventional methods does not include the effects of unremitting[15, 16] radiation, prevailing in the furnace. The first row of convection section is usually submitted to such situation which leads to incredibly high heat transfer rates. Monrad's[17, 18] empirical relation for convection coefficient is presented in eq. 3.3.

$$h_c = \frac{2.845 G^{0.667} \times T^{0.3}}{D^{0.33}} \qquad \dots (3.3)$$

where $G =$ mass flow rate $\dfrac{kg}{sec. \, M^2}$

$T = \qquad °K$

$D = \qquad$ outside dia of tube, cms.

3.2.3 Lobo-Evans Method

Working of a furnace is checked by this method. Despite it looks as an empirical one, it incorporates all the factors that influence radiation. In Hottel's equation 's' factor includes some of the factors that aid radiation, however the effects of radiation constituents and emissivity factor are not properly augmented. The check by Lobo-Evans method consists of the following steps:

(a) α-factor is evaluated,

(b) Area of wall that carry the tubes is evaluated

(c) With the known characteristics of fuel and air-fuel ratio, partial pressure (p) of carbon dioxide and water vapor is found out. Fig. 3.5a.

(d) Assuming adiabatic flame temperature as equal to bridge wall temperature, tube skin temperature is found out.

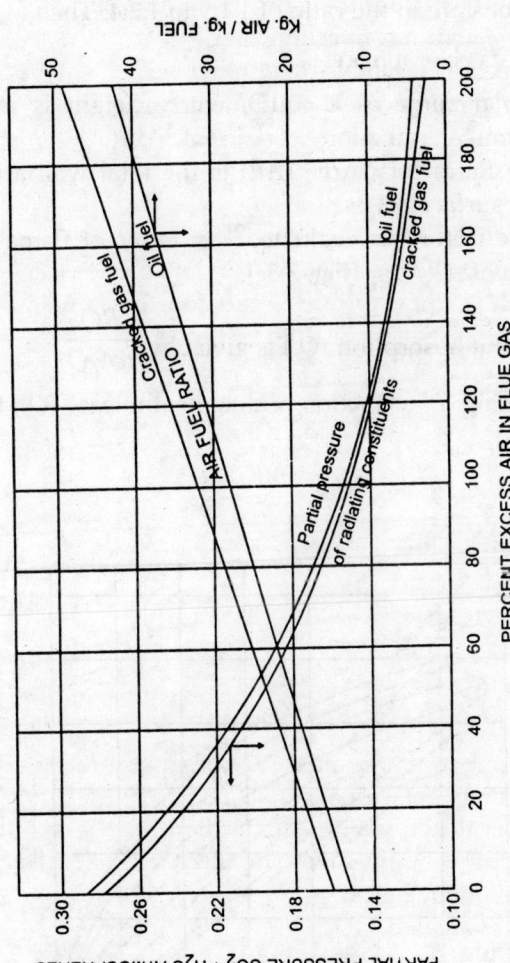

Figure 3.5a   Combustion Data on Cracked-Gas and Oils Fuels
Lobo and Evans, (Trans. A.I. ch. E.)

(e) Flame emissivity is related to partial pressure of radiant constituents (gases) and mean radiant beam length (L) (Fig. 3.5b).

Mean length of radiant beam is dependent on the geometry of the furnace.[20]

For rectangular type of furnace where length: breadth: height; dimensions are in the ratio of 1:1:1 to 1:2:4; Then

$$L = 1/3 \sqrt[3]{\text{Vol. of furnace}}$$

for circular furnaces L = 1D, where height is many times diameter.

Effective Refractory Area (AR) is the total wall area less the effective surface αA_c.

(f) Exchange factor[14] is evaluated[21] in terms of flame emissivity and ratio of $AR/\alpha A_c$ (Fig. 3.5c).

(g) Rate of heat absorption (Q_r) is given by $\dfrac{RQ}{\alpha A_c Q}$;

by assuming 'R' a fraction of total heat release (Q), this can be calculated.

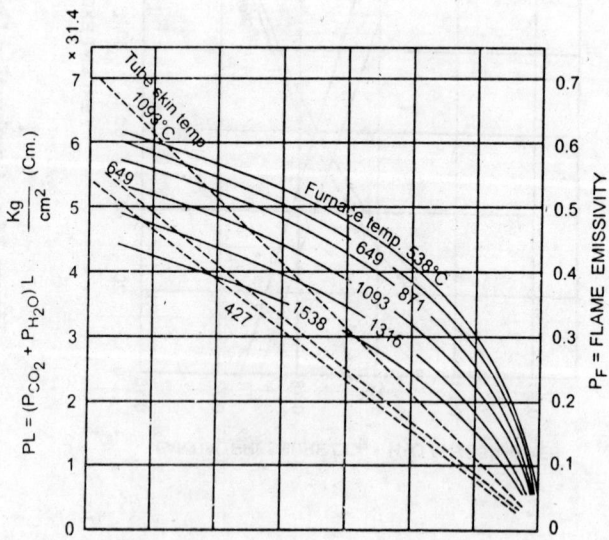

Figure 3.5b   Flame Emissivity Lobo and Evans, (*Trans. A.I. ch. E.*)

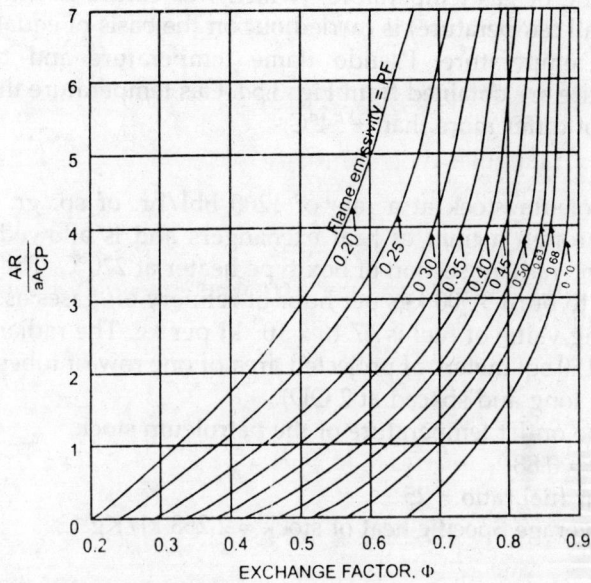

Figure 3.5c Over-all Exchange Factor Lobo and Evans, (*Trans. A.I. ch. E.*)

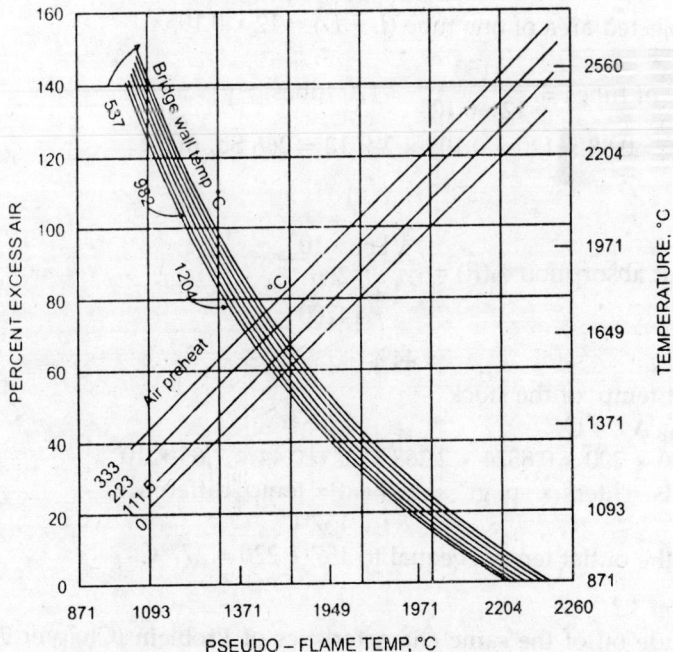

Figure 3.5d Pseudo Flame Temp. for Cracked-Gas
Fuels Lobo and Evans, (*Trans. A.I. ch. E.*)

Checking of gas temperature, (which was earlier assumed as the bridge wall temperature) is carried out on the basis of equation (3.1). The gas temperature, Pseudo flame temperature and tube skin temperature are obtained from Fig. 3.5d. Gas temperature thus found should not differ more than ± 54°C.

Problem 3.1

A petroleum stock at a rate of 1200 bbl/hr. of sp. gr. 0.8524 is passed through a train of heat exchangers and is allowed to enter directly the radiant section of box type heater at 220°C. The heater is designed to burn 3,500 kgs per hour of refinery off gases as fuel. The net heating value of fuel is 47.46×10^3 kJ per kg. The radiant section contains 150 sq. meters of projected area of one row of tubes (10.5 cm OD, 12 m long and spaced at 2 OD).

Find the outlet temperature of the petroleum stock,

Data $\alpha = 0.88$

    Air fuel ratio = 25

    Average Specific heat of stock = 2.268 kJ/Kg°C.

Solution

Total heat liberated = $47.46 \times 10^3 \times 3,500$

                    = 1.66×10^8 kJ per hour

Projected area of one tube $(L \times D) = 12 \times 0.105$

No. of tubes = $\dfrac{150}{12 \times 0.105}$ = 120 tubes

$\alpha A_c = 0.88 \times 120 \times 0.105 \times 2 \times 12 = 266$ Sq. m.

Heat absorption %(R) = $\dfrac{1}{1 + \dfrac{\sqrt[25]{\dfrac{1.66 \times 10^8}{266}}}{14200}}$

                    = 44%

Outlet temp. of the stock:

$MC_p \Delta t = Q$

$1200 \times 200 \times 0.8524 \times 2.268 \times \Delta t = 0.44 \times 1.65 \times 10^8$

(bbls × liters × sp. gr. × sp. heat × temp. difference)

                    $t = 157°C$

So the outlet temp. is equal to 157 + 220 = 377°C

Problem 3.2

Crude oil of the same characteristics of Problem (Chapter 2). No. 2.1 is to be heated in pipe still heater. The crude enters the convection section at 100°C and then enters the radiant section at a rate of 1000 bbls/hr. Maximum temperature of the crude at outlet is 350°C.

(a) Find the area of convection section and radiant section. Estimate vol. of furnace.
(b) Calculate the pressure drop in the furnace.
(c) What is the cross over temperature?
(d) Show the arrangement of tubes in radiant section.

Assume 30% Heat transfer in convection
    18% Heat losses.
Sp. gr. of crude = 0.8570
Sp. heat of crude = 0.67 cal/gm°C
Fuel net heating value: 13,000 Kcals/Kg.
Radiant heat transfer rate: 17×10^4 kJ/sq. meters
                            (on projected area basis)
Average sp. heat of flue gases: 1.28 kJ/Kg°C

Solution

Total heat liberated by fuel = Heat to raise the temp.
                               of crude + latent heat + losses.
Heat for raising the temp. of crude from 100°C to 350°C
    $= 1000 \times 200 \times 0.8570 (350 - 100) (0.67)$
    $= 297 \times 10^5$ Kcals or (1243.46×10^5 kJ) — (a)

Latent heat is to be calculated on the basis of percentage vaporisation obtained from the TBP curve of this crude. Fractions, whose boiling point is below 350°C will vaporise. Hence the latent heat for such fractions has to be found out as shown in Table 3.3

(a) From Table 3.3 latent heat 94.22×10^5 Kcals — (b)
    Heat duty $(a + b) = 391.22 \times 10^5$ Kcals/hr.
with losses 18% total =
    heat duty $(Q)_c = 461.61 \times 10^5$ Kcal/hr.
                  or
        1932.77×10^5 kJ/hr.
Amount of fuel required

$$= \frac{461.61 \times 10^5}{13,000} = 3552 \text{ Kgs/hr}$$

Feed outlet temperature in convection is roughly calculated as
    Total temperature difference × % Heat transfer in convection
    $= (350 - 100) (0.30) = 75°C$
    i.e. $100 + 175°C$.

The outlet of the feed in convection section is 175°C, % evaporation upto this temperature from 100°C is 11.7. This amount of crude is kept in liquid state by applying pressure. The pressure required is obtained from the nomogram Fig. 2.2a (Chapter 2).

Convection Section

Sensible heat from 100°C to 175°C	=	89.1×10^5 Kcals
Latent heat (11%) Vaporisation	=	42.28×10^5 Kcals
Total heat	=	131.38 Kcals
		or 551.8 kJ.

Remaining heat is absorbed in radiant section

Total heat	:	391.22×10^5
Convection heat	:	131.38×10^5
(Qr)	=	259.84×10^5
		or 1091.33×10^5 kJ.

$$\text{Area of tubes} = \frac{Qr}{0.15 \times 15 \times \pi \times 12 \times 10^4} = 91 \text{ tubes}$$

(length assumed 15 meters Dia 0.15 m)

Volume of furnace—(l)

Appropriate heat density per cubic meter

$= 12 \times 10^4$ Kcals.

$$\text{So, vol. of furnace} : \frac{\text{Total heat}}{\text{Heat density}} = \frac{1932.77 \times 10^5}{12 \times 10^4}$$

$= 1610$ cubic meters.

Arrange tubes as

30 + 30 + 31 i.e. 30, 30 tubes on side walls and 31 tubes on top wall

Spaced at 2 *OD* and providing excess space for two

more tubes on side walls

Wall height: 32 × 0.15 × 2 = 9.6 m

Width of furnace : 31 × 0.15 × 2 = 9.3 m

Length of tube : 15 m.

In the roof of furnace a length of 2.0 meters is reserved as opening for convection section inlet, this space is utilised for superheating purpose.

So total width : 2.0 + 9.3 = 11.3

Furnace volume checked by this arrangement

$= 9.6 \times 11.3 \times 15 = 1595$ meters^3

which is comparable with figure (l).

In convection section, the area is calculated on the basis of average convection heat transfer coefficient. Convection coefficient is evaluated by knowing the flue gas velocity—

Amount of fuel burnt: 3552 Kgs/hr.

Amount of air used: 3552 × 25 Kgs/hr.

Total amount of gases: 92,352 Kgs/hr.

Assuming fuel + excess air as combustion products.

$$\text{Flue velocity (mass)} = \frac{92{,}352}{\text{Area of convection duct}}$$

$$= \frac{92{,}352}{20 \times 15 \times 60 \times 60}$$

$$= 0.884 \text{ kgs per Sq. meter. Sec.}$$

convective heat transfer $(Q_c) = A_c U_c (LMTD)$

$$A_c \text{ (convective heat transfer area)} = \frac{Q_c}{U_c(LMTD)}$$

Calculation of LMTD : See Fig. 3.6.

LMTD:　$\Delta t_1 =$　gas temp. inlet convection　$-$　outlet temp. of crude
　　　　　　　　(t_1)　　　　　(T_1)

　　　　$\Delta t_1 =$　gas temp. outlet　$-$　inlet temp. of crude (100°C)
　　　　　　　　t_2　　　　　T_2

$$LMTD = \frac{\Delta t_1 - \Delta t_2}{2.303 \log \dfrac{\Delta t_1}{\Delta t_2}}$$

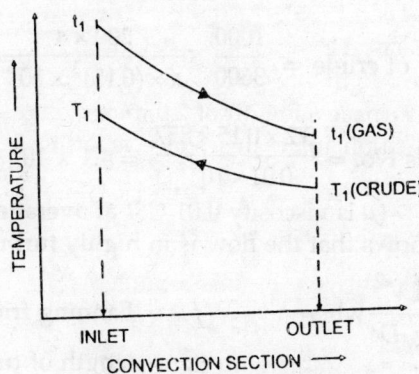

Figure 3.6　LMTD Calculation.

Gas temp. at inlet to the convection section is calculated as shown in part (C) of this problem.

So : $\Delta t_1 = 695.6 - 175° \approx 520$

$\Delta t_2 =$ the difference between outgoing temperature of gas and incoming crude should be atleast 200°C to maintain good draft.

$$LMTD = \frac{520 - 200}{2.303 \log \dfrac{520}{200}} = 334.8°C$$

U_c is obtained as $85.86 \dfrac{KJ}{hr.\ Sq.\ m°C}$

(Nelson p. 613)

Area in convection zone $= \dfrac{5518 \times 10^5}{334.8 \times 55.86} = 1920$ Sq. meters.

No. of tubes $= \dfrac{1920}{\pi \times 0.15 \times 15} = 388$ tubes

(Area of tube πDL)

(b) Pressure drop in convection section (P_c) pressure drop in radiant section (P_r) + Pressure to be imposed to restrict vaporization (P_l)

Total pressure drop due to friction

Mass velocity of crude at

inlet conditions $= \dfrac{1000 \times 200 \times 0.8570 \times 4}{3600 \times \pi \times (0.15)^2}$

Velocity of crude $= \dfrac{1000}{3600} \times \dfrac{200 \times 4}{\pi \times (0.15)^2 \times 10^5} = 3.2$ M/sec.

Reynolds No. $= \dfrac{3.2 \times 0.15 \times 857}{0.01 \times 10^{-4}} = 4.1 \times 10^8$

(μ is viscosity 0.01 CSt at average temp.)

Reynolds no. shows that the flow is in highly turbulent region.

$P_c + P_r = \dfrac{4fLV^2}{2g_c D}$ where

f : Fanning friction factor (.005)

L : length of tubes

TABLE 3.3 From TBP Analysis.

Fraction	B.P.C °C	Vol.%	API	Average mol. wt.	Latent heat kcal/kg	Sp. gr	Total heat (× 10⁴) Kcal/
Losses	—	0.47	—	—	—	—	—
—LSR	C₅— 93	3.23	69.6	80	—	0.7035	—
Cat. Reforming feed	93–176.7	11.71	46.9	120	65	0.7932	120.748
Heater oils	176.7—226.6	30.61	34.8	200	58	0.8508	302.098
Furnace oil	226.6—326.7	13.69	31.3	230	55	0.8692	292.668
Light FCU Feed	326.7—365.5	9.05	29.0	260	42	0.8816	226.680
Heavy FCU Feed	365.5—543.3	26.97	28.1	—	—	0.8865	—
Reduced Crude	>543.3	4.22	20	—	—	0.9340	—
							942.196 × 10⁴

Latent heat required for vaporising of crude from 100–350°C (65.06%) vaporisation.

$$V : \text{velocity of liquid}$$
$$g_c : 9.81 \text{ m/sec}^2$$

$$= \frac{2 \times 0.05(3.2)^2 \times 15(91 + 272)}{0.15 \times 9.81} = 375.7 \text{ meters}$$

or 32.2×10^4 Kgs/Sq. meter.

Total pressure at which liquid is to be pumped into the furnace:
$P_c + P_r + P_f$

(To simplify two phase flow, it has been assumed that under a pressure of P_f, the crude evaporation may be neglected, hence liquid volume is taken into consideration).

Back pressure required in convection and radiant sections:

The outlet temperature of convection section is 175°C% evaporation upto 175°C from 100°C = 11.7%. This 11.7% of crude should be kept in liquid state by imposing pressure i.e. operating under pressure. The back pressure required in convection section obtained from the nomogram Fig. 2.2a (Chapter 2), is equal to 11.2×10^4 kg/Sq. meter.

Similarly for radiant section (nearly evaporation takes place from 175°C – 350°C) is equal to 28×10^4 kgs/M^2.

So the total pressure in the furnace

$$= 11.2 \times 10^4 + 28 \times 10^4 + 32.2 \times 10^4$$
$$71.4 \times 10^4 \text{ kgs/Sq. meter.}$$

(c) Cross over temperature:

This is obtained by heat balance in the radiant section:

Total heat developed–Heat absorbed in radiant section–losses in radiant section = Heat going to convection section

Assuming 1% heat losses in radiant zone,

i.e. $1932.77 \times 10^5 - 19.32 \times 10^5 - 1091.33 \times 10^5$

$$= mc_p \Delta t.$$
$$= 92.352 \text{ (kgs/hr)} \times 1.28 \times \Delta 1 -$$

m = total flue gas
c_p = Sp. heat
Δt = difference in temp.

i.e. Δt = 695.6°C i.e. 665°C above room temperature

(d) Arrangement[22] is shown in Fig. 3.3a.

3.3 DISTILLATION OF PETROLEUM

Distillation is a separation technique used for separation of soluble liquid mixtures into individual components. Petroleum being a mixture of hydrocarbons has a boiling range of – 160°C (methane) to + 1000°C or more (pitch) i.e. to say a mixture of gas, liquid and solid, requires an effective and economic distillation to process into a number of cuts of small boiling range. These cuts are later processed and tailored to suit the requirements of consumers. Modern refinery techniques have meticulously laid the way to recover as many fractions as possible from crude, discarding the least possible in view of the binding situation i.e. dearness of crude.

The fractions in demand are in fact the property of a country, (although most of the fractions are common and in good demand, though not to same extent), hence a refinery should have the facilities for such fractions; this in turn makes each refinery a complete entity.

It may be mentioned, a country like the USA is deeply involved in production of more gasoline; frequently converting 70% of crude to gasoline. Whereas in a country like India, the picture is different, the accent being on middle distillates. This necessity has made Indian refineries to exalt the production of more middle distillates. It is true that the nature of crude plays an indecorous role in displaying the characteristics of its fractions, often defeating the aims of refiners. Present and future demands, future expansion, conservation of stocks etc. are some irksome spells, and a refiner has to give deep thought for all anticipated situations.

The basis of refinery distillation[23] design rests completely on TBP tests. Distillation of crude mainly takes place in two stages. First stage distillation is carried out at atmospheric pressure, hence the name 'Atmospheric Distillation unit' (ADU) is conferred on it. The undistillated portion of crude, called reduced crude is further distilled under reduced pressure in a second unit known as 'Vacuum distillation unit' (VDU). The maximum pressure in an atmospheric column seldom reaches two atmospheres and at the top of the column the pressure is only few centimeters of mercury above atmospheric pressure. These two columns differ from conventional towers in practice in a peculiar and conspicuous way by not providing any reboiler. This has been overcome by heating the feed to maximum permissible temperature only once and allowing it to flash in towers. The maximum temperature allowed in topping operations is 375°C. Higher temperatures are not permitted due to degradation of crude by thermal cracking.

When the crude contains a good amount of soluble gases, to avoid load on ADU, a preflashing or topping column is employed. Preflashing is also useful when crude has to be transported to a long distance. Light-ends-free crude gives no problem in transportation. A crude containing less than 6% light ends (gases) usually offers no problem in transportation. Preflashing is conducted at 100°C under a pressure of 3-5 atmospheres to remove these light ends.

3.3.1 Arrangement of Towers

Distillation or flash vaporisation is conducted in two units in series, when it is known as two stage distillation unit. Seldom do stages exceed three. These towers are classified into three distinguished types[24]:

Figure 3.7 shows the arrangements of these types *viz.*,

   (a) Top tray reflux,
   (b) Pump back reflux, and
   (c) Pump around reflux towers.

3.3.1.1 Top Tray Reflux

This tower has reflux at the top tray only and the reflux is cooled and sent into the tower. Fig. 3.7a. This defies reflux on any other plate.

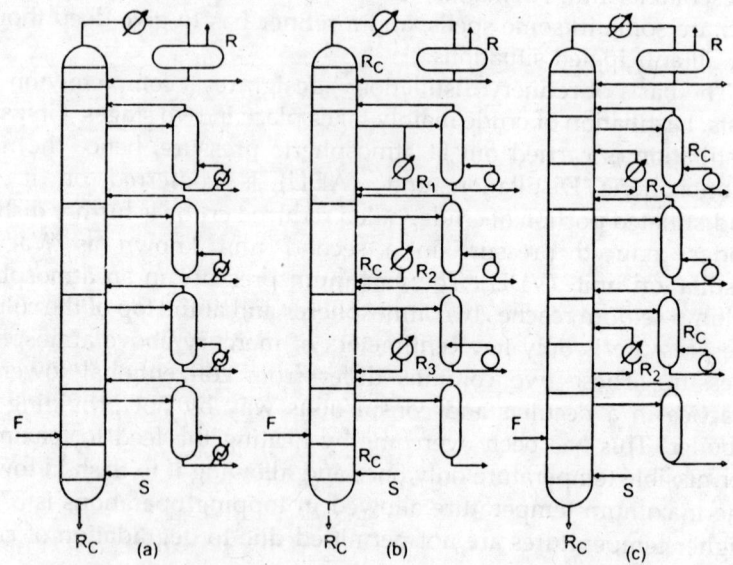

F – FEED, S – STEAM, R – REFLUX CONDENSER, R_1, R_2, R_3 – REFLUX
R_C – REDUCED CRUDE

Figure 3.7 Arrangement of Distillation Towers
a–Top Tray Reflux, *b*–Pump Back, *c*–Pump Around Reflux.

Consequently, heat input to column is through crude (bottom) and removal is at the top. This creates always 'a build up of vapor' necessitating larger tower diameters. Obviously the reflux is not proper and quality of fractions is not satisfactory. Economic utilisation of this heat is also not possible. However, the unit is simple in design and operation. But quality of fractions cannot be sacrificed for obvious reasons. Hence in practice this type of arrangement is not preferred.

3.3.1.2 *Pump back reflux (Fig. 3.7b)*

In this arrangement reflux is provided at regular intervals. This helps every plate to act as a true fractionator, due to the fact that there is always good amount of liquid. The tower is uniformly loaded, hence a uniform and lesser diameter tower will do. The heat from external reflux can be utilised, as it is at progressively higher temperatures. However, the design and operation of such towers are costly, but provides excellent service. Most of the refineries are based on this arrangement.

3.3.1.3 *Pump around reflux (Figure 3.7c)*

In this arrangement, reflux from a lower plate is taken, cooled and fed into the column at higher section by 2 to 3 plates. This creates a local problem of mixing uneven compositions of reflux and liquids present on the tray. To overcome these ados, designers treat all the plates in this zone as one single plate, the result of which is reflected on the height and number of plates in the column. Usually this pump around is not placed at more than two sections in a column.

Figure 3.7b shows a distillation unit with four side draws using pump back reflux. The number of side draws in an atmospheric column may go upto eight.[25] Side stream strippers are provided to all fractionation units, to ensure quality and close control of products to specification limits.

3.3.1.4 *Design aspects*

Design of atmospheric column is based upon experience, intuition and emphericism. Because of the unpredictable nature of crude, this type of design has been developed. Crude, even though it contains innumerable components, they fall into small close boiling cuts and so individual separation is not possible; hence the design is significantly based upon the TBP and EFV data. Modern towers operate with

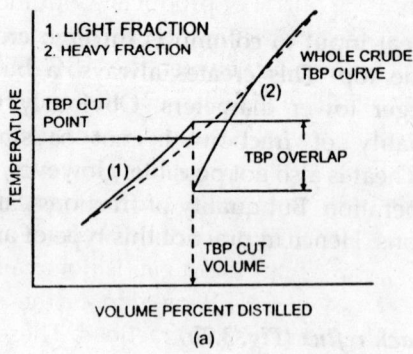

Figure 3.8a   T.B.P. Overlap and Cut Point.

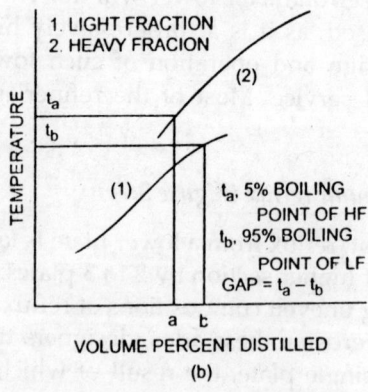

Figure 3.8b   ASTM Gap.

remarkable accuracy and match with any critically designed tower. Adoptability of resembling crudes and marginal adjustments in quality of cuts are some of the flexibilities of these towers. Further, control by conditions and quality of reflux, along with other operational parameters contribute much towards desired product pattern.

Edmister,[26] Packie,[27] Watkins[28] have furnished valuable literature for designing atmospheric and vacuum towers. Packie's classical work showed the trails of chemical engineering principles, though expressed in a different terminology. Accordingly, the design is based upon two terms, namely, "Degree of separation" and "Degree of Difficulty of Separation." Degree of Difficulty of separation is more or less connected with the purity of products. With increasing closeness of boiling points of two cuts, separation becomes a difficult task. So it is expressed as the difference between the ASTM 50% boiling points of two successive cuts. Obviously larger difference renders separation easier. Degree of separation is encountered in separation of two close boiling cuts, irrespective of the attainable purity. It is otherwise closely resembling relative volatility. ASTM gap is defined as the difference between 5% boiling point of heavy fraction and 95% boiling point of preceding cut. When ASTM gap is not available TBP overlap may be taken into account. TBP overlap is the simple difference between FBP and IBP of successive fractions. These are explained in Fig. 3.8a and b.

Packie's method is based upon these gaps and overlaps. The number of plates in a particular section depends upon the gap and reflux ratio. Separation capability is denoted by 'F' factor; given as[29]:

F = reflux ratio x number of plates in that section. This F is related to ASTM 50% difference of fractions (successive) and the gap of these fractions, as shown in Fig. 3.9.

EFV curve obtained at normal[30, 31] pressure should be converted to flash zone pressure; from this corrected EFV curve, above-mentioned gap and 50% difference should be found out, for all the fractions. In practice a good quantity of steam is utilised during distillation, steam being inert causes the reduction in partial pressure of hydrocarbons in the tower. Reduction of partial pressure of hydrocarbons may contribute to a temperature drop[32] of 10-20°C in distillation columns and side stream strippers[33].

The hot crude, where it is flashed in column is called flash zone and significance of flash zone is immense due to the following facts:

(a) Increase in flash zone pressure increases draw temperatures.

(b) Increase in over-flash decreases the side draw temperatures from the second draw onwards. [Overflash is that portion of total vapor leaving the flash zone boiling above the nearest side draw fraction; but never included in that fraction. Over flash allowance is kept to an extent of 2% total crude processed in column. This maintains a good pool of liquid and reflux on plates. The outlet of this overflash is mostly from the bottom of the column].

(c) Increase in steam in flash zone decreases the product temperature.

(d) Pressure in flash zone is reflected throughout the column in the form of plate temperatures.

232

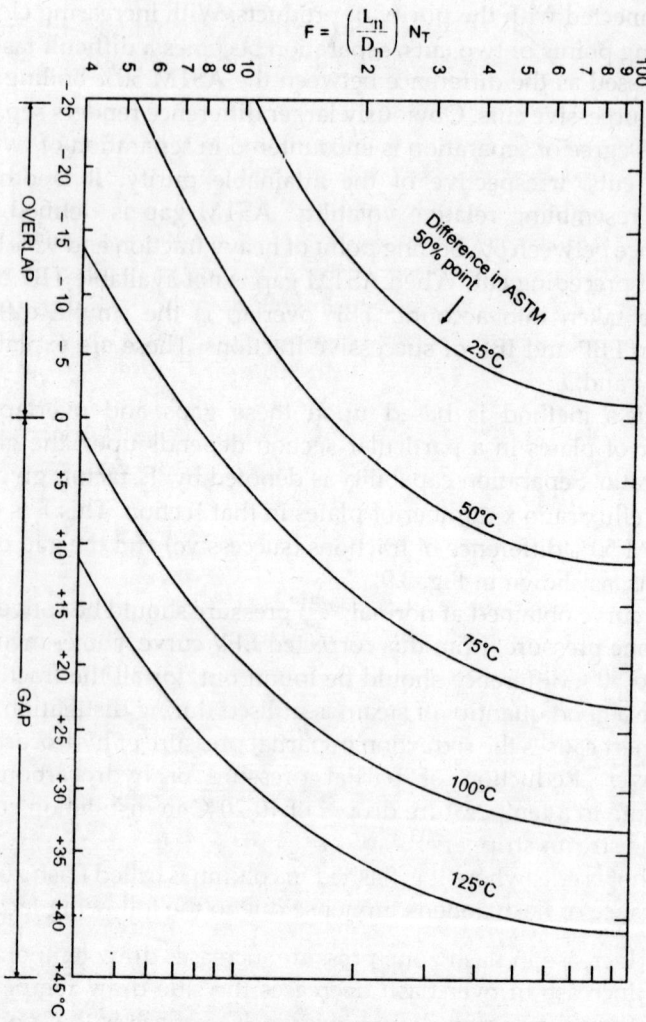

$$F = \left(\frac{L_n}{D_n}\right) N_T$$

Figure 3.9 F-Factor and ASTM Gaps
From Petroleum Refinery Distillation, Second Edition, by R.N.
Watkins, Copyright (c), 1979 Gulf Publishing Company, Houston,
Texas used with Permission.

$$\frac{L_n}{D_n} = \text{Reflux ratio} : N_T = \text{No of plates}$$

Increase in this pressure demands more quantity of steam to maintain the designed pattern. It may be stated that while distilling heavy crudes a reduction in flash zone pressure furnishes a good yield of products, without this one has to be content with the yields of products as shown by EFV curve at that pressure.

ASTM gaps followed in practice

Light naphtha to heavy naphtha	12 to 18°C
Heavy naphtha to light distillate	15 to 30°C
Light distillate to middle distillate	2 to 6°C
Middle distillate to first draw	2 to 6°C

Degree of fractionation is measured by the gap or overlap in crude distillation when 95% b.p. point is higher than the 5% b.p. point of the heavier component it is termed as overlap.

Steam strippers are necessary to strip off the lights from the cut fraction to maintain the specifications. The procedure is same as in the case of finding out the no. of plates. Usually the side steam strippers have 5 to 6 plates. The following illustration gives the procedure for finding out the fraction leaving the tray.

Problem 3.2

In a refinery side stream operation the fraction to be collected is diesel. The diesel entering the side stripper is 3500 bbls per hour, the 50% point of the cut is 275°C and contaminated with kerosene whose mid b.p. is 145°C. If the stripper is having 8 plates find the actual amount of diesel coming out of the stripper, if ASTM Gap is 18.6.

Solution

First find the F factor from Packie diagram for side stream strippers at the difference of mid bps of the streams i.e. 275 – 145 = 130° which is equal to 4.6. Dividing by no. of plates 4.6/8 = 0.55. The fraction which is likely to come out of the stripper is 55% of the total feed entering the stripper.

= 55% of 3500 = 2630 bbls per hour.

3.3.2 Atmospheric Distillation Unit (ADU)

All distillation towers are plate type towers. The arrangement of plates and tray hydraulics are not discussed here. All types of conventional trays like bubble-cap plates,[34] ballast trays, valve trays, sieve plates[35, 36] etc. are in service. Usual dimensions and operating data are given below for an Atmospheric column:

Tower diameter	5 to 8 metres
Number of plates (depends on no. of draws)	25 to 40
Maximum allowed pressure drop per plate	0.015 Kg/cm^2

Allowed pressure drop between flash zone to top of tower	0.5 to 0.55 Kg/cm^2
Pressure drop from furnace outlet to flash zone	0.3 to 0.4 Kg/cm^2
Pressure at the top of the tower	1.20 to 1.4 Kg/cm^2 (abs)
No. of plates required for separation	
Light naphtha to heavy naphtha (80°-130°C)	6 to 8 plates
Heavy naphtha to light distillate (180°C)	5 to 6 plates
Light distillate to middle distillate (250°C)	4 to 6 plates
Middle distillates to gas oil (diesel) (330°C)	3 to 4 plates
Flash zone to first draw	3 plates
Flash zone to bottoms	3 plates
Tower top temperature	100-110°C
Reflux drum pressure	1.1 to 1.15 Kg/cm^2 (abs)
Steam rate per barrel of crude	4 to 5 Kgs
Reflux ratio for light fractions	2 to 3
Reflux ratio for heavy fractions	1.5 to 2.5

Top of the ADU is fitted with reflux condenser. Gases at equilibrium with light naphtha will escape, leaving a mixture of two layers comprising of light naphtha and water. Water is drained of from this condenser and light naphtha is sent for stabilisation purpose. This reflux condenser usually operates at slightly higher pressure than the atmospheric pressure to condense the vapors.

3.3.3 Vacuum Distillation Unit

Reduced crude from the bottoms of ADU is further distilled under reduced pressure to get the remaining (or as many) fractions. The bottoms of vacuum distillation column is known as 'Goudron or residuum'.

Design of this tower is more empirical than ADU. The purpose of this tower is clearly defined even before designing. Major alterations are very difficult. Crude certainly dictates its destiny, that is, whether the stock is best for lube oils or best for cracker feed. Number of side cuts are limited to three, may occasionally go upto four. The operation of the tower is more costly than ADU and rests on the economic production of steam.

Amount of steam in demand is dependent on the extent of vacuum. So even the design of tray, plays its own role. A plate that gives least pressure drop is well preferred. Steam with vacuum, separates the fractions with a good economy rather than any of them alone. Vacuum alone is used when highly emulsifying crudes are handled, and in such cases, the vacuum is maintained below 15 mm Hg. This type of operation is very costly and hence not practised. Most of the towers operate at a flash zone pressure of 30-40 mm Hg. Which gives a top pressure of 20-25 mm Hg. This leaves an overall pressure drop of 10 to 15 mm from flash zone to top plate; this signifies the importance of low pressure drop plates. It is clearly seen, that by placing the partial condensers directly over the head of the column, saves additional piping and minimises the pressure drop in a vacuum column:

Vacuum tower operating data

Temperature of reduced crude in flash zone	350-400°C
Pressure in flash zone	30-40 mm
Pressure for long residuums	40-60 mm
Top pressure	12-15 mm
Temperature at top	225-250°C
Trays between two draws	3 to 4
Steam (at 370°C) rate	0.3 to 5 Kg per barrel
Over flash	2%

Vacuum distillation tower is designed for a limited number of products, and a tower contemplated for one operation may not be suitable for other products. The main products of commercial interest are:

1. Cracking stocks,
2. Fuel oil mixtures (Heating oils),
3. Lube oils,
4. Bitumen.

Towers for the production of cracking stocks require less elaborate fractionation. The temperature of the feed stock may cross even 400°C. All cracking stocks (catalytic) must be free from detrimental metals like vanadium, nickel etc. to avoid catalyst poisoning. Metals and other sulfur, nitrogen bearing compounds may not be that much hazardous, when hydrocracking operations are in progress. To satisfy such requirements, the vacuum tower has to give only one top product and one side product. Bottoms of tower are meant for asphalt making. Towers for such operations are closely packed; if trays are used, wire mesh screens are inserted between plates. This is very essential, at least around draw plate, to stop the metallic bodies being carried into main product. Towers employed for heating oils do not require much fractionation. However, metallic bodies should be stopped from mixing the main product which alleviates ash content. Two side stream products are usually obtained with difference in gravity to serve light and heavy duty operations. Such towers are preferably fitted with grids or packings. In this case the primary importance goes to the specifications of products rather than mode of operation. When cracked fuel oils are primarily aimed, the long residuum requirement is high, accordingly a moderate vacuum of 60 mm at flash zone is conductive.

A New type of packing has been designed by 'KOMSOMOLSK' Refinery to achieve low pressure drop and carry out dry distillation which is shown in Fig 3.10b.

Towers devoted for the production of lube oils or waxy distillates employ 3-4 side draws. These are equipped with trays of minimal pressure drop. These towers may be fed with reduced crudes, bright stocks or other lube bearing stocks for close fractionation. Viscosity and flash point specifications are closely guarded. Raw neutral oils produced must have viscosity 77 SUS, and a flash point of 180- 200°C. Heavier cuts usually have, 140 SUS and flash point 220°C. In all the towers, draw plates are fitted with separators. Lube towers are profoundly fitted with cascade trays. When bottoms are the desirables, deliberately light fractions are allowed to seep into this fraction. This enhances the plasticity of bitumen which enables easy processing.

Bitumen is graded by the penetration index. Different applications require different penetration quality, an index in the range of 85 down to 100 is commonly desirable. API gravity, in the range of 5 to 8 is recommended. Though bitumen is mainly separated under the existing conditions of tower, the desirable qualities may not be well anticipated. Further processing of bitumen is always necessary. Vacuum distillation may also be conducted in two stages for effective separation, if speciality oil products are ventured. Fig. 3.10a, b describes the arrangement for different operations.

3.3.4 Overhead Corrosion in Distillation Unit

Corrosion in distillation unit is mainly due to hydrogen sulfide and chloride ion. High sulfur and salt containing crudes generally create these problems. Though prior to distilling, desalting of crude is practised, ineffective desalting brings this problem. Slight amount of hydrogen sulfide causes a protective film as iron sulfide, but excess of this leaves sulfide ion and chloride ion as threatening corrosion contributors. P_H of this system is directly proportional to the partial pressure of these components.

Chloride ion is mainly contributed by hydrolizable calcium, magnesium chlorides. If crudes can be desalted to a minute extent of one kg of salt per 1000 bbls; perhaps resulting acid concentration may not exceed 0.6 Kg per 1000 bbls, and though it seems to be not much, still it can offer problems under those operating conditions.

Prevention of this type of corrosion is accomplished by sending required amount of ammonia into column. It is also seen that the presence of high amount of ammonium chloride can lead to severe troubles. Even though ammonium chloride is fairly soluble in water and carried away but a concentration above 70 ppm would lead to the appearance of crystals. The amount of ammonia admitted depends largely on P_H of H_2S. The system can be written as[37]

$$NH_3 + H_2S \rightleftarrows NH_4^+ + HS^-$$

$$2NH_3 + H_2S \rightleftarrows 2NH_4^+ + S^{--}$$

i.e. 2 moles of NH_3 can combine completely if second ionisation constant of H_2S is also taken into consideration. But, it is not necessary, as the ionisation constant is very small.

Ionisation constant (K) for such system can be written as

$$K = \frac{\left(NH_4^+\right)(HS)^-}{(NH_3)(\text{Partial pressure of } H_2S)}$$

From which the partial pressure of H_2S can be found out. The system resembles the situation where NH_3, H_2S and H_2O are present.[38] Therefore, it is seen that 20 to 30% of H_2S, should be neutralised leaving excess of H_2S[39] to form the buffer system $H_2S — NH_4OH — NH_4HS$. That leaves the system at P_H 5.5 to 6.5. Modern practice is to use organic[40] amines instead of ammonia. Amines are more basic than ammonia and serve without any side effects. These react usually to form soluble compounds with chloride and sulfide ions, thus tendency of crystallisation does not appear. It may be even significant to introduce caustic, if the concentration of sulfur is high.

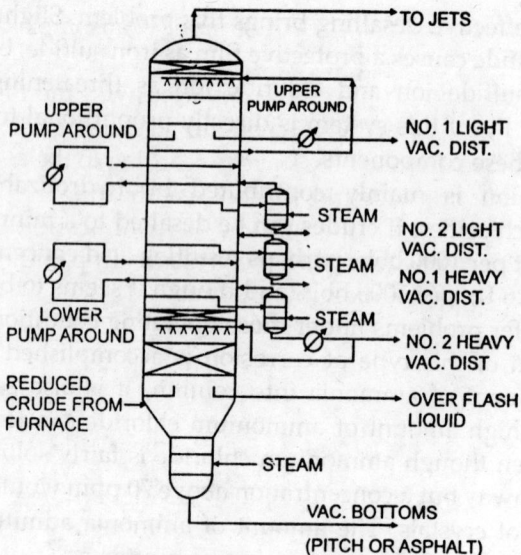

VACUUM TOWER FOR LUBES (PUMP AROUND)

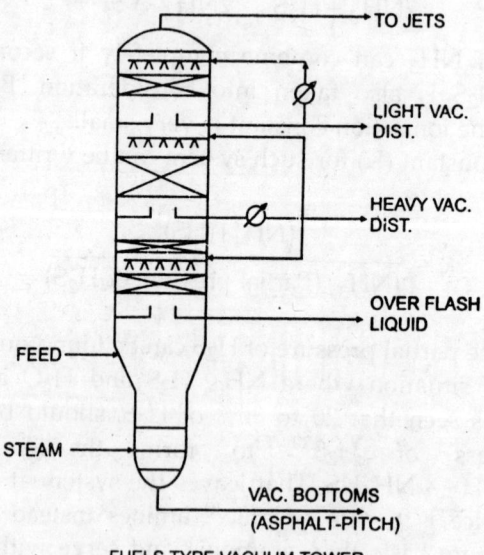

FUELS TYPE VACUUM TOWER

Figure 3.10a Vacuum Distillation Column:
a-Lubes—Type, *b*-Fuels—Type

Vacuum column of AWT, D=7000 mm, Komsomolsk Refinery of Rosneft Joint-stock Co.

All trays in a column have been replaced with the PETON cross-flow regular packing without changing technological piping and supporting constructions of the trays. According to the task given, the column should work with and without feed heating in the furnace at residual pressure in the upper part 54-80 kPa in conditions of "dry" distillation. Proceeding from this, the pressure drop in the column has been designed to be not more than 8 kPa.

There has been achieved:

- the pressure drop no more than 4 mm Hg at residual pressure in the upper part up to 17,29 kPa

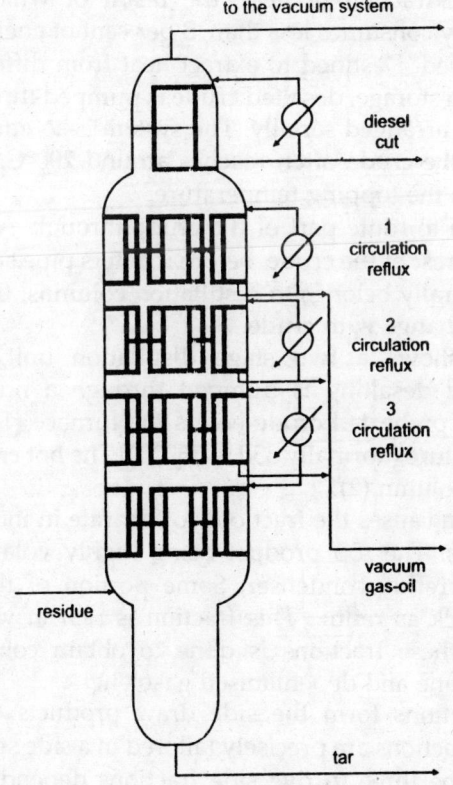

Figure 3.10*b*   Vaccum Distillation

yield and quality of products at the design level. Yield of vacuum gas-oil when operating without heating is 10-12%, with heating up to 19-20% on crude.

The creation of deep vacuum by the injecting vacuum-creating system has made it possible to carry out "dry" distillation of an atmospheric residue after modernization of PETON packing at the top of the column.

3.3.5 Topping Operation

In practice, separation of different fractions from crude is done by simple fractionation.

Refinery being a main energy source, has to avail as much less energy as possible during processing of crude. Bountiful combinations of heat exchanging systems were explored for effective and economic extraction of heat, the result of which a critically designed refinery consumes less than, 3 per cent of energy equivalent of crude processed. Destined to extract heat from different fractions before they reach storage, desalted crude is pumped through different heat exchangers arranged serially. The system is so effective that the temperature of the crude often reaches around 200°C, some reports say almost up to the topping temperature.

It is advisable to route part of the crude through exchangers and mingle with the rest of the crude, before it enters pipe still heater. The exchangers generally belongs to distillation columns, through which hot fractions exchange with crude.

Figure 3.11 shows a two stage distillation unit. Crude after dehydration and desalting is pumped through a number of heat exchangers. The preheated crude enters the furnace (1) to attain the topping temperature (normally 330 to 360°C). The hot crude is flashed in atmospheric column (2).

The distillation causes the fractions to separate in increasing order of boiling points. The top product being highly volatile has to be condensed in a reflux condenser. Some portion of the condensed fraction goes back as reflux. This fraction is rich in volatiles hence stabilisation of these fractions is done to obtain components like butane and pentane and depentanised gasolene.

All other fractions form the side draw products of distillation column. These fractions are precisely tailored in a side stream stripper (3). There shall be three to five side fractions depending upon the design. The fractions are usually classified as heavy naphtha, kerosene, diesel.

241

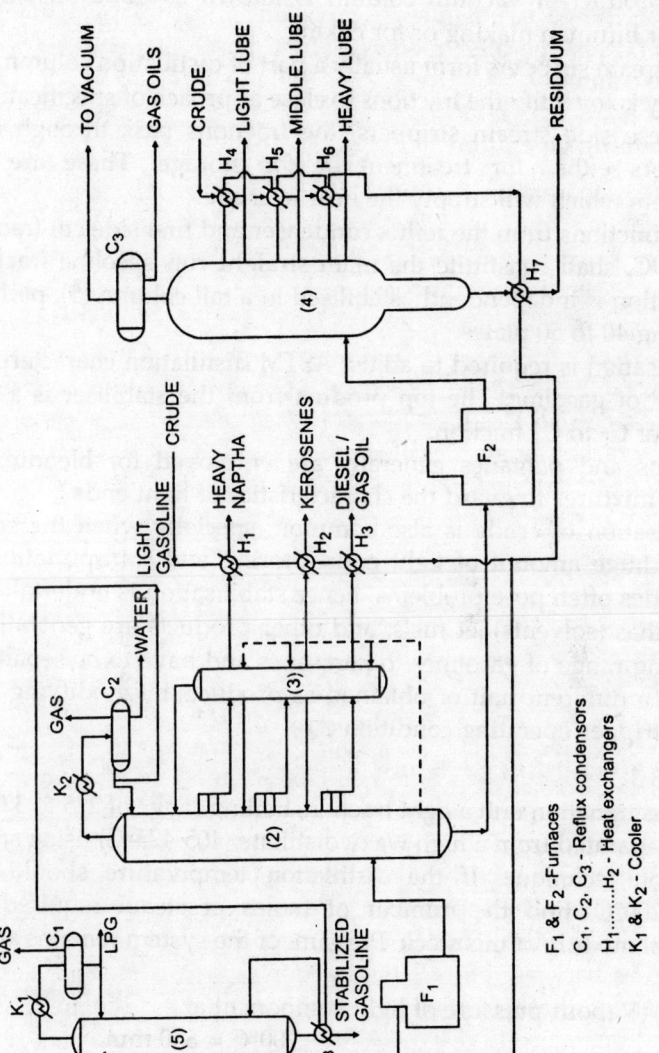

Figure 3.11 Two Stage Distillation Unit with Stabiliser.

F_1 & F_2 - Furnaces
C_1, C_2, C_3 - Reflux condensors
H_1......H_7 - Heat exchangers
K_1 & K_2 - Cooler

Bottom product of atmospheric column is now again routed through a furnace to reach a temperature of 350 to 400°C and is allowed to flash in a vacuum column (4). Vacuum gasoil/heavy diesels, lube oil cuts/pressure distillates shall be the side cuts. The bottom product of vacuum column is known as goudron, and is meant for bitumen making or for coking.

Side stream strippers form usually a part of distillation column and their duty is to rectify the fractions to close approach of specifications. From these side stream strippers, the fractions pass through heat exchangers either for treatment or for storage. These are the exchangers which will supply the heat to crude.

Light fractions from the reflux condenser and first side cut fraction from ADC, shall constitute the main straight run gasoline fraction. This fraction is independently stabilised in a tall column (5), perhaps containing 40 to 50 plates.

Stabilisation is required to adjust ASTM distillation characteristics and RVP of gasoline. The top product from the stabiliser is a rich mixture of C_2 to C_4 fraction.

Butanes and pentanes generally are employed for blending in gasoline mixtures to guard the characteristics of light ends.

Stabilisation of crude is also common, specially when the crude contains huge amount of light ends. Storing and transportation of such crudes often pose problems, hence stabilisation is undertaken.

Naphthas (solvents) jet fuels, and other products are generally in the boiling range of gasolines to kerosenes and have to be separated either in a different unit or obtained as products by modifying side stream stripper operating conditions.

Problem 3.3

In a fractionation unit a light fraction, boiling range of 225 to 325°C, is to be separated from a high waxy distillate (405-425°C) using steam distillation technique. If the distillation temperature should not exceed 200°C, find the number of moles of steam required for equimolal mixture of this stock. Pressure of the system remains at 760 mm.

Data: (a) Vapour pressure of light component at
$$200°C = 650 \text{ mm}$$
Vapour pressure of heavy component at
$$200°C = 35 \text{ mm}$$
  (b) Mol. wt. of light fraction = 225, $c_p = 0.47$
  Mol. wt. of heavy fraction = 435, $c_p = 0.42$

Solution

    (i) Assuming steam does not condense and the mixture is at 200°C.

$$\frac{\text{No. of moles of steam}}{\text{No. of moles of light component}} = \frac{\pi - p}{p} = \frac{760 - 650}{650}$$

where, π =   total pressure of the system

               p = pure component vapour pressure

No. of moles of steam to distill light component

$$= \frac{110}{650} = 0.169$$

No. of moles of steam to distill heavy component

$$= \frac{760 - 35}{35} = 20.7 \text{ or } 375.6 \text{ kgs}$$

Steam Condenses

At 200°C latent heat of steam = 769.5 (kcal/kgm)

Heat content of the mixture from a temperature of 30°C to 200°C

$= m_1 \, cp_1 \, dt_1 + m_2 \, cp_2 \, dt_2$

$= ML$; and $(dt_1 = dt_2)$

$= (225 \times 0.47 + 435 \times 0.42) (200\text{-}30) = M \times 769.5 \text{ or } M = 63.7 \text{ kgs}$

So total steam requirement of 63.7 + 375.6 = 439.3 Kgs.

Problem 3.4

In a flash vapourisation experiment of benzene-toluene of equal amount, is vapourised at 295 mm pressure and 65°C temperature. Find the vapour composition. If the separation is conducted at 1300 mm pressure and 110°C, find the composition.

		Vapour pressures	(mm)
		65°C	110°C
Benzene	—	440	1850
Toluene	—	150	750

Solution

Mol. wt. of benzene   = 78

Mol. wt. of toluene   = 92

1st case

$$\text{Moles of benzene in vapour} = \frac{440 \times 0.5}{295} = 74.6\% \text{ approx.}$$

$$\text{Moles of toluene in vapour} = \frac{150 \times 0.5}{295} = 25.4\%$$

2nd case

$$\text{Moles of benzene in vapour} = \frac{1850 \times 0.5}{1300} = 71.2\%$$

$$\text{Moles of toluene in vapour} = \frac{750 \times 0.5}{1300} = 18.8\%$$

Problem 3.5

A distillation column is fed with feed at its bubble point. A reflux ratio of 4 is used. From the operating data given below, calculate number of plates, feed plate and its temperature required to obtain the following compositions:

DATA:

	Feed (moles) (F)	Overhead moles (D)	Bottom (W)
Pentane	60	59.5	0.5
Hexane	20	0.5	19.5
Heptane	20	—	20.0
	100	60.0	40.0

Basis: Reflux ratio 4

Solution

Mole fractions of these components in feed, overhead and bottoms:

	Feed	Overhead	Bottom
Pentane A	0.6	0.992	0.012
Hexane B	0.20	0.008	0.488
Heptane C	0.20	—	0.500
	1.00	1.000	1.000

Operating lines

Below feed point: (Reflux ratio) = L_n/D = 4

Reflux = 4 × overhead = 4 × 60 = 240 moles (LN) product

Total vapour = 240 + 60 = 300 (V_m)

Liquid flow downwards: $L_n + F$ = 240 +100 = 340 moles (L_m)

Vapor (V_m) : [(L_m-W)] 340 - 40 = 300

Operating lines for each component (below feed plate)

$$Y_m = L_m/V_m x_{m+1} - \frac{W}{V_m} x_w = \frac{340}{300} x_{m+1} - \frac{40}{300} x_w$$

Y_m stands for vapor composition corresponding to X_{m+1}.

For
$$A : Y_A = 1.133\, x_A - 0.134 \times 0.012$$
$$B : Y_B = 1.133\, x_B - 0.134 \times 0.488$$
$$C : Y_C = 1.133\, x_C - 0.134 \times 0.5$$
Set-I

Above feed plate, the composition of vapor is given by,

$$Y_N = \frac{L_n}{V_n} x_{n+1} + \frac{D}{V_n} x_D \qquad (V_n = V_m)$$

For
$$A : Y_A = \frac{240}{300} x_A + \frac{60}{300} x_D$$
$$A : Y_A = 0.8 x_A + 0.1984$$
$$B : Y_B = 0.8 x_B + 0.0016$$
$$C : Y_C = 0.8 x_c$$
Set-II

Starting from bottom composition, corresponding vapor composition is calculated. From the vapor composition using the Set-I eqns. for individual components the liquid composition in the above plate is calculated. This procedure continues till the feed plate composition is obtained. Above feed plate Set-II eqns. are used to calculate the compositions.

It is understandable that the mixture should always boil between the BPs of high boiling component and low boiling component. At any temperature in between these limits, the equilibrium constants (K) for each of these fractions are found out and to match $\Sigma Kx = 1$, number of trials, for K are necessary.

TABLE 3.4 Equilibrium Composition of the Bottom Plate.

Component	X_w	Ks at* 79.5°C	X_w	Y_w	X_{w-1}
A	0.012	3.3	0.0396	0.039	0.04
B	0.488	1.39	0.678	0.677	0.65
C	0.500	1.57	0.285	0.284	0.31
			1.0026	1.000	1.000

* Consult Chemical Engineer's Handbook, Perry.
** Elements of Fractional Distillation, Robinson and Gilliland.

Y_w is corrected to $\Sigma Kx_w = 1$; x_{w-1} is liquid composition in the above plate; obtained from eqn. of Set-I**

$$Y_A = 1.133\ X_A - 0.0016 \therefore 0.039 = 1.133\ X_A - 0.0016$$
$$\text{or } X_A = 0.04$$

	X_1	K at 72.3°C	$\Sigma KX_1=Y$	Y_1^*	X_2	K at 66.7°C	ΣKX_2	Y_2
A	0.04	2.8	0.112	0.111	0.099	2.5	0.2475	0.247
B	0.65	1.15	0.7475	0.744	0.715	0.95	0.6792	0.679
C	0.31	0.47	0.1457	0.145	0.186	0.40	0.0744	0.074
			1.0052	1.000	1.000		1.001	1.000

	X_3	K at 61.7°C	Y_3	X_4	K at 54°C	Y_4	X_5
A	0.219	2.0	0.438	0.388	1.66	0.645	0.571
B	0.657	0.8	0.525	0.520	0.64	0.333	0.351
C	0.124	0.3	0.037	0.092	0.24	0.022	0.078
			1.000	1.000		1.000	1.000

• Adjusted value for $\Sigma KX = 1$

Feed-l Composition approximately matches fifth plate

	X_5	K at 46.6°C	Y_5	X_6	K at 41.7°C	Y_6	X_6
A	0.571	1.42	0.81	0.764	1.2	0.909	0.888
B	0.351	0.5	0.175	0.217	0.41	0.088	0.108
C	0.078	0.19	0.015	0.019	0.145	0.003	0.004
			1.000	1.000		1.000	1.000

	K at 38.4°C	Y_7	X_8	K at 38°C	Y_8	X_9	37°C
A	1.08	0.96	0.952	1.03	0.983	0.98	1.01
B	0.36	0.039	0.047	0.35	0.017	0.02	0.345
C	0.125	0.001	0.001	0.122	—	—	—
		1.000	1.000		1.000	1.000	

Y_9 composition is A 0.993 which is approximately equal to designed overhead composition

$$B\ 0.007$$
$$C\ \underline{}$$
$$1.000$$

Hence nine plates are required for this operation.
Feed plate is number 5.
Distillation temperature ranges from 79.5°C to 37°C.

Problem 3.6

A crude distillation is a column having a capacity of 10,000 bbls per day. Because of ineffective desalting the chloride (Cl^-) ion concentration at the top of the column is found to be 5 ppm. If the crude contains volatile sulfide (H_2S) upto 0.7 gms/liter. Find the amount of ammonia required for neutralisation. (One barrel is approximately 200 litres) API gravity of crude is 20.

Solution

5 ppm Cl^- ion is equivalent

$$= \frac{5 \times 10000 \times 200}{1000 \times 1000}$$

$$= 10.0 \text{ kgs/day}$$

$$H_2S \text{ content} = \frac{0.7 \times 1000 \times 200}{1000} = 1400 \text{ kgs/day}$$

$$\text{or } \frac{0.7}{34} \frac{\text{gm. moles}}{\text{liter}}$$

P_H of the solution to be maintained at 6
System consists of H_2S-NH_4OH ($NH_3 + H_2O$)
The first ionisation constant of H_2S at 25°C = 9.81×10^{-8}
P_H of the buffer is given by equation

$$P_H = P_{Ka} + \log \frac{\text{Salt}}{\text{Acid}}$$

where Ka is ionisation constant, Salt is ammonium hydrogen sulfide; and Acid is H_2S.

i.e. $6 = -\log (9.81 \times 10^{-8}) + \log \left(\frac{\text{Salt}}{0.7/34} \right)$

$$\log(\text{Salt}) = 6/ + \log (9.81 \times 10^{-8}) + \log \left(\frac{0.7}{34} \right)$$

NH_4HS Salt = 2.0197×10^{-3} gm. mol. per litre
Ammonia in salt = 0.343 gm/litre
So ammonia required per day = $0.0343 \times 10^4 \times 200/10^3 = 68.6$ kg

so total requirement of ammonia = ammonia required for chloride ion
+ 68.6

$$= \frac{10 \times 17}{35.5} + 68.6$$

$$= 73.4 \text{ kgs/day}$$

3.3.6 Calculation of Plate Temperatures

It has been reported in the literature that the slope of a flash curve is
practically independent of pressure. This makes easy for drawing
flash curve at any pressure, once flash curve at a known pressure is
available. Either the percentage intersection of TBP and flash curve or
50% boiling point of flash curve at that pressure be known available.
The effect of diluent (steam) on the overhead product and its
temperature (draw) could be better understood by the following
illustration.

Problems 3.7

An atmospheric column is operating at 810 mm Hg. The top
product light naphtha has a final boiling point of 140°C. Reflux on the
top plate is maintained at 2.5:1. The condensate is found to contain 6
Kgs of water per every 100 Kgs of hydrocarbon. From this calculate
the draw temperature of tray and partial pressure of hydrocarbon.

(Assume Mol. wt. of HC as 130)

From this data it is understood that the 100% boiling point on flash
curve is 140°C at normal pressure.

1000 Kgs of HC

Total product of HC = 1000 + 2.5 × 1000
= 3500 Kgs

$$\text{Rates} = \frac{3500}{130} = 26.93$$

Moles of steam: $\dfrac{.06 \times 3500}{18} = 11.67$

Total moles = 38.60

Partial pressure of HC = $\dfrac{26.93}{38.60} \times 810 = 565.2$

At this pressure 100% EFV boiling point shall be (Fig. 2.4).
The draw temperature of the plate = 126.5°C

In commercial operations, the draw temperatures, are found to be slightly less than the calculated ones for middle distillates where the effect of steam is not commendable. Draw temperatures are approximates taken as

$$0.88 \times (5\% \text{ point on TBP cut})$$

The sound method of calculating the temperatures at the draw plate is by heat balance in that section; however this is also a complex thing.

3.4 BLENDING OF GASOLINES

Blending, an important operation in refinery, is a physical process in which accurately weighed quantities of two or more components are mixed thoroughly to form a homogenous phase; the components mixed being similar or dissimilar in nature.

Most of the products obtained from distillation columns are invariably blended with fractions obtained from other units to help in keeping the wastage minimum and increasing the quantitative production. Almost all products from gas to lube oil are not only blends of fractions but additives too. All such blends shall be formulated to have the required properties conforming to the specifications.

3.4.1 Blending Processes[41]

There are two ways by which effective blend can be produced; one is by batch blending the other is continuous blending. Batch blending starts with mixing known amounts of components in a tank mixer. The mixer shall be fitted with an agitator and other accessories like pressure gauge, liquid level indicators etc. Some times agitation by blowing air is convenient. Suction blending is quite alluring specially when toxic materials like lead, biocides are blended. Mixing in tanks furnishes some advantages like inserting cooling or heating coils in tanks.

3.4.1.1 Line Blending

This is widely used in refineries as well as in blending plants. All components to be blended are pumped simultaneously into a common header at rates specified as per the formulations (Fig. 3.12). Rate of flow is proportioned by control values operated by pneumatic or electric relay system. Signals received are corresponding to flow

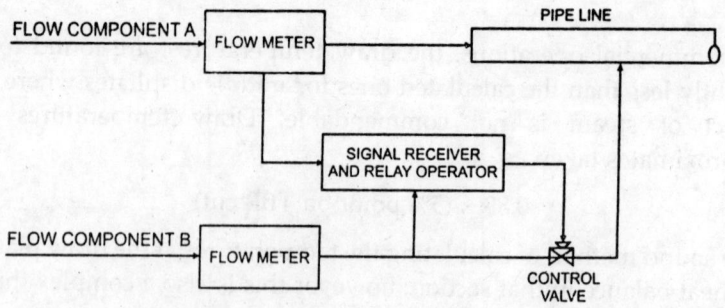

Figure 3.12   Schematic Diagram for Blending of Gasolines.

rates, and these can accurately modulate the flow rates by adjusting valve openings.

The long pipe line through which all these proportioned components travel acts as a mixer to produce the blend. Additives can also be injected into the system.

3.4.1.2   Gasoline Blending

Different gasolines like alkylated, reformed polymer, cracked, straight run etc. are blended along with various additives to boost the performance value of gasoline; however the blend should faithfully respond to the specifications. The two important inviolable properties on which blends are critically constituted are vapour pressure and octane number.

Vapour pressure of a mixture can be estimated by Raoults Law. But scant information on the molecular composition of a blend does not permit it; and laborious experimentation for evaluating molecular composition is also not wise.

The following problem on blending of gasoline exposes the labour involved in such blending operations.

Problem 3.8

In a refinery, the following gasolines are available as shown in Table 3.5. It is desirable to make a blend with these, guarding RVP and octane number as specified. Formulate the blend as per specifications (Table 3.6).

Solution

A suitable blend has to be tried by trial and error methods. An initial blend composition is chosen arbitrarily that nearly satisfies

TABLE 3.5 Data of Components.

Component	B.P. °C	at 15°C sp. gr.	Mol. wt.	O.N.	Component cuts	B.P. °C	Sp. gr. at 15°C	Mol. wt.	Octane No.	Vapor pressure (mm) 37.8°C
A	35-180	0.7133	99	61	a_1	35—50	0.6521	75	70	720
B	65-200	0.8118	113	60	a_2	51—75	0.6833	86	40	380
C	120-150	0.7846	125	95	a_3	76—110	0.7188	100	70	160
D	78-180	0.7870	106	84	a_4	111—180	0.7686	130	60	35
					b_1	65—75	0.6845	90	40	300
					b_2	76—110	0.7175	100	70	150
					b_3	111—180	0.7666	125	60	32
					b_4	180	0.7784	150	50	4
					d_1	75—110	0.7844	90	90	160
					d_2	111—180	0.7904	120	80	35

A = Straight run
B = Straight run
C = Alkylated stock
D = Reformed stock

TABLE 3.6 Specification of Finished Product.

Temperature	% distillation
60°C	10 IBP. 42°C
115°C	50
175°C	90
Octane No.	
(Research)	Minimum 65
R.V.P.	0.7 Kgs/Cm^2

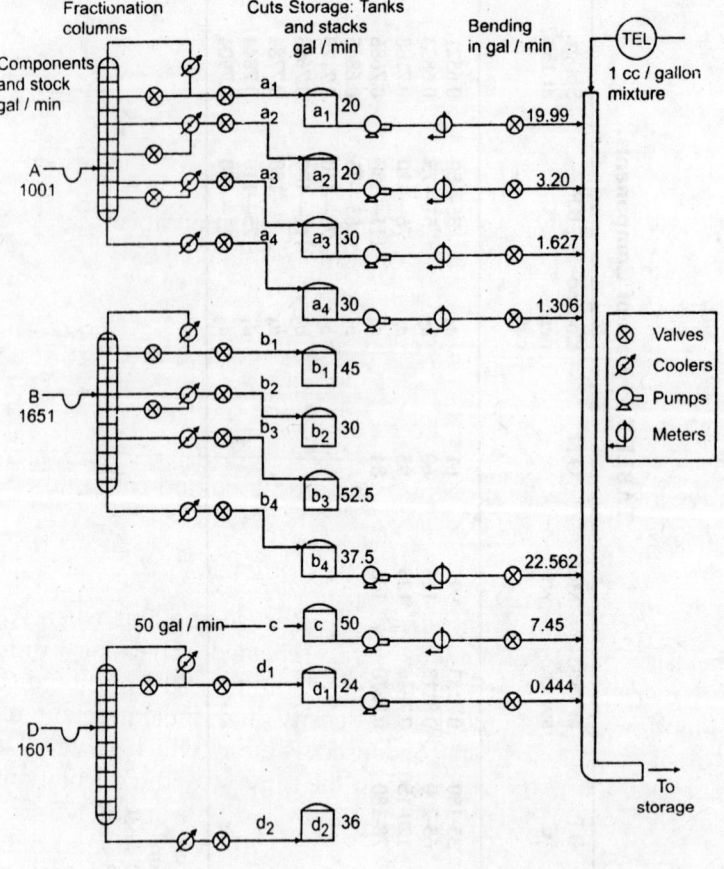

Figure 3.13   Pipe Line Blending of Gasoline.

octane number and RVP. The equilibrium composition of liquid and vapour are theoretically found out from equilibrium constants (K) at a particular temperature. As shown in specifications, 10% of the stock has to be distilled at 60°C. The problem is to first find out the equilibrium composition of liquid with vapour, at that temperature for 10% distillation; the molar ratio of vapour to liquid is found to be 0.13/0.87. Now finding the K values for these components which depend upon the vapour pressures, the vapour composition is found out.

Say in the case of straight run gasoline a fraction 35 to 50°C is named as component (a_1), has an equilibrium constant 2.1 at 60°C. Assuming a feed composition 0.45 mole fraction in the mixture, its vapour composition[42] is given by

$$X_A = \frac{X_{AF}}{L + (1-L)K}; \ P_A/P. = K$$

Where = X_A : Eq. concentration of A with vapour.
X_{AF} : 0.45
X_E : Residual liquid composition at 60°C.
L : Fraction of liquid residual.
$P_A/P =$: K
P_A : Vapour pressure.
P : Total pressure (760 mm)

$$X_A = \frac{0.45}{2.1 \times 0.13 + 0.87} = 0.3937$$

Like this for all components[43] the residual liquid composition (X_E) is found at

$$i.e. \ 0.3937 \times 0.87 = 0.3425$$

for all components a_1, a_2, a_3, a_4, b_4, c are found out when $X_E = 1$. Otherwise the calculations are repeated choosing different compositions of the feed mixture. The volume of the portion distilled is found out by knowing the density and mole fraction of the individual component in condensed vapour, which is very nearly equal to 10% of the distillation. In the same way the calculations are repeated at different temperatures 115°C and 175°C and corresponding percentage amounts of distillates are found out. The RVP of the final mixture, satisfying the distillation pattern is thus evaluated. The blend may slightly differ in octane number, and it is to be boosted by lead as shown in Table 3.7 and Fig. 3.13.

TABLE 3.7 Blend-Composition-Properties.

	Feed X_1 mol. fraction	Vol. of feed X_1 (cc)	K at 60°C	$V/L = \frac{.13}{87}$ X_A	$X_A =$ 0.87X_A	Vol. 0.87 X_A	K at 115°C	$V/L = \frac{58}{42}$ X_A	0.42X_A	Vol. 0.42X_A
a_1 :	0.45	51.75	2.10	0.3937	0.3425	39.38	7.20	0.0979	0.0411	4.73
a_2 :	0.066	8.30	1.00	0.0660	0.0574	7.23	4.50	0.0217	0.0091	1.15 –
a_3 :	0.030	4.21	0.60	0.0316	0.0274	3.85	2.70	0.0151	0.0063	0.89
a_4 :	0.020	3.38	0.10	0.0226	0.0197	3.33	0.37	0.0315	0.0132	2.23
b_4 :	0.303	58.38	0.01	0.3477	0.3025	58.29	0.095	0.6377	0.2678	51.68
c :	0.121	19.48	0.10	0.1370	0.1192	19.19	0.41	0.1839	0.0772	12.43
d_1 :	0.01	1.15	0.60	0.0105	0.0092	1.05	2.7	0.0050	0.0021	0.24
Total :	1.00	146.65		1.0091		132.32		0.9928		73.35
Distillation %						9.8%				49.98%

(Contd.)

TABLE 3.7 : (Contd.)

K at 175°C V/L = $\frac{0.92}{0.08}$ X_A	$X_A =$ $0.87X_A$	Vol. $0.08X_A$	R.V.P of the feed (mm)	Octane Number of the feed
16.0	0.0304	0.0024	0.28	324	0.45 × 70
12.0	0.0059	0.0004	0.06	25.08	0.066 × 40
5.8	0.0055	0.0004	0.06	4.8	0.030 × 70
0.95	0.0209	0.0016	0.28	0.7	0.020 × 60
0.31	0.8296	0.0663	12.79	1.20	0.303 × 50
1.40	0.0884	0.0070	1.13	4.23	0.121 × 95
5.8	0.0018	0.0001	0.02	1.6	0.01 × 90
Total	0.9825		14.62	361.0	64 + ICCTEL/gallon
Vol %			90.04%	0.7 Kgs/cm²	

Generally refiners depend upon two components butane and pentane as the vapour pressure adjustors. For easy estimations depentanised gasoline shall be regarded as low volatile component, whose vapour pressure is more or less constant at given temperature. By adding highly volatile butanes of known amounts the vapour pressure can be easily adjusted. In practice blending values are assigned to each component and the value for butane is taken as 100. With respect to butane other components are assigned the blending values depending upon the volatility. The vapour of the mixture can be derived by law of additives.

Problem 3.9

Different gasolines are to be blended along with *n*-butane. Find the resulting vapour pressure of blend from the data given below:

	Blending Value at 38°C	Volume % in blend
Alkylated gasoline	5.0	10
Cat. cracked gasoline	27.3	20
Straight run gasoline	30	60
n-butane	100	10

Solution

Contributed value of each fraction

Alkylated gasoline: $\dfrac{5 \times 10}{100} = 0.5$

Cat. cracked gasoline $= 27.3 \times \dfrac{20}{100} = 5.46$

Straight run gasoline $= 30 \times \dfrac{60}{100} = 18.0$

n-butane $= 100 \times \dfrac{10}{100} = 10.0$

Total value of blend : 23.96

Octane value of a blend follows strictly the additive principle (by volume); hence a marginal adjustment may be tried with TEL or TML whenever a blend falls short of the value two to three units.

257

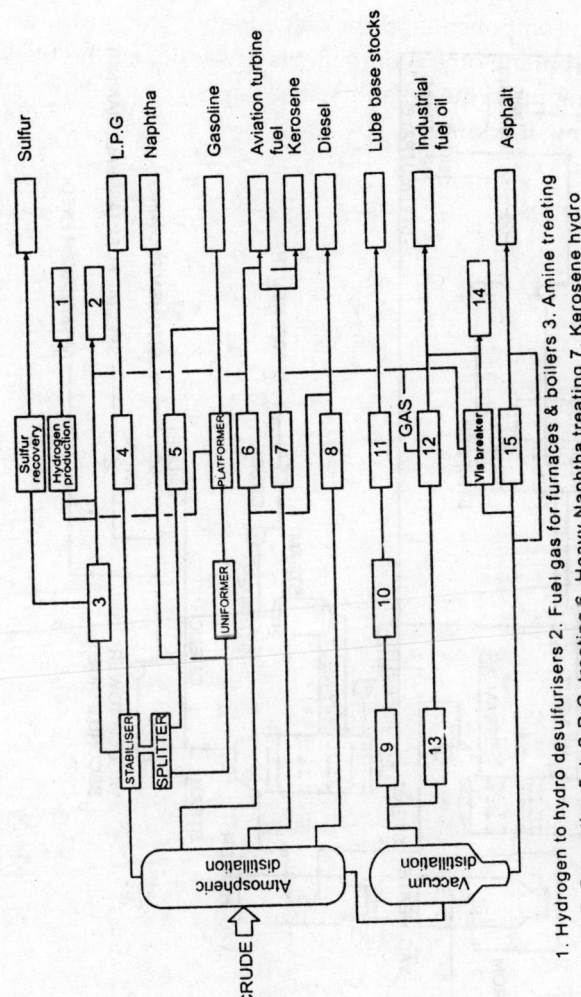

Figure 3.14 Layout of a Modern Petroleum Refinery.

1. Hydrogen to hydro desulfurisers 2. Fuel gas for furnaces & boilers 3. Amine treating
4. L.P.G. treating 5. L.S.R.G. treating 6. Heavy Naphtha treating 7. Kerosene hydro
desulfuriser 8. Diesel hydro desulfuriser 9. Furtural extraction 10. M.E.K. dewaxing
11. Lub oil hydro finishing 12. Thermal cracker 13. Vacuum distillate hydro desulfuriser
14. Refinery fuel oil for furnaces and boilers 15. Bitumen air blowing

258

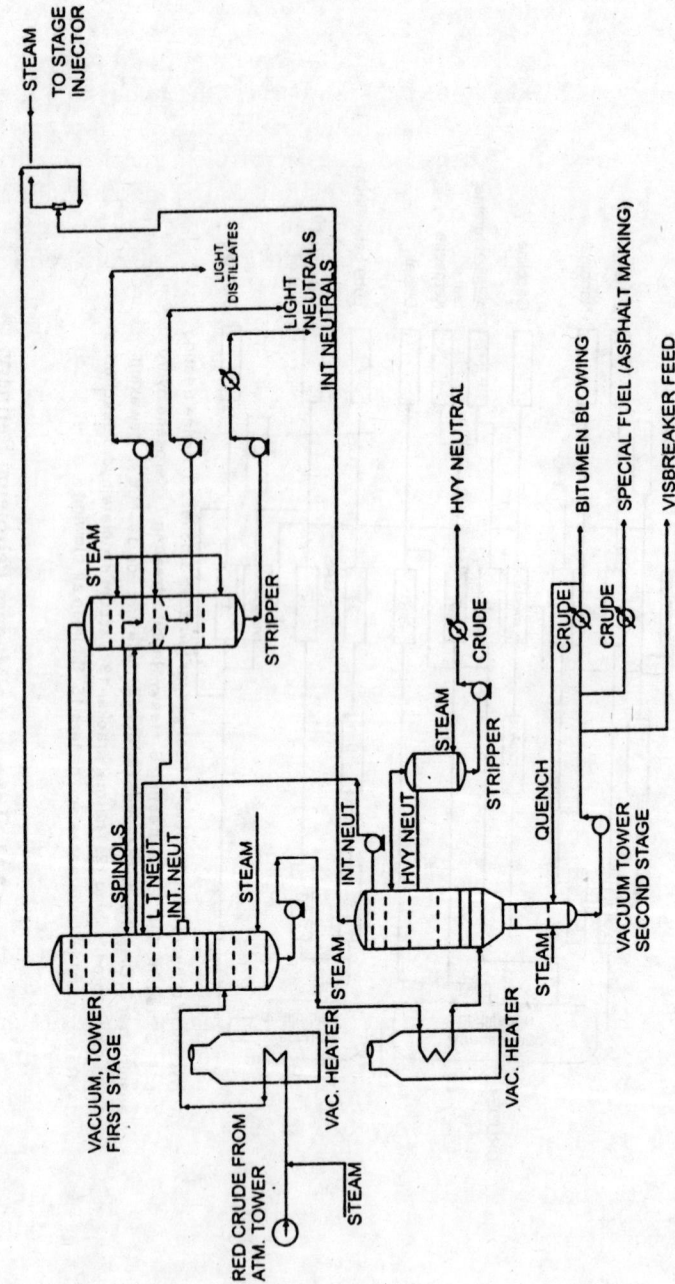

Figure 3.14 Layout of a Modern Petroleum Refinery (*Contd.*).

3.4.2 Integrated Refineries

Present trend is to install a refinery at the consumer point rather than at the place of oil production. This can harvest the riches to a great extent, as the products are immediately sold, thus not much with holding of investment. Further, the basic petro-chemical industries expedient for other necessities can be instituted. The modern refinery gingerly incorporates on an intergrated circuit, all necessary treatments, waste disposals storage facilities etc. A modern refinery lay out is shown in Fig. 3.14.

Modern refinery is a much cleaner industry compared to many other industries. However, even if a low sulfur crude is handled the off gases containing sulfur, shall be high. For example a 3 million ton plant processing 0.10% sulfur crude can easily give rise to pollution problems (as the off gas contains roughly 30% of sulfur intake). The tendency, of recovering sulfur is increasing, as it forms a cheap by product, specially for a country like ours.

Disposal of acid sludges, clays, water, gas etc. all require prudent planning.

Fire hazards though uncommon, cannot be overruled. Fire fighting equipment should be in the reach of all units.

Most of the refineries in our country, like elsewhere, are coastal based; this helps in an unfaltering way the disposal operations.

PROBLEMS

1. (a) A petroleum stock at a rate of 1200 bbl/hr. (sp. gr. 0.8524) is passed through heat exchangers before it is fed into radiant section of a box type heater at 200°C. The pipe still heater burns refinery off gas at a rate of 3,250 kgs/hr. The refinery off gas has a heating value of 11,300 Kcals/kg. In the radiant section of furnace, there is one row of tubes of 12 metres long of 10.5 cm. O.D. spaced at 2; O.D. The projected area of tubes is 150 sq. metres.

   Find the outlet temperature of oil

   (Air fuel ratio 25)

   Sp. heat 2.66 kJ/kg°C   (Ans. 342°C)

   (b) Find the volume of the furnace (combustion space only). Comment on the heat density in this furnace as well as heat generation capacity with respect to projected area of tubes.

   (Ans. 560 cubic metres.)

(c) Calculate the area of duct. Assume a reasonable mass flow rate of gases.

2. A petroleum stock at a rate of 1000 bbl/hr. is passed through convection section of box type heater. In this process the temperature of the feed is raised from 30°C to 180°C. Assuming an average flue gas velocity 2 metres per sec. at an average temperature of 625°C. Find the heating area required.

   Data :

   O.D. of the tube 10.05 cm.

   Mass flow rate of gases 100 kg/sec. sq. meter.

   Sp. heat of stock: 0.52 kg cals/kg.°C.

3. Carbontetrachloride to be steam distilled. The ambient pressure is 775 mm Hg. Calculate how many K. Cals are required to distill 100 kgs. of CCl_4, starting with a mixture of 50 kgs of water and 500 kgs of CCl_4 at a temperature of 25°C.

   (Use Cox chart to get the boiling point of this mixture try between 65-68°C).

   Latent heat of vaporisation of CCl_4 : 46.5 gm. Cals/gm.

   (Ans. 13504 kg cals)

4. The vapour pressures for Toluene and Benzene are given below at different temperatures. Assuming that mixture obeys Raoult's Law construct the B.P. diagram at 760 mm Hg total pressure and compare it with 900 mm Hg total pressure.

   Problem 4: (Data)

T°C	Vapour Pressures	
	Benzene (A)	Toluene (B)
80	760	—
82.2	811	314
87.7	957	378
93.3	1123	452
99.0	1310	538
107.2	1520	635
109.2	1620	687
112.0	—	760

5. A Butane-pentane mixture containing 40% butane is distilled at 90°C at 10 Kg/cm^2 pressure. Find the composition of condensate. Equilibrium vaporisation constant

   For butane   1.32

   For Pentane   1.02

   $\left(\begin{array}{l} \text{Ans. 0.82 Butane} \\ \quad\ \ \, 0.18 \text{ Pentane} \end{array} \right)$

6. A 2 mt refinery distills crude that contains chlorides and sulfur to an extent of 700 ppm and 2000 ppm respectively. If PH at the top of the column is maintained at 6; to avoid corrosion—
   (a) find out the requirement of ammonia,
   (b) Blend of naphtha from the same refinery has the following overhead composition (in stabiliser). Suggest the methods for prevention of corrosion in stabiliser.

   No =10 ppm; SO_2=50 ppm; H_2S=0.001%
   Other sulfur 0.0003%
   Compounds
   Chloride concentration = 100 ppm.

REFERENCES

1. Gary, J.H; Glenn, E. Hand Werk. *Pet. Refining*, Marcel Dekker, N. York, p. 31.
2. 7th WPC 1967, Vol. IB p. 16.
3. [3, 20, 24] Nelson, W.L. *Pet. Refining Engg.* 267, 238-239, 594.
4. *HCP*, Sept. 1972, p. 188, 189.
5. Lapedes, D.N. *Encyclopedia of Energy*, McGraw Hill; 1976, p. 557.
6. *CEW*, January 1979, p. 55.
7. Dikshit, R.D. Datta, P. & Chalihar, S. *CEW*, Jan. 1979, p. 81.
8. Deen, H.E., Kaestner, A.M., & Stendahl, C.M., 7WPC, Vol. 4, p. 43.
9. Purnaprajna, V. & Shroff, A.C., 2nd LAWPSP, Symposium IIT Bombay.
10. Mekler, L.A. & Fairall, R.S. *Pet. Ref.* Dec. 1952, p. 151.
11. Wilson, Lobo & Hottel. *Ind. Eng. Chem.* 24, (1932), p. 486.
12. Frank L.E. Evans., Equipment Design Handbook, Vol. 2 GPC, Houston, p. 14.
13. Lobo & Evans., *Trans. A.I.Ch.E.*, 35, (1939), p. 743.
14. Kern., Process Heat Transfer, McGraw Hill p. 690.
15. Campbell, O.F., *Pet. Ref.* Jan. 1950, p. 109.
16. Morrow, R., *Pet. Ref.* 1957, Feb. p. 159.
17. Monrad, *Ind. Engg. Chem.* 24, 1932, p. 505.
18. Sergio Menicatti & Luigi Cappiello, *HCP*, Sept. 1972, p. 226.
19. Scott, R., *Trans. Inst. Chem Engrs*, 1950, 28, 4, p. 205.
21. Mathis, H.M., Schweppe, J.L. & Wimpress, R.N. *Pet. Ref.* April 1961, p. 17.
22. William, F. Bland & Robert, L. Davidson, Petroleum Processing Handbook McGraw Hill p. 4-2.
23. Parekh, R.H., *Pet. Ref. Nov.* 1952, p. 99.
25. Prater & Boye, C.D., *Oil and Gas J.* May 2, 1955, p. 72.
26. Ludwig, E.E., Applied Process Design for Chemicals and Petro-chemical Plants (GPC), Vol. 2, 1979, p. 65.
27. Packie, J.W., *Trans. A.I.Ch.E*, Vol. 37, 1941, p. 51.
28. Watkins, R.N., *HCP* Dec. 1969, p. 93.
29. Watkins, R.N., *Petroleum Refinery Distillation* (GPC) 1973, p. 5.
30. Van Winkle, M., *Pet. Ref.* 33, 1954, p. 171.
31. Van Winkle, M., *HCP* April, 1964, p. 139.
32. Winn, F.W., *Pet. Ref.* 33, 1954, p. 131.
33. Atkins, G.T. & Wilson, G.W., *Pet. Ref.* 33, 5, 1954, p. 145.

34. Norman, W.S., Absorption, Distillation and Cooling Towers, Longmans Green & Co. Ltd., 1962, p. 177.
35. New Turbogrid Trays, Emeryville, Calif. Shell Development Corp. *Pet. Ref.* Nov. 1952, p. 105.
36. Rein Hard Billet, Distillation Engineering, Heyden & Sons, London, 1979, p. 350.
37. Fred. C. Riesenfeld & A.L. Kohl, Gas Purification, Gulf Publishing Co. Ed. p. 123.
38. Proce, of NPL Conference. Teddington, Middlesex, U.K. 11-12. Sept. 1978, p. 69.
39. Hausler, R.H. & Coble, N.D., *HCP*, May, 1972, p. 108.
40. Little, R.S., Baun., W.H. & Anerousis, J.P, *HCP*, May, 1977, p. 205.
41. Kenneth A. Kobe, John. J. McKetta, Advances in Petroleum Chemistry and Refining, Vol. 5, p. 137.
42. Norman, W.S., Absorption, Distillation and Cooling Towers, Longmans p. 128.
43. Rao, B.K.B., *C.I.D. (CP&E)* Aug. 1975, p. 15-16.

4

Treatment Techniques

4.1 FRACTIONS—*IMPURITIES*

In general all petroleum fractions require some kind of treatment in consensus with the specifications. Such treatments widely vary from fraction to fraction, and sometimes a chain of treatments is unavoidable. The rigorousness of treatment is proportional to the corrosive and indign constituents in fractions. Fractions from naphthenic crudes are obviously subjected to such drastic treatments. All fractions contain impurities, which can be classified under two headings namely mechanical and chemical.

4.1.1 Physical or Mechanical Impurities

The sources for mechanical impurities are plenty; during mining, crude may be contaminated with dust, mud and moisture and processing operations may bring catalyst dust, moisture or solvent droplets etc. Indubitably, all these have to be separated first.

In case of gases, dust particles, moisture droplets and oily matter are such physical impurities. These can be removed by a solvent wash; cheap solvent for such wash is water. Oil matter can be dissolved by passing the gas through gas oils or any convenient fraction. After water wash, drying is usually resorted to. When the gases are treated for removal of chemical impurities, solutions are employed. In such cases separate water-spray washing of gas may not be necessary, unless quantum of the suspended impurities is very high. Physical impurities from liquid fractions can be easily separated by subjecting to settling in towers of convenient sizes and then filtering. Moisture is removed by using suitable dehydrating agents.

4.1.2 Chemical Impurities

These exist in all fractions to varying extents. Most of them exhibit close boiling range in par with the fraction, which renders separation by distillation impossible. As an example, impurities in LPG streams are mercaptans and unsaturates of C_3 and C_4, all boil at the same temperature. In fact, the unsaturates may not eventually cause any inconvenience in burning, as they aid the flame propagation very much, hence their presence may not be a bar in such mixtures. But rich amounts of these unsaturates are undesirable, due to the fear of impending gum formation (polymers).

Sulfur, nitrogen and oxygen compounds are major impurities in all fractions. Of all the objectionables, sulfur heads the list. In gaseous form sulfur is present, other than mercaptans, as hydrogen sulfide, carbonyl sulfide, carbon disulfide and sulfonil chloride. The spree escape of these components into LPG streams due to close boiling points has been perceived, and as a result, number of treatment techniques have been devised.

Nitrogen and oxygen compounds are slightly different in their reactions with ordinary chemicals and sometimes require special procedures.

Removal of sulfur compounds that cause damage is known as sweetening. Usually acidic sulfur can be removed by washing with any amines or caustic. Doctoring is principally based upon the following criteria.

1. Oxidising mercaptans to disulfides.
2. Physical extraction.
3. In-situ destruction of sulfur bearing compounds.
4. Catalytic conversions in presence of hydrogen.

All the above methods are in practice, although the last mentioned method is universally acclaimed. Depending upon the commercial or chemical importance of a product the severity of treatment also increases.

4.1.2.1 Oxidising Mercaptans to Disulfides

This is, perhaps the oldest sweetening technique involving conversion of active mercaptans to inactive disulfides. Disulfides in gasolines are found to be harmless compared to sulfides; however for pollution considerations both are equally harmful. Important sweetening reagents are chlorides and hypochlorites of copper and iron and lead dioxide and lead sulfide. The reagents react with mercaptans in the following way:

Reaction

$$4RSH+4CuCl_2 \rightarrow 2R_2S_2+4CuCl+4HCl$$

Regeneration

$$4CuCl+4HCl+O_2 \rightarrow 4CuCl_2+2H_2O$$

$$Na_2PbO_2+2RSH \rightarrow (RS)_2Pb+2Na(OH)$$

$$(RS)_2Pb+S \rightarrow R_2S_2+PbS.$$

The reagents are often regenerated back. Regeneration of copper chloride is achieved by blowing air in contact with hydrochloric acid. In case of lead doctoring, the resulting lead disulfide is converted to lead sulfide by additions of fresh sulfur. Evidently for a fraction that contains free sulfur, this lead process is remarkable. Lead sulfide is converted back to lead plumbite by oxidising the slurry with air at 80°C. If free sulfur is in excess of equivalent mercaptan, fresh additions of mercaptan is essential to consume the sulfur. This is a paradoxical situation; fresh additions of sulfur are invited for completion of reactions, which may lead to improper balance in the treated stocks, often resulting in more sulfur than anticipated.

A simple reagent like lead sulfide alone can also do the same works as given by the following equation.

$$2RSH + S + PbS, Na (OH) \rightarrow PbS, (R_2S_2) + Na_2S + 2H_2O$$

The reaction is carried out in presence of air or oxygen over a catalyst bed (lead sulfide) where, thorough contact of gasoline and sulfur takes place. When gasolines contain naphthenic acids and phenols; sodium soaps shall be formed, resulting in a difficult situation i.e. the soaps hinder the precipitation of lead sulfide. Exact amount of litharge for precipitation has to be calculated. Industrial dosage is about 0.02 kg/bbl for straight run fractions and 0.5 kg/bbl for cracked stocks.

At present this process is losing its place to other processes. Carbon dioxide can also be removed simultaneously along with acid gases, thus; itself does not require any treatment.

4.1.2.2 Extraction Processes

Physical extraction of sulfur compounds is practised widely under different trade names. ADU gases are relatively free from impurities compared to off gases from crackers. ADU gases and natural gas can be treated in a single stage itself. If high sulfur-natural gas or wild gasolines are handled, it is preferred to first wash with caustic. Ethanolamines are profusely used for washing purpose, because of the ease of regeneration. Ethanol amines react with acidic components as shown below:

$$(HO\ CH_2\ CH_2)NH_2 + H_2S \rightleftharpoons (HO\ CH_2\ CH_2\ NH_3)_2\ S$$
$$(HO\ CH_2\ CH_2)NH_2 + CO_2 + H_2O \rightleftharpoons (HO\ CH_2\ CH_2\ NH_3)_2\ CO_3$$

All forward reactions proceed below 40°C where as backward reactions occur at about 105°C. 20-30% aqueous amine solutions are employed for this purpose. Fitzgerald[1a] et al. presented the solubility data on H_2S and CO_2 in amines. The equilibrium concentrations of CO_2 and H_2S at loading conditions in MEA, DEA[1b] solutions are available in literature. The idea of concentrations (equilibrium) is prerequisite for designing such absorbers. Extraction of H_2S and mercaptans from naphtha cuts by using caustic soda solutions were dealt by Dunstan[2] and the data is presented below:

Component		B.P.°C	% Removed	B.P. of corresponding disulfide°C
H_2S			100	—
Ethylmercaptan		38.9	97.1	152.6
n-propyl	-do-	67.8	88.8	—
i-propyl	-do-	59.0	87.2	175.5
n-butyl	-do-	98.2	63.2	—
i-butyl	-do-	87.8	62.8	220
i-amyl	-do-	117.8	33.0	

When gas is washed by wet solutions, drying of gas is inevitable. Drying[3a] agents like dehydrated salts, alumina, silica-gel, activated charcoal are in service. Ethylene glycols are the modern solvents employed for drying paraffins of lower order[4]. Paraffins have the capacity to form hydrates. A survey of various processes in operation is presented in Table 4.1.

4.1.2.3 Destruction of Sulfur Compounds by Sulfuric Acid[5]

One method of in-situ destruction of sulfur compounds is by sulfuric acid. Acid of above 93% strength is normally used for drastic

TABLE 4.1 Survey of Different Processes for Treatment.

	Name of the process (1)	Chemicals involved (2)	Constituents removed (3)	Operating conditions (4)	Regeneration (5)	Ref: (6)
1.	Girbotol	Ethanol amines-monodi	H_2S, CO_2 from Natural, cracked gases	Reaction temp. 25°C 20% concentration	Backward reaction 105°C	Chyuan-Chung Chen & Andrews—Ng HCP, April, 1980, p. 122
2.	Shell solutiser	Tripotassium phosphate	H_2S, from Natural gas	35°C	105-110°C Air blowing	
3.	Alkacid BASF	Sodium, potassium salts of diethyl/methyl amino acids, propionic-acetic acids	H_2S, CO_2 from Natural gas liquid hydro carbons			
4.	Benefield Corporation	Hot potassium carbonates containing benefield reagents	CO_2, H_2S, COS Natural gas, synthesis gas	High pressures, above 8 Kg/cm² Any convenient temperature		
5.	Cat Carb	Potassium salts	CO_2 (NG, H_2, NH_3 or Sour gases)			

(Contd.)

TABLE 4.1.: (Contd.)

(1)	(2)	(3)	(4)	(5)	(6)
6. Adip	Dipproponal amine	COS, H$_2$S, CO$_2$ mercaptans, acid gases (NG, SG, Refinery gas)	Wide flexibility in operation conditions	P: 20-25 Kg/cm^2 T: 35°C–38°C	
7. Unisol Atlantic	Methanol + KOH	CO$_2$			
8. Mercapsol	Cresols+ naphthenic acids	Mercaptans (gas, gasoline)			
9. Tannin solutizer	Organic salts+ Potassium phosphates	H$_2$S, CO$_2$ (gas)			
10. Vetrocoke	Sodium potassium arsenite sol. (K$_3$AS$_2$O$_3$)	CO$_2$ (NG, RG)			
11. Battersea	Aromatic amines (alkyl salts)	SO$_2$ (gas)			
12. Fluor Econamine	Aqueous sol. of primary alkyl amine/diglycolamine (HO-C$_2$H$_4$-O-C$_2$H$_4$-NH$_2$)	H$_2$S, CO$_2$ (NG, SG, RG)	Sol. 40-60% P: upto 80 Kg/cm^2	Steam strippings P: 80 Kg/cm^2 T: 180°C	
13. Purisol	N-methyl pyrrolidone	H$_2$S/acidic constituent Gas)	Pressure 80 Kg/cm^2 T 30°C		I & EC Vol. 62/1970 P 37-43

(Contd.)

TABLE 4.1. : (Contd.)

	(1)	(2)	(3)	(4)	(5)	(6)
14.	Rectisol	Methanol	CO_2, H_2S, NH_3, COS, HCN Gum formers (gases)			Oil & Gas J. March 20, 1967 (116-118)
15.	Solexol	Dimethyl ether of Polyethylene glycol	SO_2, CO_2 NG, SG			HCP April 1987, 35
16.	Sulfonil (Shell)	Na_2CO_3 Vanadium Organic nitrogen Compounds in water	H_2S, COS, CO_2 Mercaptans (NG, SG & LNG)	>80 Kg/cm^2	Air blowing	
17.	Locap gasoline sweetening	Locap catalyst conditioned with sodium sulfide	Sweetening of gasoline	T: 70°C P: 5 to 10 Kg/cm^2	Total re-placement of catalyst	HCP, Sept. 1972, p. 211
18.	Fluor solvent	Anhydrous organic compounds, propylene carbonate (Diglycolamine)	O_2 & H_2S from NG, SG	30-40% sol. P: upto 80 Kg/cm^2	Steam stripping at 180°C and 80 Kg/cm^2	
19.	SNPA-DEA	Diethanol amine (modified)	H_2S+CO_2 NG	P: 50-75 Kg/cm^2 T: Low (30°C)		HCP, April 1973, p. 101

269

(Contd.)

TABLE 4.1. : (Contd.)

(1)	(2)	(3)	(4)	(5)	(6)
20. Takahax	Alkaline sol. (P_H 8.5) Sodium 1,4 naphthoquinone, 2 sulfonate as redox catalyst	(NG and SG)			HCP, April 1973, p. 110

treatments. Olefins and higher unsaturates are readily attacked by sulfuric acid of 75% strength. Next to sulfur compounds, oxygen and nitrogen compounds are also significantly removed by sulfuric acid. Though the process is still continued for limited operations, its usefulness for lighter fractions is questionable. Oxygenated compounds like sulfur compounds are attacked with same vigour. Phenols form condensation products with naphthenes during acid treatment. About 50% of oxygenated compounds are eliminated in this process. Non-basic nitrogen compounds are more resistant to sulfuric acid. Gasolines from high nitrogen crudes are recommended to be treated with 25% sulfuric acid. The attack of sulfuric acid is least on paraffins and lower ones are not affected at all. Relatively, solubility of particularly hydrocarbon increases with molecular weight and strength of acid, further more complex molecules are readily oxidised. In refining practice 93% acid is the weakest that can be economically employed.

The action of sulfuric acid on mercaptan is shown below:

$$RSH + H_2SO_4 \rightarrow RSHSO_3 + H_2O$$
$$RSH + RSHSO_3 \rightarrow (RS)_2 SO_2 + H_2O$$
$$(RS)_2 SO_2 \rightarrow R_2S_2 + SO_2$$

Sweetened gasolines are likely to be converted back to sour gasolines by decomposition of disulfides. Thiophenes are very slow to react and require fuming acid. Sulfoxides and sulfones are soluble in 75 93% acid. Carbondisulfide is immune to this treatment. It is found with fuming acid and at high temperatures any hydrocarbon can be reacted. The quantum of acid and its strength required, for treating different fractions is given below:

Natural gasoline	1 Kg/bbl of 75% acid
	Temp : <30°C
	Time : <1 minute
Straight run gasoline	1-2 Kg/bbl of 75% acid
	Temp : <30°C
	Time : 1 to few minutes
Kerosene	3 to 6 Kg/bbl of 93% acid
	Temp : <40°C
	Time : about 30 minutes
Lube Oils	30 Kg/bbl of 98% acid
	Temp : 75°-80°C
	Time : 1-4 hours
Transformer Oils	30-40 Kgs (of oleum 20%) per bbl.

Temp : 80°-90°C

Time : in hrs.

With increasing strength and temperature, acid sludge also increases. Disposal of these sludges forms a major obstacle in generous adoption of this process. With modern technical advancements, material wastage is prevented; hence sulfuric acid treatment is replaced by other processes wherever it can be.

4.1.2.4. Catalytic Desulfurisation

Of late, sweetening operations are carried out by catalysts. In such reactions whole molecule bearing sulfur is not removed, as seen in extraction or acid treatment, but only the sulfur atom is picked up. Thus material loss is negligible, with possible recovery of sulfur.

The first catalytic process for removal of organic sulfur was introduced by Carpenter and Evans[6]. Using sulfides of nickel. The success of this has led to many catalysts. Still search for better catalysts is going on. Iron[7] oxide catalysts with 5-15% Cr_2O_3 were used for hydrogenation and hydrolysis of carbonyl sulfide from gases (manufactured). Such catalysts operate at a temperature of 300-400°C at 1-25 kg/cm^2 pressure. Removal of carbonyl sulfide is almost complete. Copper-Chromia-Alumina Catalysts, Huff Catalysts, (Copper-Cr-Vanadium oxides) promote conversion of organic sulfur compounds to H_2S, which is retained in the form of metal sulfide, on catalyst. Cobalt, molybdenum catalysts (12-13% Co) supported on bauxite or on fullers earth are new in this field. Holmes Maxted[8] process employs molybdates of copper, iron, zinc and cobalt. Shell hydrodesulfurisation technique which is carried out in presence of hydrogen, uses sulfides of tungsten and nickel. Cobalt molybdenum catalysts are in fact the best used hydrode-sulfurising catalysts. The process is described in detail in the next chapter.

Treatments required for different fractions are elaborately discussed below.

4.1.3 Production and Treatment of LPG (SNG, NG)

LPG and natural gas are most widely used gaseous fuels in domestic and industrial circles. Some gases like LPG is supplied in the form of vapor-liquid mixtures in cylinders of convenient size, while other gases may be directly supplied through pipe lines in gas phase itself. All these gases are to be given certain treatments, before consumer gets them. In case of LPG, the gas is mainly obtained from distillation unit, hence relatively pure. Oddly, natural gas displays variable compositions from mine to mine, and even in the same mine from time to time. All gases are likely to contain external impurities and these have to be separated.

Production of LPG

Production of LPG is mainly based upon the principle of separation. C_3-C_4 fractions can be separated from the rest of fractions in many ways, however at present the following methods are used in industry.

(a) Distillation at low temperature
(b) Absorption and desorption
(c) Compression and expansion
(d) Combined methods

Of all these methods, (a) and (b) are the most popular ones in the industry.

Distillation is based upon the boiling points or relative volatility, while absorption is intimately connected with the selective absorption capacity of oils for certain fractions.

The absorption method is most convincing and presented here (Fig. 4.1*a*).

A stream of combined gases consisting of gases from ADU, isomerisation and alkylation units/refinery off gas (catalytic/cracker/hydrocracker/stabiliser) is first given a water wash to remove the suspended impurities. The gases are then passed into dryer (1) containing any suitable adsorbent to remove moisture. Issuing gas from dryer is compressed to 10 atmos pressure and chilled and sent into the absorber (2). Absorber always maintains a cold lean oil circulation. Methane, ethane, hydrogen and oxides of carbon, hydrogen sulfide, nitrogen and its compounds are usually not soluble in this cold oil; hence they leave the absorption unit without any change. Heavier gas fractions; C_3 and C_4 compounds are carried over with the fat oil. Small amount of light gas absorbed in oil may be separated by decreasing the pressure, as is done in deethaniser (3). Deethaniser bottoms are taken into another stripper called debutaniser (4). The top fraction consists of a mixture butane and propane. Bottom product contains solvent and other hydrocarbons higher than C_4 fraction. When natural gasoline or virgin gasoline is the feed, C_5, C_6 components shall be accompanying and these fractions are absorbed by lean oil. Some portion of this lean oil is cooled and sent into deethaniser and remaining to absorber.

274

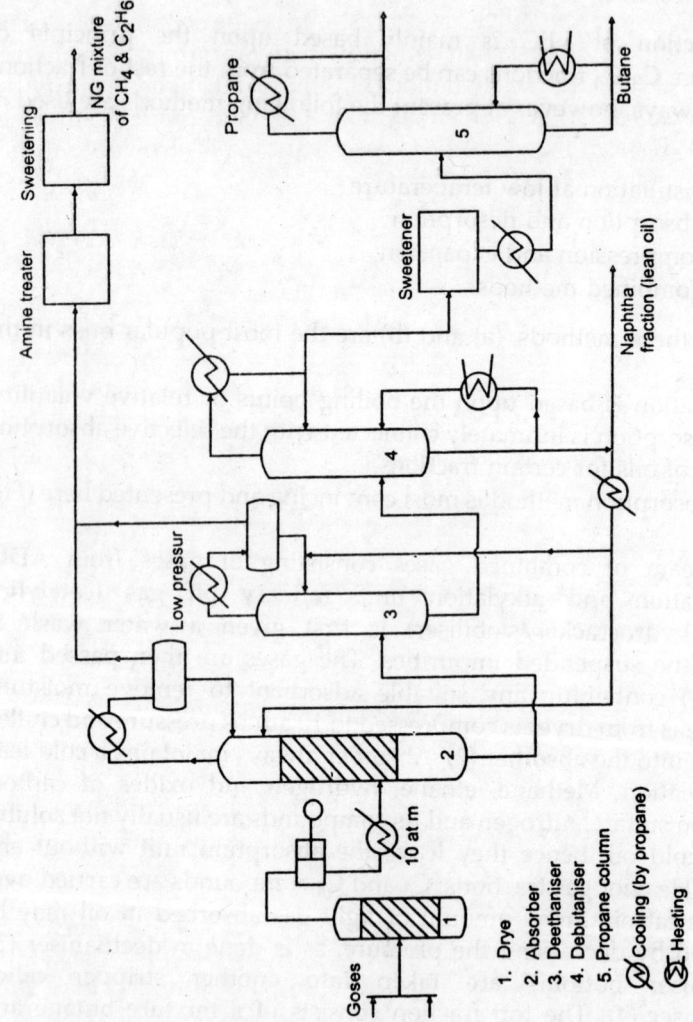

Figure 4.1a LPG Production by Absorption Technique.

Recovered C_5, C_6 fractions are blended with gasoline. C_3-C_4 mixture is sweetened by any of the given techniques, and later it is chilled and fed to propane column (5). Propane, butane mixture is sold as LPG, part of butane is always reserved for blending with gasoline. Oil absorption gives very pure hydrocarbons. A purity of 98% for propane and 95% for butanes at a recovery of 70-80% is reported. If cracked gases contain much unsaturates, propane separation is done by other techniques, one such being low temperature fractionation.

Low temperature fractionation is employed, when purity of products is very much essential, like polymer grades, ethylene, propylene etc. and the quantity of gases handled is also very much.

4.1.3.1 LNG Technology

Technology on liquefaction of natural gas was largely developed and progressed after the induction of cryogenics into this field; however the stupendous investment (to go down to such low temperature; methane - 160°C) for low temperature—technology has been the real problem for conservative application. Separation of these gases at low temperature demands special alloys to combat the embrittlement of metals.

LNG, LPG technologies do not differ to a great extent; the over doing of cooling to much needed methane in LNG processing is little bit expensive. Improved cascade cooling to liquefy methane, has been the main feature of cryogenics in the industry. Turbo expander technology, the newcomer in this field has been acclaimed and has gained momentum.

In India first LPG plant based on 2.21 million Cubic metres of natural gas available from Naharkatiya Oil field, has been commissioned (August '82) at Duliajan. This plant is designed to produce 60,000 tons of LPG and 12,000 tons of natural gasoline per year, using sophisticated turbo expander technology; incidentally, this becomes the first plant in Asia working on this know-how.

4.1.3.2 Turbo Expander liquefaction

The underlying principle in this design is quite simple: the gas is compressed in two or three stages and expanded in turbines; the cooled gas is permitted to exchange heat with incoming gas. Thus the inability prevailing in Joule-Thomson cycles to recover and utilise mechanical energy has been successfully overcome in turbo expanders. Simplified natural gas processing is shown in Fig. 4.1b. In each expander, as condensation of liquids should not take place in

276

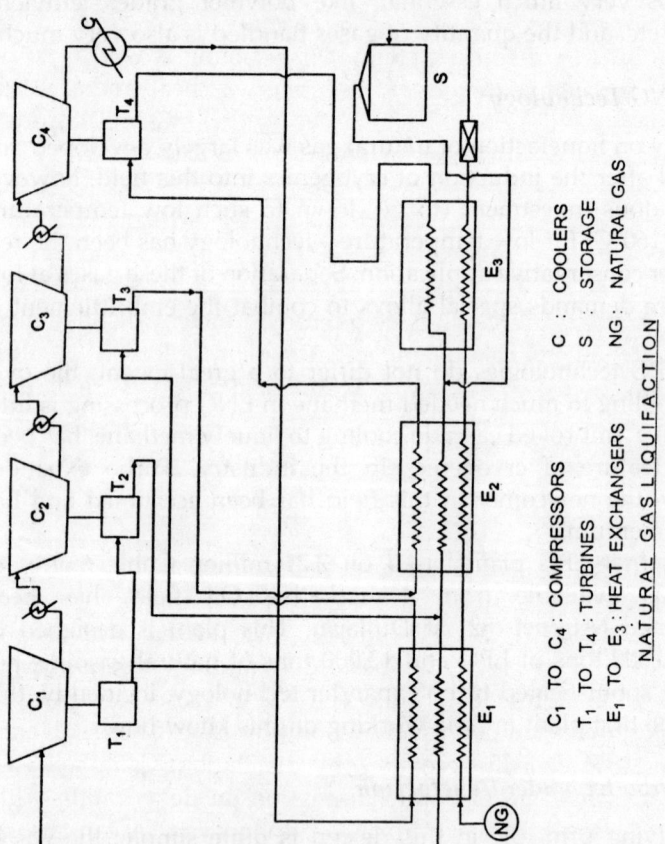

C₁ TO C₄: COMPRESSORS
T₁ TO T₄: TURBINES
E₁ TO E₃: HEAT EXCHANGERS

C : COOLERS
S : STORAGE
NG : NATURAL GAS

NATURAL GAS LIQUIFACTION

Figure 4.1b Natural Gas Liquefaction.

any engine, the liquids are taken from heat exchangers, and further fractionated.

Purification of natural gas is essential before such cycle is attempted.

Absorber gases, can be directly fed into fuel system, without much purification. Removal of acid gases and sulfur compounds will suffice. If desirable, methane, ethane, ethylene fractions can be separated by low temperature distillation or by hypersorption. Low temperature fractionation depends upon the efficient production of such low temperatures.

In this connection cascade system of gas cooling is worthy to be mentioned. Propane being produced in refinery is cheap and acts as a good refrigerant and the starting material for this system. As much low temperature as —160°C is achieved by cascade cooling. The principle of which is: Ethylene at 19 atm. Pressure[9] is liquefied by propane evaporation at 40°C. Propane is liquefied at 18 atm. pressure by water cooling at 50°C. Ethylene at -100°C can condense methane at 35 atm.

Separation of ethane and ethylene mixture is done by distillation at low temperature. Ethane-ethylene mixture is distilled at 21 atm. pressure in a 60 plate column, where liquid propane acts as coolant in condenser. The temperature of the reboiler is kept around -7°C, in these columns.

4.1.4 Dehydration of Gases

Dehydration of natural gas or light petroleum fractions like naphtha, gasoline etc. is done by passing the gases through absorbers containing ethylene glycols (Di, mono, tri). Mixed with slight ethanolamines, they serve as best dehydrating agents, even comparable with lithium chloride. Each Kg of these solvents can absorb 3 kgs of water to depress dew point by 25°C. These solvents after use are economically recovered by distillation. Campbell and Laurence[3b] presented an elaborate discussion on dehydration with these solvents. Molecular sieves for NG drying Type 3A can absorb water and H_2S but not carbon dioxide.

4.1.5 Sweetening Operation for Gases

Ethanolamine Treatment

Figure 4.2 illustrates the sweetening operation in detail. Feed gas containing acid components is first compressed to 10 to 15

278

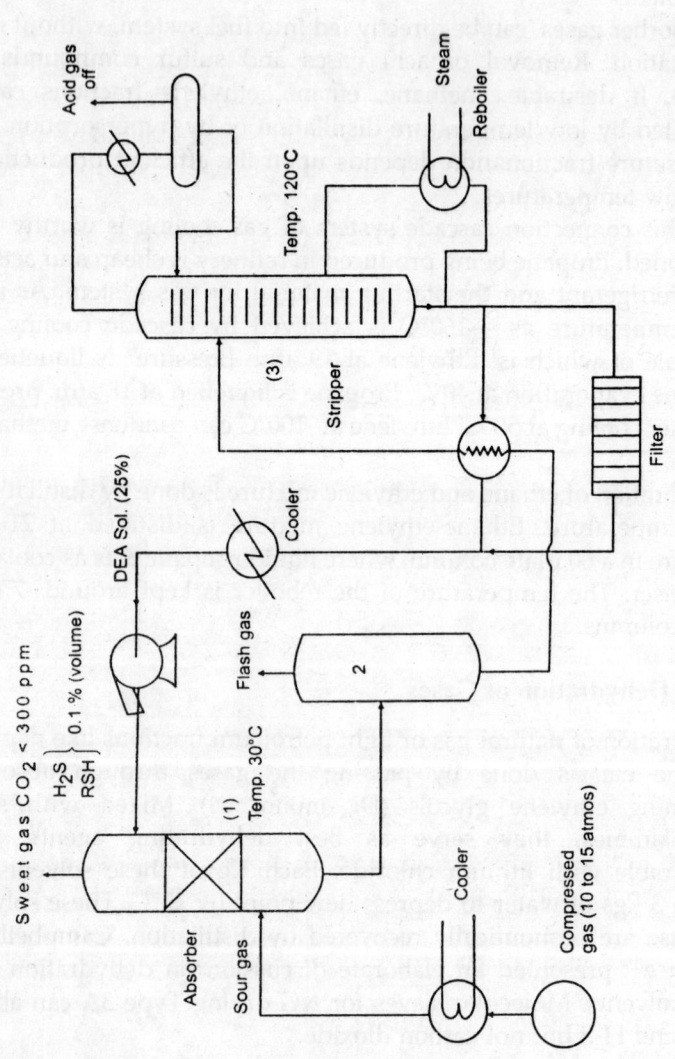

Figure 4.2 Amine Treatment for LPG.

atmospheres and then cooled by cold water. The cold gas is admitted into absorber (1). The absorber may be plate type or packed type containing 20 theoretical stages. Ethanol amine solution (30% in water) is distributed from the top of the tower and gases are fed counter currently. Undesirable constituents of gas are absorbed by the solution, and sweet gas leaves from the top of absorber. The fat solution is flashed in flash tower (2); here, some of the dissolved gases escape due to low pressure. The solution is later fed into a stripper (3) where the stripping of acid gases takes place. The stripper is again a 20 tray-column, operating at 105 to 120°C. Heating is done by steam coils. The amine solution from the stripper is filtered if necessary and sent back to the absorber after cooling to 30°C. The off gases from stripper are sent for sulfur recovery. Sulfonil, Vetrocoke, SNPA-DEA processes also work on the same lines as amine plants.

4.1.5.1 Stretford Operation (Figure 4.3 illustrates this)

Complete removal of hydrogen sulfide is contemplated here, partial removal of accompanying organic sulfur is also achieved.

The treating solution consists of a mixture of sodium carbonate, sodium vanadate, anthroquinone disulfonic acid and arsenic oxide. Sodium carbonate can remove all carbondioxide, when it exists in small percentages.

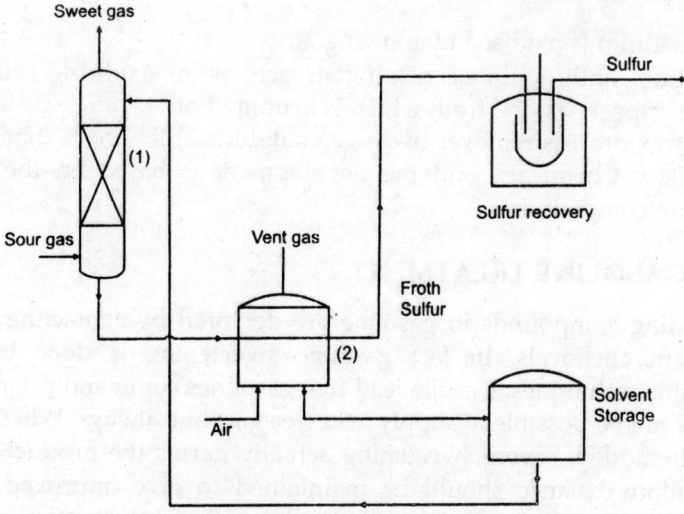

Figure 4.3 : Stretford Process.

In the contactor (1) (a packed bed absorber) sour gas comes into contact with circulating solution. Hydrogen sulfide is first removed because of its reaction with 5-valent state vanadium. Hydrogen sulfide is completely oxidised to elemental sulfur in one operation itself. Sulfur liberated in the reactions must be freed from solution. This can be done by submitting the fat solution to frothing by blowing air through tank (2). Sulfur is removed as a floating layer. This layer is later filtered and processed for sulfur. The settled liquid is sent to storage to begin with the operation. The operation is moderately simple and continuous.

When air is blown into fat solution hydroquinone is oxidised back to original disulfonic acid, which keeps the activity of vanadium, by restoring its five valence state.

Air blown separation technique for getting elemental sulfur is contemplated in other processes like, Giammarco Vetrocoke and Takahax. Beavons Process for sulfur dioxide and sulfur is based upon hydrogenation at moderate pressure and temperature in presence of a catalyst.

$$H_2O + COS + S \rightarrow H_2S + CO_2$$

In Stretford Process, hydrogen sulfide is oxidised by and sodium vanadate by sodium carbonate

$$Na_2CO_3 + H_2S \rightarrow Na\,HS + Na\,H\,CO_3$$

$$NaHCO_3 + NaHS + Na\,VO_3 \rightarrow S + Na_2VO_5 + H_2O + Na\,CO$$

Vanadium is oxidised by blowing air.

Sodium anthroquinone disulfonate acts as an oxidising catalyst. Sulfur appears as fine froth which is skimmed off.

Sulfrex process removes hydrogen sulphide at less costs (Shell Oil and Dow Chemicals) and the development is based on the Iron chealate compounds.

4.2 GASOLINE TREATMENT

Offending compounds in gasoline are doctored by contacting with different chemicals. In fact gasoline sweetening is done by all available techniques. Despite lead free gasolines command premium, it may not be possible to supply lead free gasoline always. When TEL is to be added, severe sweetening actually harms the product, so a meticulous balance should be maintained to give improved lead susceptability. Odourous methyl and other mercaptans, and thiophenols are kept at less than 0.002 ppm in gasoline to give a negative doctor test, for jet fuels, the figure is kept at 20 ppm to give a satisfactory copper strip corrosion test (3 hrs at 50°C).

4.2.1 Copper Chloride Process (Figure 4.4)

Gasoline free of sulfur and hydrogen sulfide is first passed through moisture remover (1). Moisture is removed by passing through dehydrated salts like $CaCl_2$ or $NaCl$. The feed is heated by exhaust steam to a temperature of 40-60°C and sent into mixer (2). Part of circulating slurry consisting of copper chloride and clay (-200 mesh) in water is mixed up thoroughly with the feed. Oxygen is admitted into this mixer containing slurry and gasoline. This mixer acts as a reactor when air is sent in. The floating layer consists of gasoline and is taken to water washing system (3). Water removes the trapped particles of catalyst and acid. After settling the gasoline phase is routed through another dehydrator (salt bed) to storage tank.

The efficient circulation of slurry and its contact with fresh oxygen usually determine the success of operation.

4.2.2 Inhibitor Sweetening

Caustic stripping followed by addition of good oxidation inhibitor, converts sour gasoline to sweetened product. Mercaptans are removed by caustic and oxidation. Caustic is regenerated by blowing air and recycled back along with fresh additions. Inhibitor retards the oxidation of gasoline, while it permits caustic to react. Air inhibitor sweetening is desirable to cracked gasolines, while air-solulizer sweetening is suitable for all gasolines.

Extraction of mercaptans is greatly facilitated by solubility promoters such as i-butyric acid, methanol, cresols, alkyl phenols and naphthenic acids.[10]

The improvement in extraction coefficient with the promoters is shown below:

$$\text{Extraction coefficient} = \frac{\text{Mercaptan in caustic layer}}{\text{Mercaptan in oil layer}}$$

Treating Solutions[11]

	10% NaOH	36% KoH +26% cresols	36% KoH +37% cresols
Methyl Mercaptan	1,000	3,500	12,000
Ethyl Mercaptan	200	1,000	2,750

282

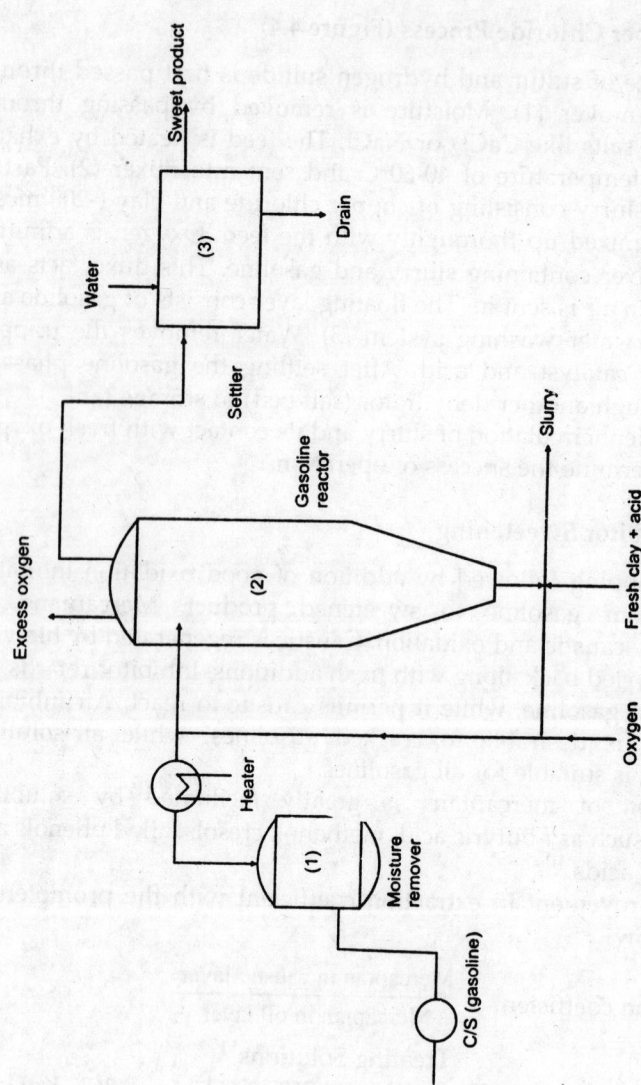

Figure 4.4 Sweetening by Copperchloride Process.

4.2.3 Caustic and Methanol (Unisol Process)

The treatment is performed in a counter current extraction tower, as shown in Fig. 4.5a. The extraction tower is provided with two inlets one at the top and the other at the middle for separately introducing solvents, *viz.*, caustic and methanol respectively. Sour gasoline comes into contact with these two down flowing solvents. Methanol removes some nitrogen compounds in addition to mercaptans. The temperature during operation is kept around 40°C. The spent solvent is recovered back in two distillation columns, one for caustic and the other for methanol. Methanol steam-stripper separates methanol and mercaptans. The recovered streams are again proportioned to the extractor. This process can successfully eliminate 60% of sulfur compounds while mercaptan removal is complete. Caustic to methanol ratio is maintained around two and consumptions are found to vary. Electrolytic mercapsol process adopts a slightly different regeneration technique. In this an electrolytic cell is used to convert mercaptides into disulfides.

4.2.3.1 Dualayer Process

In Dualayer process, highly concentrated potassium cresylate solution is employed as solvent. Like methanol, cresols and salts of fatty acids

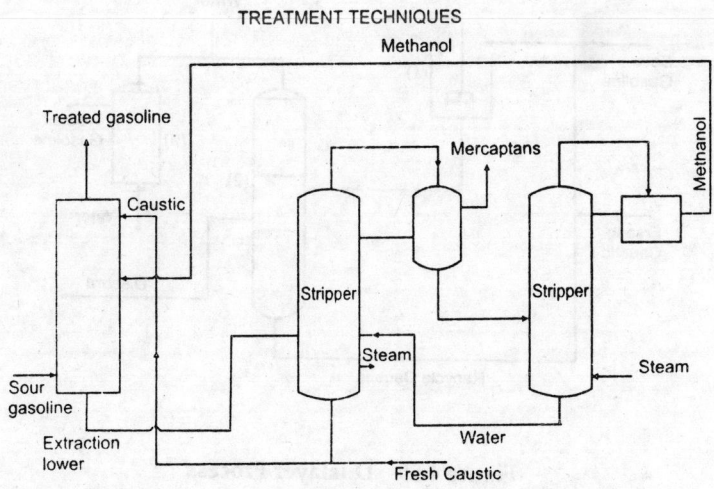

Figure 4.5a   Caustic Methanol Treatment for Gasoline.

dissolve, freely mercaptans. Higher boiling mercaptans are less readily extracted than the lower boiling. As cracked stocks contain a higher proportion of low boiling mercaptans, treatment of these stocks is easy.

Dualayer process for treating gasolines and gases, is similar to Unisol process. Here the contact between Dualayer agent and sour gasoline is brought out in at least two stages at a slightly higher temperature than Unisol process (i.e. 50°C).

Figure 4.5b shows the outlines of the process. Sour gasoline is thoroughly mixed with fresh and recycled caustic in a mixer (1). The mixer is now allowed to settle in a column (2). Three separate layers are formed, top most layer will be gasoline and the bottom layer will be caustic for recycle. Middle layer contains organic salts and impurities, hence to be discarded. The treated gasoline is washed with water and the aqueous phase is separated from gasoline by settling in a column (3). The remaining moisture may be removed by passing through any drying agents. Steam stripping of the extract layer separates out mercaptans. The advantages accrued in this process are less losses and less sediment-formation, even after a substantial period.

Nalifining is another process, that employs acetic anhydride and caustic rinse to convert compounds of sulfur and other impurities into

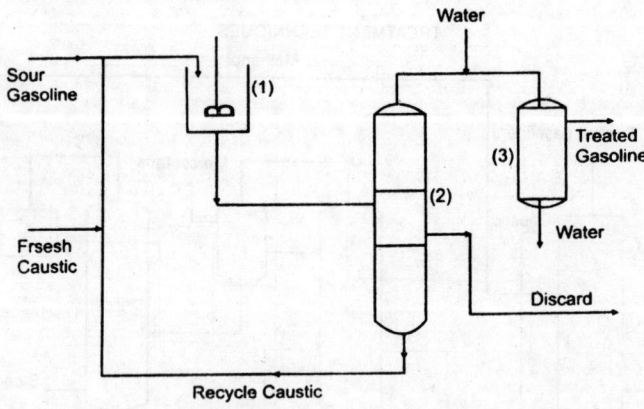

Figure 4.5b Dualayer Process
(1) Mixer (2) Column Settler (3) Water Washer and Settler ... ...

less objectionables. The anhydride reacts with OH ions to form esters; SH ions and nitrogen are converted to thiaester and amides respectively. Final caustic rinse is required to neutralise corrosive acetic acid.

4.2.4 Lead Doctoring of Gasoline (Fig. 4.6)

In this method of treating charge stock (c/s), *i.e.*, sour gasoline is thoroughly agitated in Mixer 1 with a solution of caustic plumbite. Lead becomes lead sulfide during reaction as shown below:

$$2RSH + Na_2 Pb O_2 \rightarrow Pb (SR)_2 + 2 NaOH \qquad (a)$$
$$Pb(SR)_2+S \rightarrow PbS+R_2S_2 \qquad (b)$$

This lead alkyl sulfide (a) is reacted with fresh sulfur (Mixer 2) to free lead sulfide from disulfide as shown in the above reaction (b). Lead sulfide is converted back to sodium plumbite by blowing air into the spent solution obtained from the settler (2).

$$PbS + Na (OH) + O_2 \rightarrow Na_2 Pb O_2 + Na_2 SO_4 + 2H_2O$$
$$\text{(excess)}$$

The regenerated solution is ready for treatment. Lead consumption as is evident from the reactions, is negligible, but alkali consumption is high.

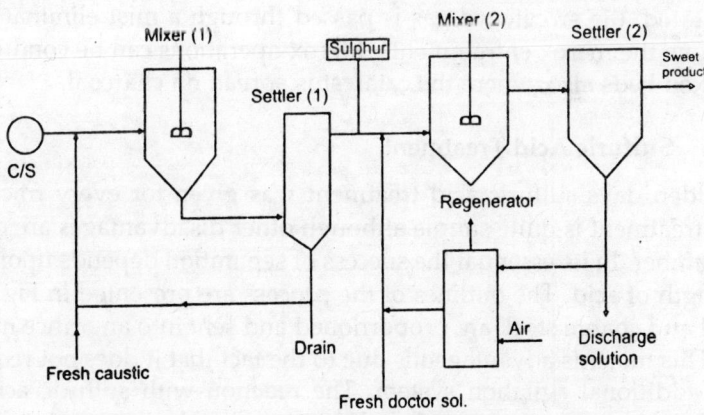

Figure 4.6   Lead Doctoring of Gasoline.

Operating Data.

Ratio of Doctor sol. to gasoline 1 : 5	Sulfur addition and Stirring 5 to 15 min.
Time of contact 1/2 min. to 1 minute	Air regeneration temp. 100°C Pressure 1 to 5 kg/cm^2

There is one difference in this doctoring, that is the sulfur in organic phase (gasoline) is converted to a disulfide and is allowed to go into the stream, as disulfide is considered to be less harmful than sulfide. The calculated amount of fresh sulfur should be added otherwise sulfur will increase in the end product.

4.2.5 Merox Sweetening (Figure 4.7)

Mercaptan oxidation (Merox) process is most appealing as it is suitable for treating LPG, gasolines and kerosenes. This process has shown commendable abilities in ablating mercaptans in the form of harmless disul fides. All modern refineries in India are bent upon using this process. C/S and merox solution (catalyst cobalt salt) in caustic solution are contacted in extractor (1). Sweetened product from the top of extractor goes into merox separator (3) where a fresh gust of air comes in contact with merox solution. Merox is regenerated here and the solution is transferred to a settler (4). The phases are separated in this horizontal settler (4), and the heavy merox is obtained from the bottom of the tank. This merox being regenerated is fed back to extractor along with fresh merox. The extract containing mercaptans is oxidised by blowing air in towers (2). Disulfides are separated in settler (5) by decantation and regenerated merox is collected for circulating into extractor.

A reduction of 90% sulfur is possible in this operation. When LPG is treated, the sweetened gas is passed through a mist eliminator to remove the merox entrainments. Merox operations can be conducted in fixed beds also, where the catalyst is spread on charcoal.

4.2.6 Sulfuric Acid Treatment

In olden days sulfuric acid treatment was given for every fraction. The treatment is quite simple although other disadvantages are quite in number. In its essential the success of separation depends upon the strength of acid. The outlines of the process are presented in Fig. 4.8. Acid and charge stock are proportioned and sent into an orifice mixer (1). This mixer is advantageous due to the fact that it does not require any additional agitation system. The reaction with sulfuric acid is exothermic; hence care must be taken to see that the temperature does not exceed 40°C. As a precautionary measure cold water sprays may be kept in reserve. After the reaction is over the contents are sent into a settler (2) where a settling time of 5 to 6 hours is aptly given. Gradually, the acid sludge separates out and concentrates at the bottom. Sulfur, unsaturates and metallic bodies are all attacked by the

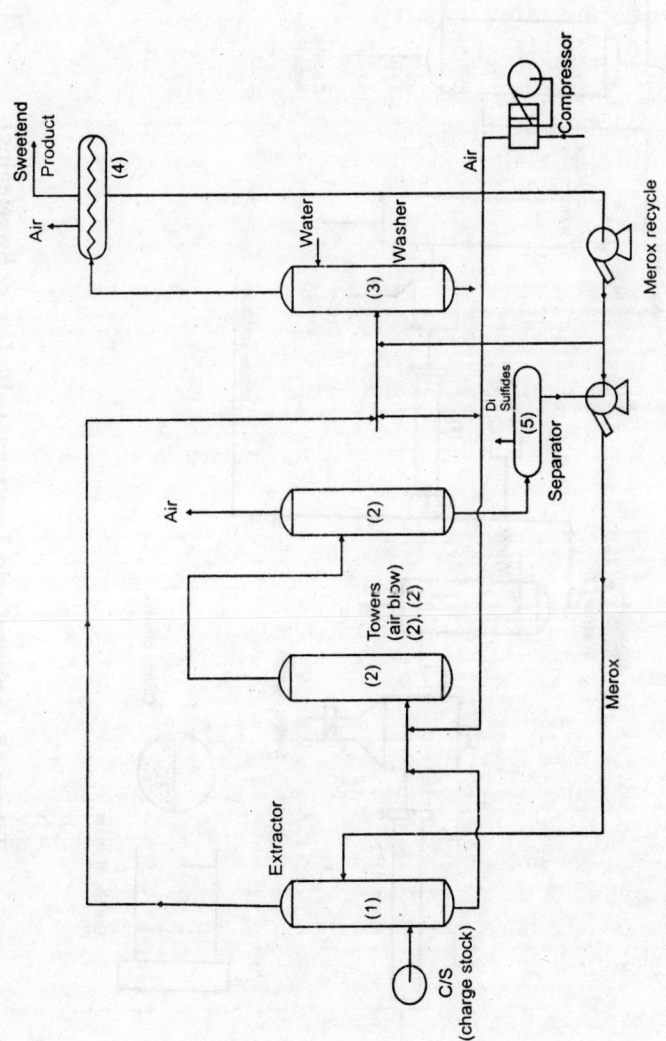

Figure 4.7 Merox Process.

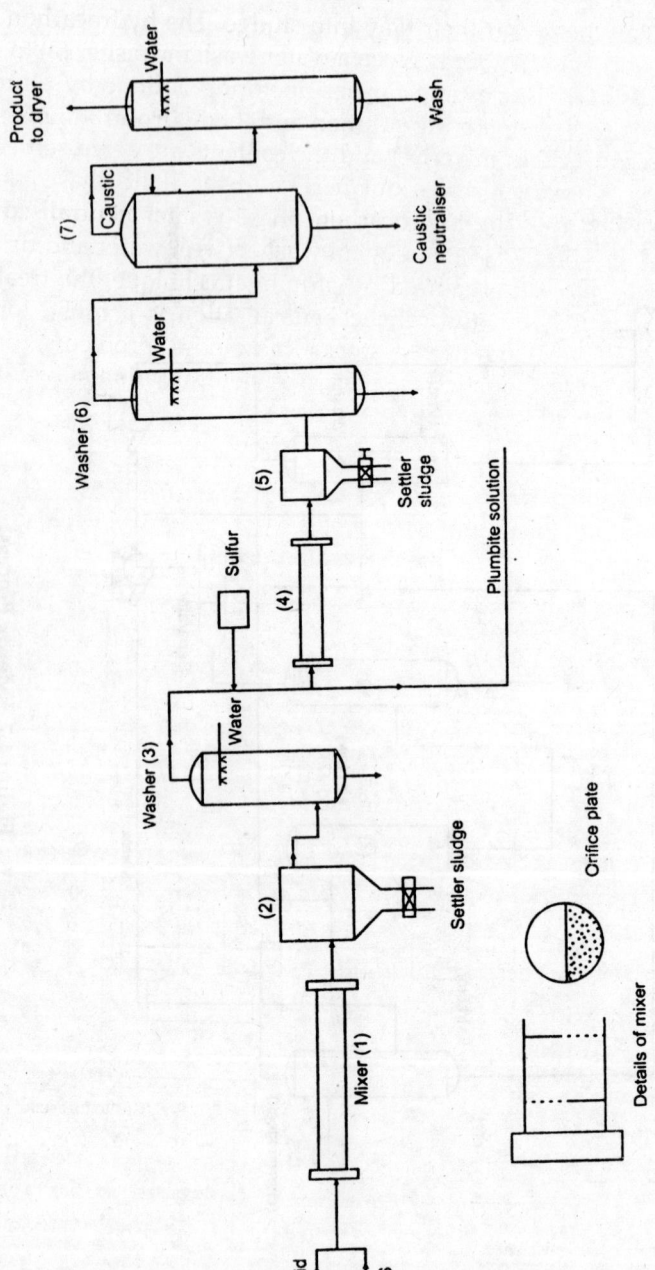

Figure 4.8 Sulfuric Acid Treatment (with Doctor Sweetening).

acid and all these find their way into sludge. The hydrocarbon layer from the top of the settler is given a water wash in washer (3) to scrub of acid residues. Some time further doctoring is done by plumbite-sulfur mixers. The doctoring solution and the hydrocarbon phase are thoroughly mixed in mixer (4) and the contents are again settled in a settler (5). Sludge separates out here, and the light phase is again water washed (6). The hydrocarbon phase is later neutralised with caustic in a contactor (7); and again washed with water and dried.

If doctoring is not required by plumbite technique, the treatment may be omitted. But the caustic neutralisation is a must, for acid treated stocks. Some heavy stocks may pose problems, due to emulsifying characteristics; in such cases, neutralisation may be given up; and clay contacting may be applied.

4.2.7 Catalytic Desulfurisation

Desulfurisation technique is a predecessor to hydrodesulfurisation. All fractions lighter than diesels can be desulfurised. Desulfurisation

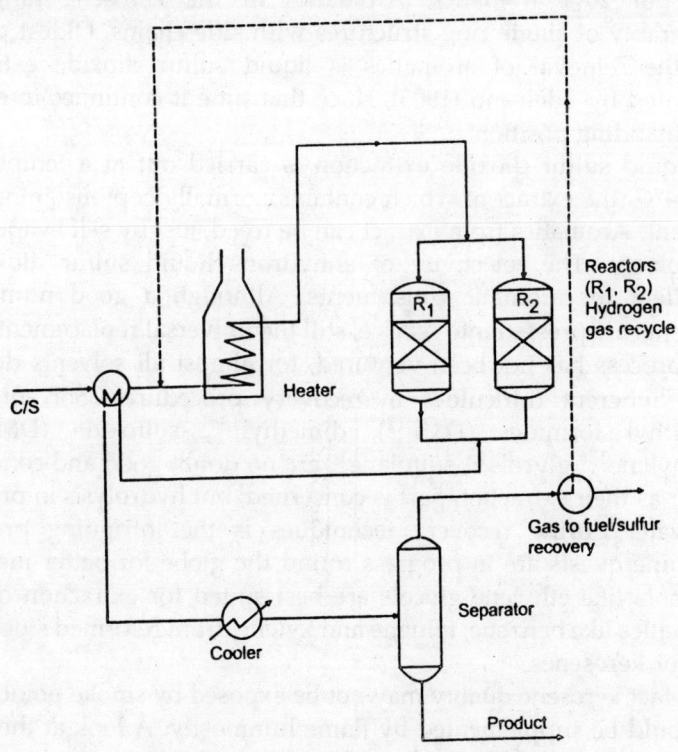

Figure 4.9 Catalytic Desulfurisation.

is conducted at higher temperatures (upto 400°C); hence the feed should be vaporisable without any thermal degradation. The reaction is brought out in catalytic reactors. Any suitable catalyst ranging from synthetic nature to natural bauxite can be employed. Often the gases are vented out. But recycling is essential when hydrogen gas is a co-stream.

Desulfurisation as followed in Autofining process satisfactorily operates even without hydrogen recycle. Depending upon the nature of catalyst, 90-95% of sulfur can be removed. Catalytic desulfurisation operation is shown in Fig. 4.9.

4.3 TREATMENT OF KEROSENE

For kerosenes (illuminating oils) a single treatment technique like desulfurisation is not sufficient; as the illuminating quality of kerosene is not improved in this operation. Smoke point improvement is achieved by removing smoke causing ingredients, namely aromatics. Formulations of good kerosene touch a maximum limit of 20% aromatics. Aromatics in the kerosene range are preferably of single ring structures with side chains. Oldest process for the removal of aromatics is liquid sulfur dioxide extraction invented by Edeleanu (1905), since that time it continues to enjoy a commanding position.

Liquid sulfur dioxide extraction is carried out at a temperature of -14°C, the extract of which contains normally copious amounts of solvent. Aromatics from extract can be freed, just by self evaporation of solvent. The selectivity of anhydrous liquid sulfur dioxide is excellent for aromatic constituents. Although a good number of solvents are pressed into service, still the universal replacement of the old process has not been ventured, for almost all solvents do have their inherent difficulties in recovery procedures. Solvents, like dimethyl formides (DMF[12]), dimethyl[13a], sulfoxide (DMSO[14]), diethylene[13b] glycols[15], sulfolane[16] are no doubt good and conducive as far as their extraction part is concerned; but hydrolysis in presence of water during recovery techniques is the intriguing problem. Genuine quests are in progress round the globe for better methods. Solvents like ethylene glycols are best suited for extraction of light aromatics like benzene, toluene and xylene, from reformed stocks, but not for kerosenes.

In fact kerosene quality may not be exposed by smoke point alone, it should be supplemented by flame luminosity. A look at the Table 4.2 would reveal that glycols may not be good solvents for improving smoke point, but the treated stocks posses a good luminosity compared to conventional solvents. Hence, while extraction of smoke causing ingredients is desirable, at the same time retention of illumination-contribuents like naphthenes is foremost. Different treating processes for kerosene are shown in Figs. 4.10a, b.

4.3.1 Liquid Sulfurdioxide Extraction of Aromatics

Figure 4.10a illustrates the process. Kerosene is first deaerated by exposing to vacuum. Any moisture in kerosene is removed by passing through a tower of adsorbent (1), (usually calcium chloride serves the purpose). The charge is then passed through a heat exchanger (2), cooled by raffinate and further chilled by liquid sulfurdioxide (3) to bring down the temperature to -8°C. The cold stock is fed into extractor (a) where counter current extraction takes place. Liquid sulfurdioxide being heavier, the extract is obtained from the bottom of the extractor. Raffinate and extract are separately processed for pure streams in two sets of evaporators. Raffinate stream through a heat exchanger (4) fed with treated and hot kerosene enters first unit of the evaporation system (b_1). The evaporators resemble more or less distillation units, operate in series, under decreasing pressure. First evaporator (b_1 -1) is operating above the critical pressure of sulfur dioxide, but below critical temperature. In evaporators, pressure ranges from 6 to 15 atmospheres. This permits the temperature during operation to reach as high as 80 to 110°C without vaporising sulfur dioxide. Heating is mainly done by steam in an outside boiler (5), and in this process some sulfur dioxide is expelled in vapor phase (or in liquid phase). The raffinate is now taken to low pressure evaporator (b_1-2) where most of the sulfur dioxide is vaporised and taken for compression system. The column is usually atmospheric and the temperature here will be 20°C more than that of the first evaporator. Condensation of sulfur dioxide does not take place, hence only gaseous sulfur dioxide escapes. The bottoms of the evaporator through reflux (6) heater are now fed to third stage (C_1) unit where the pressure is maintained at 25 mm to 190 mm mercury, and that keeps the temperature around 15°C. Sulfur-dioxide from this unit is recovered by vacuum system, (d_1) and then sent into the compression system (d_2) along with other streams of sulfur dioxide. Sulfur dioxide is compressed to above critical pressure and then cooled by a train of heat exchangers (e) fed with cold water at 20°C, which liquefies sulfurdioxide. The liquid sulfurdioxide is stored in tank (f) and self evaporation can bring the temperature of sulfur dioxide to -10°C.

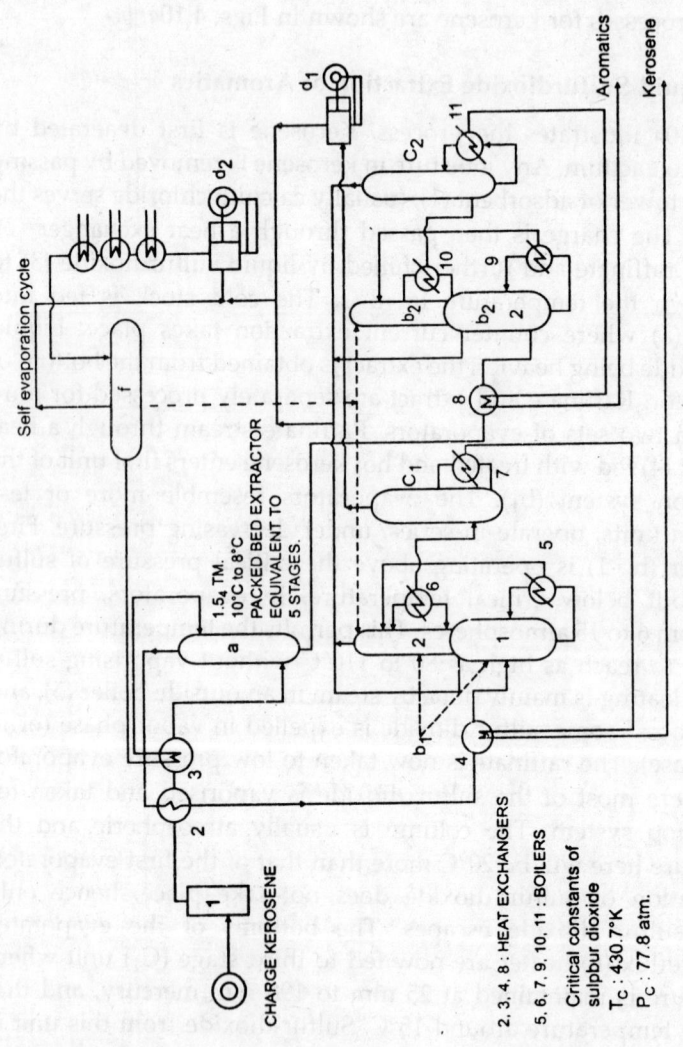

Figure 4.10a Liquid SO₂ Extraction Process.
2, 3, 4, 8: Heat Exchangers; 5, 6, 7, 9, 10, 11 are Boilers

Self evaporation cycle

Aromatics

Kerosene

2, 3, 4, 8 : HEAT EXCHANGERS
5, 6, 7, 9, 10, 11 : BOILERS

Critical constants of sulphur dioxide
T_C : 430.7°K
P_C : 77.8 atm.

1.54 TM.
10°C to 8°C
PACKED BED EXTRACTOR
EQUIVALENT TO
5 STAGES.

CHARGE KEROSENE

TABLE 4.2 Smoke Point and Luminosity Index.

Solvent	Temperature of extraction °C	Solvent to feed ratio	Improvement in smoke point		Luminosity Index*	Calorific value kJ/gm at 30°C
			Without dilution (%)	With dilution (%)		
DMF	25	2	11	10(10%)		
	40	2	14	12(10%)	(32)	46.85
	60	2	12	11(10%)		
DMSO	30	2	12	13(5%)		
	50	2	12	13(5%)	(38)	44.17
	70	2	12	13(5%)		
Ethylene-Glycol	30	4	4	4(5%)		
	50	4	5	5(5%)	(40)	
Diethylene-Glycol	30	4	6	6(5%)		
	50	4	6	6(5%)	(42)	37.16
Raw kerosene	(smoke point=18)				(13)	46.85
Tetralin	(smoke point=9)				(0)	46.85
i-octane	(smoke point=35)				(50)	47.44

* Luminosity is determined at highest smoke point elevation only.

294

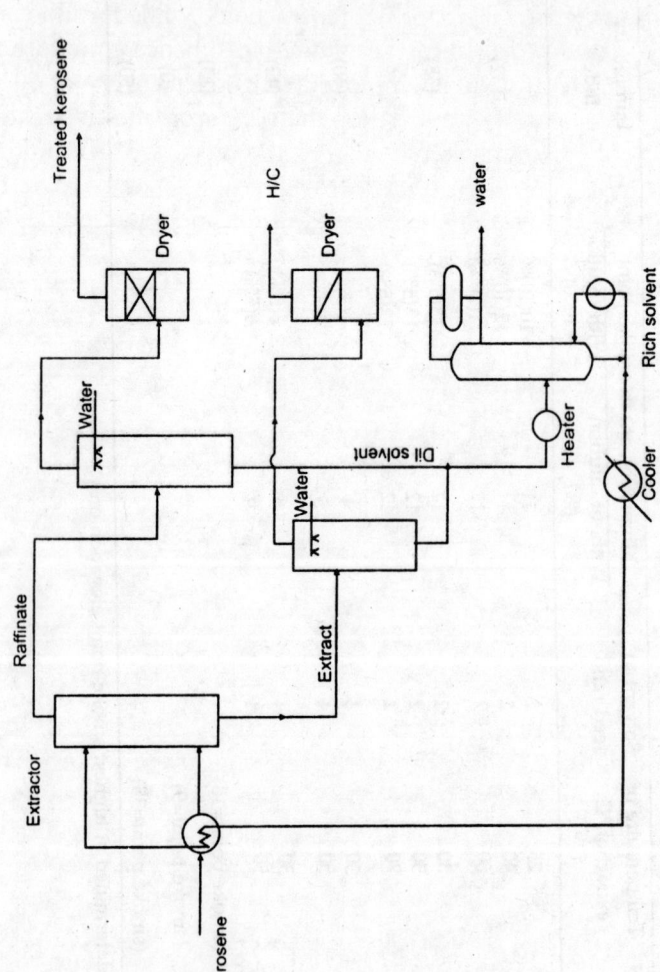

Figure 4.10b Improvement of Smoke Point with Solvents (DMF & DMSO).

The same type of evaporation cycle is executed for recovery of sulfurdioxide from extract. The entire operation should be conducted in moisture free situation.

Depending upon the stock, the amount of solvent to feed varies. When smoke point elevation of kerosene is done then liquid sulfur dioxide to kerosene ratio is kept in the range of 1 to 1.5, and when diesel stocks are used a ratio of 1 to 1.4 is preferred. Sulfur dioxide extraction is quite convincing for lighter fractions, while for lube oils the solvency power of solvent is very much less, hence a mixture of benzene and sulfur dioxide is employed. A good recovery of 90 to 95% aromatics is possible leaving less than 2% aromatics in treated stocks.[17] A smoke point elevation of 10 to 15 units is frequent in this operation. With lube oils the treating temperature is kept above the pour point of the oil. The loss of solvent is considerably low (about 0.2%), and that really maintains the economy of process.

Information on liquid sulfur dioxide treatment of lubes and kerosenes is presented below[31a].

	Kerosene	Lubes
SO_2 vol.	100%	180%
Treating temp.	$-10°C$%	40 to 60°C
Raffinate yield	75%	63%
Sulfur in feed	0.15%	
Sulfur in finished product	0.026%	
Smoke point (Feed)	24 mm	
Smoke point (Finished product)	45 mm	
Feed viscosity index	—	55
Product viscosity index	—	95

Snam Progettis[18] new aromatics extraction process uses N-formylmorpholine solvent. This has been suggested to be quite competitive with present processes, has a greater flexibility to feed stocks and keeps abreast with the developments. Extraction of aromatics by other solvents like DMF, DMSO, glycols is presented in Fig. 4.10b. It is apparent from the flow sheet, that solvent from the rich phase is separated by water washing and later by distillation. The hydrocarbon phases obviously contain some moisture and have to be dried separately; steam stripping may not be satisfactory because of the close boiling ranges of solvent and stock. Though extraction of aromatics is favoured, the recovery of solvents without suffering hydrolysis, has not been reported; hence it stands as an obstacle in the way. Operating data of solvents DMSO, DMF and glycol, on improvement of smoke point are shown in Fig. 4.11a to f.

Properties of solvents used for elevation of smoke point of kerosene are listed in Table 4.3.

4.4 TREATMENT OF LUBES

By providing a thin film of oil between sliding surfaces, lubrication increases mechanical advantage, efficiency, and the life of machine. Another function unnoticed by many, is the disposal of heat developed in frictional movements. A carefully chosen lubricant must

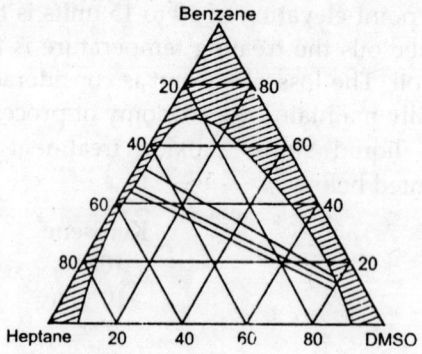

Figure 4.11a **Mutual Solubility Envelope Showing Tie-lines for Extract and Raffinate when using Undiluted DMSO**

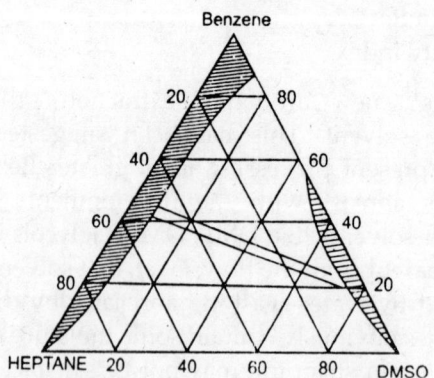

Figure 4.11b **The Mutual Solubility Envelope has Greater Area when DMSO is Diluted with 10 wt. % Water**
Phase Diagrams for the System DMSO and Heptane, Benzene

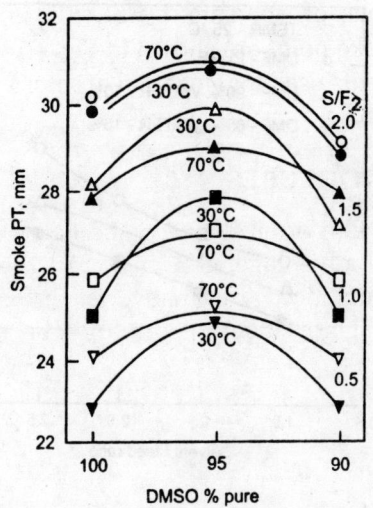

Figure 4.11c : Dimethyl Sulforide as Solvent for Improvement of Smoke Point (Effect of Temperature (30°C and 70°C) S/F Ration and Dilution).

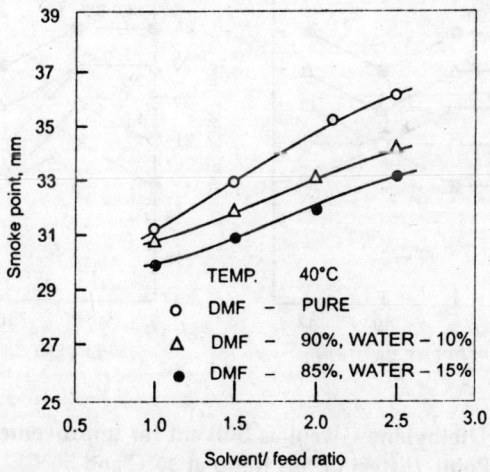

Figure 4.11d Dimethyl Formamide as Solvent for Improvement of Smoke Point (at 40°C) (Effect of S/F ratio, with and without dilution).

maintain fluidity and viscosity at the conditions of operation like temperature, load, shocks etc. Truly a single oil cannot contain all the requirements for different conditions and operations thus necessitating several formulations for specific functions. The basic

298

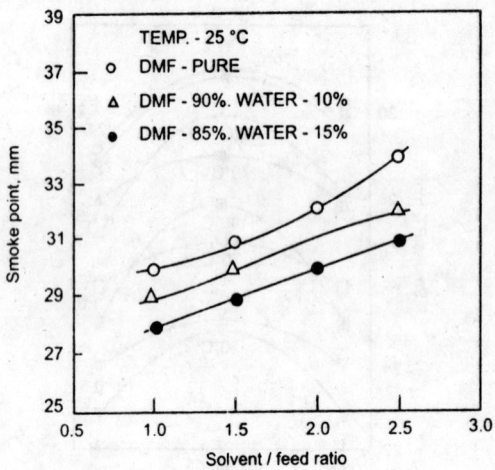

Figure 4.11e Dimethyl Formamide as Solvent for Improvement of Smoke Point at 25°C (Effect of S/F Ratio with and without Dilution).

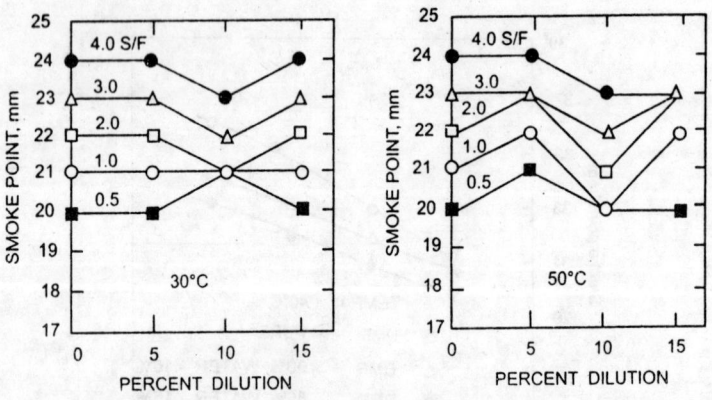

Figure 4.11f Diethylene Glycol as Solvent for Improvement of Smoke Point. (Effect of S/S Ratio at 30°C and 50°C).

nature of the oil rests upon the source of oil (base of crude); some properties of allegiance are incorporated by blending with other oils or additives. A raw lube oil having a boiling point of above 350°C is composed of the following general groups, (except purely synthetic ones).

TABLE 4.3 Properties of Solvents used for Smoke Point Improvement.

Solvent	Mol. wt.	MP°C	BP°C	Sp. gr. at 20°C	Sp. heat J/gm °C	Latent heat J/gm	Viscosity Cst at 20°C	Surface tension dynes 25°C	R.I. at 25°C	Solubility in water wt.%
1. Sulfur dioxide	64	−93	−10	1.45 at−14°C	1.338	387.76 at 17°C	−	−	−	19 at 0°C
2. Dimethyl formamide	73.1	−61	152.8	0.9489	2.09	575	0.83 at 25°C	35.2	1.4629	fair
3. Dimethyl sulfoxide	78.3	18.49	188	1.101	2.05	−	19 at 25°C	−	1.4770 at 20°C	fair
4. Ethylene glycol	−	−14	197	1.116	2.41	450.0	19.0	53	1.43	fair
5. Diethylene glycol	106.11	−	245	1.118	2.09	627.0	30.2	48.5	1447 at 20°C	fair
6. N-formyl-morpholine	115	+20	244	1.1528	1.588 at 25°C	435.0	8.13	−	1.4869 at 20°C	fair

Paraffins

Naphthenes

Aromatics (Poly, Condensed, Branched chains)

Waxes

Asphalts and resins

Inorganic constitutents (Oxygen, Nitrogen, Sulfur compounds and contaminants like metals)

All these kindred groups offer their own properties to oil without any reservation. Paraffins are credited for high viscosity index, while aromatics for low viscosity index. Comparatively, aromatics are generally more viscous than paraffins. Paraffins in higher ranges are more branched and invariably increase the pour point of oil. So evidently starting fluidity, viscosity and maintenance of viscosity at different temperatures cannot be satisfied by any single species. Naphthenes being intermediate in properties, can naturally form a bridging gap of these extremities. A high viscosity index[19] oil, thus can be produced from paraffinic-naphthenic base or paraffinic base crudes only. Wax is an undesirable constituent in any oil due to its high pour point, although its viscosity index is very high.

There are four established processes for improvement in quality of lubes as given below:

(a) Sulfuric acid treatment

(b) Clay treatment

(c) Solvent refining

(d) Hydro treatment

4.4.1 Sulfuric Acid Treatment

Acid treatment for lighter fractions was already discussed. This method is now reluctantly adopted although its disappearance is not yet complete, as some refineries practice it today also (like Assam Oil Company etc.). The main disadvantage of this process is inherent loss of material; further there is no improvement in viscosity index. However it is good for stocks containing more inorganics. Sludge disposal, material loss and tedious procedure are all the repugnant reasons for its systematic disappearance.

Industrially, acid treatment is done with 93% of sulfuric acid at a temperature of 60°C. 2 Kgs to 50 Kgs of acid are required per barrel of stock treated, covering light to drastic treatments. Treatment is done in cylindrical lead lined tanks with conical bottoms. Air is used for agitation. Required temperature is ensured by steam coils placed in the agitators. After the reaction is over the sludge is carefully removed and the oil is subjected to severe washing with water till the P_H is in neutral region. Caustic neutralisation is inevitable for naphthenic stocks. Naphthenic acids are prone to form sulfonates and these sulfonates offer considerable resistance during sludge settling. In this treatment, improvement in colour is significant as asphaltenes, resins and unsaturates are gradually depleted in the treated product. This treatment is most convincing for speciality oils like transformer oils, medicinal and white oils, where the material loss is of no significance (some times the loss goes upto 50%). Acid treatment is followed by clay treatment to avoid any emulsification difficulties during caustic or water wash.

4.4.2 Clay Treatment

This treatment is listed as one of the oldest methods and even now-a-days also it is followed in a limited scale. Various mineral clays and synthetic adsorbents are the listed materials for this treatment. Asphalts and resins are adsorbed on these clays. Resins being the colour contributors to oil, removal of these reflects in improving colour of the oil (unsaturates are preferably adsorbed over saturates). Further, clays have the tendency to adsorb some sulfur compounds and metallic bodies. Material loss is very much because of wetting the whole quantity of clay and retention of some in pores. The material adsorbed is governed by Freundlich equation and this is useful only for decolorising operations and the equation is given as[20]:

$$\log \frac{X}{M} = \log a + \frac{1}{n} \log c$$

where X = units of impurities adsorbed
  M = quantity of adsorbent
  c = concentration of impurities in equilibrium at surface
 a & n = constants

Clay treating was once a technique or decolourising and stabilising cracked gasolines and also for desulfurising virgin gasolines and kerosenes. In the treatment of light fractions the clays acts as catalyst; for heavy oils it is an adsorbent purifier.

Gasolines are also clay treated to eliminate materials inclined to give colour or form gums. The tendency of natural clays like fuller's earth or bauxite to polymerise unsaturates cannot be set aside during this treatment. For this reason, the clay treated gasolines may have to be further processed (distilled) to remove the resultant polymeric groups. The Gray, the Stretford and the Osterstrom processes are enlisted for such kind of treatment. Clay being catalyst in this case, the treatment is frequently conducted in fixed beds containing 40 to 60 mesh size particles. The charge may be in vapor or liquid state. Slurry type reactors are the new comers in this field and are condigned for cracked stocks.

Most of the clay treatment operations are conducively carried at a temperature of 125 to 250°C, desirably at a high pressure, almost in the vicinity of 20 atmospheres. These operating conditions are quite typified in Gray and Stretford operations. In other processes the pressure is still higher. In contrast, when desulfurisation of virgin gasolines is attempted, a temperature of 400°C is opted, and this helps in venting sulfides and mercaptans in the form of hydrogen sulfide. Clay treatment for cracked stocks, enriched with thiophenes is not sound as these are refractory. Due to fine deposition of coke on the clay, the activity of clay diminishes. Some refiners try to regenerate the clay by burning off the coke under controlled conditions, others discard. A ton of bauxite can usually handle 10,000 bbls of virgin gasoline, 1200 bbls of cracked gasoline or 5,000 bbls of kerosene. A comprehensive information regarding removal of sulfur compounds from naphtha cuts using different adsorbents and sulfuric acid treatment was presented by Kalichevsky and Kobe[21a] and is furnished in Table 4.4. Evidently sulfuric acid is superior in removing sulfur compounds especially like thiophenes.

TABLE 4.4 Comparison of Adsorbents and Sulfuric Acid Adsorbent used about 13.6 Kgs. per Barrel of Stock Acid used 24.5 kgs per Barrel of Stock % Sulfur Compounds Removed.

	Silicagel	Fuller's earth	Al_2O_3	Sulfuric acid 93%	Sulfuric acid Fuming
Free sulfur	8	4	4	0	0
H_2S	12	12	12	12	12
1-amyl mercaptan	69	3	17	24	100
Dimethyl sulfate	100	100	25	100	100
Methyl-p-toluene solufonate	80	67	0	77	100
CS_2	0	0	0	0	0
n-butyl sulfide	57	3	0	93	100
n-propyl sulfide	37	11	5	58	100
Thiophene	11	11	0	89	100
Diphenyl sulfoxide	100	100	0	90	100
n-butyl sulfone	100	100	12	100	100

In case of lube oils of naphthenic origin, clay treatment seems to be effective, and succeeds acid treatment. Acid treated stocks in such cases pose problems for caustic washing due to emulsification tendency. For materials having high surface area, such as Attapulgus, bauxite can adsorb acidic, resinous and asphaltic materials from heavy oils.

Clay treatment is conducted in two ways, namely contact filtration and percolation technique. After treatment the clays are regenerated by burning (controlled oxidation) carbon deposits; although it is not possible to completely remove these deposits without suffering the efficiency of adsorbent. Properties of different adsorbents are presented in Table 4.6.

Residues upto 6% are left behind even at temperatures (500-600°C) of regeneration. A close look at the percolation data given in Table 4.5 by Kalichevskey and Kobe[21b] reveals, that by increasing the ratio of oil to Fuller's earth (adsorbent), the sulfur compounds increase in treated oils after an initial depletion; hence a proper oil to adsorbent ratio has to be found out and utilised.

TABLE 4.5   Percolation Data.

	API	Sulfur %	C. Residue wt. %	Viscosity SUS (at 10°C)
Unfiltered stock	41.6	0.134	2.25	156
4.5 Kgs	48.3	0.02	0.027	87
34 Kgs	43.2	0.08	0.546	125
68 Kgs	41.9	0.13	1.49	140
225 Kgs	41.8	0.13	1.61	140

A survey of different adsorbents was made available by Felix Heinemann and Henz Heinemann[22] and is presented by Table 4.6.

Continuous contact filtration is shown in Fig. 4.12. It is based on the principle, that finely divided clay (-100 mesh size) is thoroughly mixed with the stock and allowed to stay in a tower to complete the reaction (adsorption). Then the slurry is filtered in suitable filter. Heavy oils are viscous and try to settle, this will bring unlimited snags by choking the lines and filtration equipment. Suitable measures, like addition of diluent or elevation in temperature of operation are necessary. Addition of sulfuric acid is desirable to keep the activity of clay.

TABLE 4.6 Adsorbents and Properties.

Activated alumina	Activation Temp. (appox)°C	Bulk density	% voids	Surface area M²/gm	Pore dia °A	Heat of wetting kJ/gm
Grade F-1	194	0.88	44	200	72	58.980
H-151	194			300-350		
Bauxite	233	1.33	44	27	76	79.477
Silica gel	200	0.94	42	151	43	87.843
Synthetic	194	0.82	41	602	10	121.725
Silica gel	194	0.82	41	374	15	
Fluid catalyst	194	0.64	34	372	33	
Pelleted Cat. A	194	1.00	36	280	16	
. Cat. B	194	0.79	30	420	22	
Head catalyst	194	0.56	44	126	95	99.137
Fuller's earth	194	0.56	42	119	109	
Attapulgite	600	0.68	41	236	43	
Bentonite	372					
Diatomaceous earth	194	0.32	49	4	5500	
Activated carbon		0.42	44	1397	112	39.738

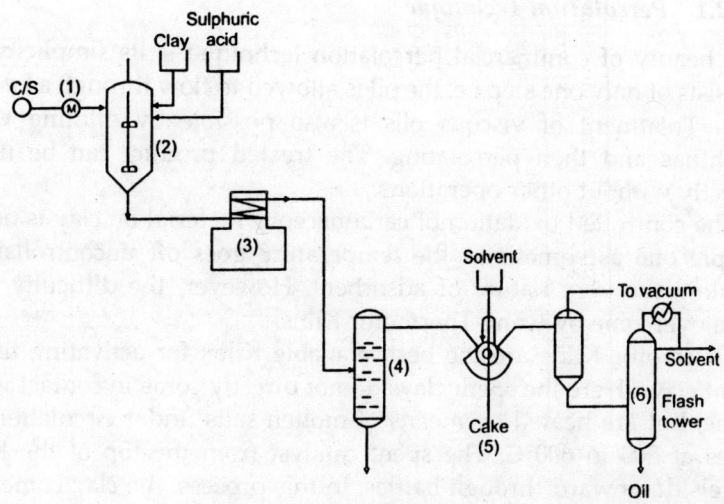

Figure 4.12   Contact Treatment.

In contact filtration or percolation the size of the adsorbent plays a significant role. Adsorption is favoured by smaller sizes, on the contrary filtration may not be easy with small sizes. The process begins with heating the charge stock by exchanging with out-going hot oil in exchanger (1). Oil and clay are now mixed in required proportion, and thoroughly agitated in mixer (2). The slurry is then routed through a convection of a heater (3) to reach a temperature of 200 to 225°C. The hot slurry is pumped into a contact tower (4). Usually an empty tower, fitted with steaming arrangement will suffice. The residence time in the tower is approximately one hour, and by that time the decolourisation operation comes to end. The slurry is now transported to filtration equipment (5). Filtration is sometimes difficult with highly viscous stocks, hence calls for fresh additions of diluent. Neutral high cylinder stocks, oils for reclamation are processed in this way. Filtrate containing diluent is distilled or flashed in a stripper (6); diluent is collected as top product which is cooled and kept ready for recycle. Treated oil from the stripper is also cooled and stored.

Though, spent clay recovery is disliked in most of the cases, it can be practised. Firstly, the cake is extracted with diluent to remove the adhering oil. Mixture of this oil and diluent is flashed to recover the diluent. The extracted cake may be regenerated by burning the carbon in a suitable kiln.

4.4.2.1 Percolation Technique

The beauty of commercial percolation technique is its simplicity; it consists of only one step i.e. the oil is allowed to flow through a bed of clay. Treatment of viscous oils is also possible by diluting with naphthas and then percolating. The treated product can be used directly without other operations.

The controlled oxidation of carbonaceous material on clay is not a simple one as sometimes the temperature goes off uncontrollably, spoiling the very nature of adsorbent. However, the difficulty has been overcome by using Thermofor Kilns.

Thermofor Kilns are the best available Kilns for activating these spent clays. Here the spent clays do not directly come in contact with flame, but are heated by means of molten salts under circulation in tubes at 500 to 600°C. The spent catalyst from the top of the Kiln travels downward through baffles. In this process, the clay comes in contact with hot tubes. This ensures burning of coke particles without fusing surface pores. With the introduction of this type of kiln, clay treatment has regained its lost importance and continued to be used in industry on a moderate scale.

4.4.3 Solvent Treatment

Solvent treatment, in refinery operations enjoys a unique position due to its multifunctional designs. Viscosity index improvement, deasphalting and dewaxing etc. are some of the major applications. A number of solvents for each of such operations are in service and more are being introduced. Each of these operations are discussed below.

4.4.3.1 Viscosity Index Improvement

A strong temptation to produce high viscosity index lubes has led to a prolific growth of solvent extraction. Solvents widely used for this affair are furfural, phenol, mixtures of cresols and propane. May other solvents like aniline, sulfur dioxide are enlisted; but without much use. A solvent credited with many irrevocable properties also is not free from controversies, as it cannot be hundred percent foolproof. General characteristics responsive towards good extraction tendencies for any solvent are

(a) High electronegativity of the hetero atom.
(b) High dipole moment.

(c) Presence of atoms of O, N, S with activity to form hydrogen bonds.

(d) Increased selectivity by substitution in β (5-ring) or m (6-ring) positions.

(e) Free rotatability of the molecule, no screening by further substituents.

(f) Accumulation of selectively acting groups.

Invariably, some of the functional groups are sacrificed and the final selection of a solvent is based on other factors as indicated below:

1. Solubility relationship
2. Adoptability to various feeds
3. Ease of recovery
4. High density (high molecular weight)
5. Less corrosiveness and good stability in presence of moisture
6. Low viscosity, less cost, less toxicity
7. Low interfacial tension
8. Low freezing point and less latent heat.

Even though high selectivity and solvency power govern extraction process, sometimes highly selective solvents are disregarded in favour of other solvents having less assuming properties. As an example, even though DMF, DMSO are good solvents for extraction of aromatics but are not considered favourably, because of accruing difficulties in recovery.

Phenol and furfural[23, 24] are very competitive solvents in field. Both these solvents accomplish the task satisfactorily; a survey on the performance of these two solvents is presented below:

1. Furfural extraction may require a higher temperature of extraction and more number of stages[25].
2. Yield of raffinate is more with phenol.
3. Phenol treated lubes have better oxidation stability.
4. VI. improvement may be same in both cases, however solvent dosage may be slightly more for furfural on comparative basis.
5. Phenol is more adoptable to different stocks and is unavoidable when naphthenic acids are more.
6. The selectivity of phenol increases with water (upto 10%). This is deemed to be a great advantage over furfural.
7. Furfural is unstable, hence requires nitrogen seal, during storage and recovery.

8. Losses can be negligible with phenol.

9. Residual solvent in treated stocks is not harmful; provided the solvent is phenol.

After extraction, the pour point of the stock increases, sometimes as high as 12-15°C. Despite this, refineries, without any doubt keep the extraction units before dewaxing units, and the result is, extraction units are destined to carry more loads. The following dewaxing operation naturally tailors the lubes to required pour point. Even with many of the above points in favour of phenol[26], furfural is able to sustain its position (in fact trying to replace) in view of pollution and toxic nature of phenol. Properties of solvents used for lube treatment are shown in Table 4.7.

4.4.3.2 Phenol Extraction of Lubes

Plate or packed bed extractors are generally employed in phenol extraction. Phenol extraction cannot be carried out at a temperature less than 50°C, because of the high melting point of phenol (40°C). Addition of water upto 10% is approved in all operations, although anhydrous phenol can do the extraction. Common operating temperatures range from 60 to 100°C. In practice a thermal gradient of 10 to 20°C[27] from top to bottom is imposed on the column; this enhances the internal turbulence due to mixing of hot solvent and relatively cold feed. Solvent to feed ratio of 1.5 to 2.5 for processing paraffinic and naphthenic stocks is conducive. An increase of V.I. by 25 to 50 Units[28a] is confirmed. Some authors had utilised the viscosity gravity constant as parameter instead of viscosity index[29].

Raffinate yields vary from 50 to 90% depending upon the solvent to feed ratio. Higher solvent ratios give better quality of raffinate but with less yields. Some of the operating data on phenol extraction is presented in Fig. 4.13 a, b and c. Phenol treated oils possess good oxidation stability with less sludge forming and carbon depositing tendencies. Number of stages required for phenol treatment is less than four with most units operating at three. In India, Barauni Refinery and Bharat Refinery (Bombay) are practicing this process. Description of the Process: (Figure 4.13a).

Solvent to feed ratio is one of the closely guarded parameters in extraction operation and for phenol extraction it lies in the range of 1.5 to 2.5 in going from paraffinic to naphthenic oils. In an extraction process number of stages play a vital role in determining the size of extractor, in phenol extraction the stages are about 4, and in general 3 stages are satisfactory. Solvent power and selectivity of phenol at

TABLE 4.7 Solvents & Properties (Lube Treatment).

		B.P. °C	M.P. °C	Sp. gravity at 20°C	Sp. heat Joules/gm°K	Latent heat J/gm	Solubility in water at 38°C
1.	Furfural	161.6	−36.6	1.162	1.757	419.137	9.0
2.	Phenol	182	41.5	1.07	2.342	445.480	15
3.	Sulfur dioxide	−10	−74	1.45	1.338	387.764 at−17°C	19 at °C
4.	Chlorex	177.8	−51.8	1.22	1.548	260.183	1.02
5.	Duo Sol	180-205	—	1.045	2.217	418.3	3.1
6.	Propane	−42	−190	0.51	2.509 (liquid) 1.966 (vapour) 1.422	422.483	—
7.	Nitrobenzene	211	5.4	1.207		296.575	0.1 at 20°C
8.	Aniline	184.4	−6.2	1.027		433.777	

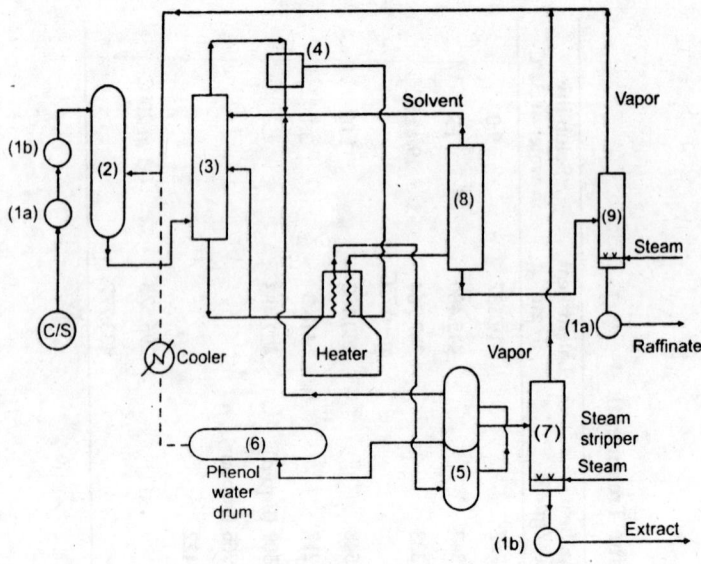

Figure 4.13a   Phenol Extraction.

lower temperatures matches with chlorex and at high temperature it is near to furfural. Hence the dual nature of the solvent at all temperatures is quite convincing.

Charge stock heated to 80-100°C by exchanging heat with raffinate and extract in exchangers (1a and 1b) is first allowed to absorb the phenol vapors in absorber (2); these vapors are delivered from the strippers. The stock then goes to extraction unit (3). This consists of a counter current extractor (packed bed or sieve plate column, or centrifugal extractors) where a continuous down flow of solvent takes place from top of the tower. Feed being lighter is admitted at the bottom of the tower. Extract and raffinate are separated as distinct layers Raffinate from the top is led into a separator (4) to settle solvent drops, if any. The collected solvent is mingled with the solvent stream that goes into extractor. Bottom product is extract, and always a part of this is circulated back to extractor as a reflux and the rest is processed for recovering solvent.

Raffinate phase contains less solvent; hence it is easily stripped off in fractionator (8) under vacuum. The stripped raffinate is again blown with steam to separate the remaining solvent from oil phase (9). If amount of solvent is small then it may be directly steam stripped.

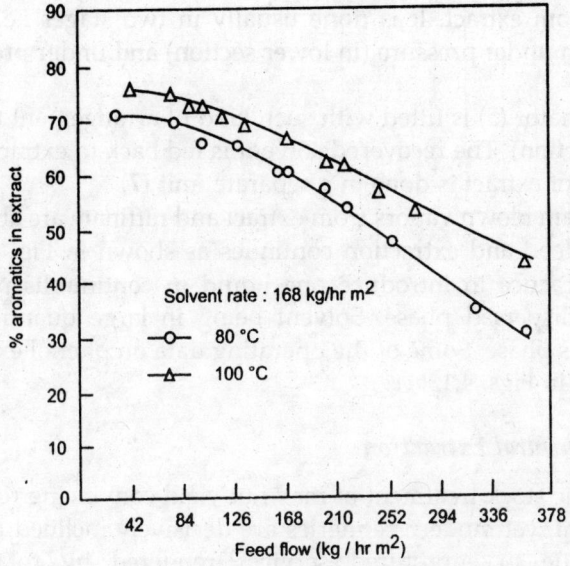

Figure 4.13b Effect of Feed Rate and Temp.
Operating Data on Phenol Extraction.

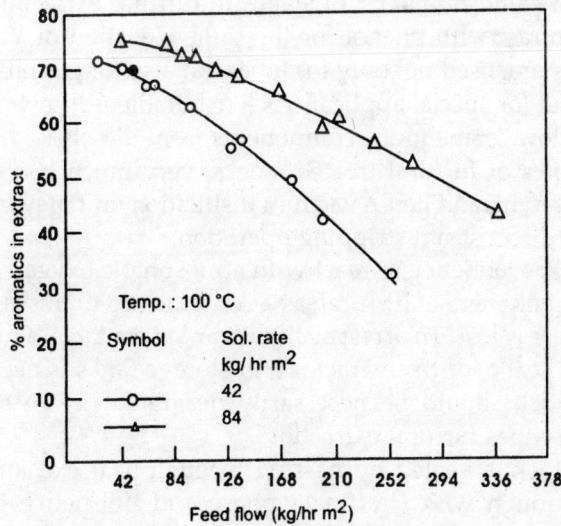

Figure 4.13c Operating Data on Phenol Extraction
(Effect of Feed and Solvent Rate).

On the contrary, difficulty may be experienced while separating solvent from extract. It is done usually in two stages i.e. carry the distillation under pressure (in lower section) and under pressure and vacuum.

Fractionator (5) is fitted with such kind of arrangement to vacuum (upper section). The recovered solvent is fed back to extractor. Steam stripping of extract is done in a separate unit (7).

The steam blown vapors from extract and raffinate are absorbed by the fresh feed and extraction continues as shown in Fig. 4.13a. It is general practice to introduce one liquid in continuous phase, and other in dispersed phase. Solvent being in large quantities forms continuous phase. Some of the operating data on phenol extraction is presented in Figs. 4.13b, c.

4.4.3.3 Furfural Extraction

Naphthenic stock treatment of may not yield convincing results with this solvent, yet modern refineries are decisively inclined to use this solvent due to enraptured services rendered by rotating disc contactor. In India, Haldia, Cochin, Madras refineries are using this process and planned refinery (Mathura) is also to follow the predecessors. A temperature range of 75 to 125°C under a gradient of 20 to 30°C makes the operation a success. Demands of furfural are slightly more, and number[30] of stages in furfural extraction are also more in contrast with phenol for the same elevation of VI. Furfural extraction is practised not only for lubes but also for gas oils, catalytic cycle oils and for special applications like butadiene[31c] extraction and removal of low cetane index components from diesels.

In raffinates of furfural treated stocks, very much less quality of solvent is present, and hence vacuum distillation unit may be skipped and opt for direct steam stripping operation.

Furfural has tendency to react with atmospheric oxygen gradually resulting in polymers of furfural; as a consequence to this change, the solvent power is lost. To arrest such unwanted contacts, a nitrogen or an inert gas seal over the extractor and storage tanks is necessary, as also feed stock should be necessarily deaerated before extraction. Fig. 4.14 describes furfural extraction.

Charge stock is heated by passing through heat exchangers HE_1 and HE_2 through which raffinate phase and finished extract pass before reaching storage tanks. This may raise the temperature of feed to 70 to 80°C. Feed then enters the extractor RDC (1) at two or three points and solvent is admitted into the upper portion. Raffinate phase

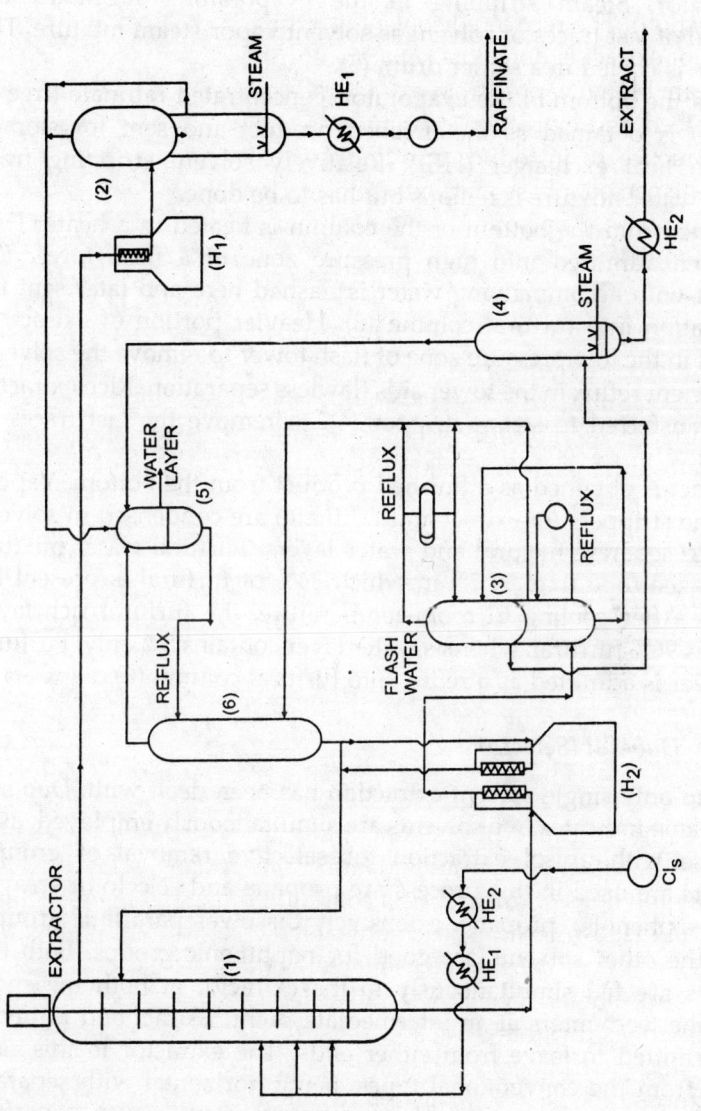

Figure 4.14 Furfural Extraction.

being lighter leaves from the top of the contactor. It is heated in a heater (H_1) and flashed in evaporator (2) working at a low pressure.

This ensures the removal of solvent contaminated with water vapor if any, the heavy stream goes into the bottom portion of the evaporator. Steam stripping in the evaporator[32] facilitates the removal of last traces of solvent as solvent vapor-steam mixture. The mixture is cooled in a settler drum (5).

From the bottom of the evaporator, concentrated raffinate (free of solvent) is obtained as the finished product and sent for storage through heat exchanger (HE_1). Relatively solvent stripping from extract water mixture is tedious but has to be done.

Extract from the bottom of the column is heated in a heater (H_2) and then admitted into high pressure zone of a flash tower (3). Solvent with accompanying water is flashed here and later sent for rectification into furfural column (6). Heavier portion of extract, is flashed in the low pressure zone of flash tower to remove the solvent. An efficient reflux in the tower aids flawless separation. Rich extract is now transferred to steam stripper (4), to remove the last traces of solvent.

Extract is obtained as a finished product from the bottom. Vapors from the strippers (of extract and raffinate) are condensed in solvent drum to separate furfural and water layers. Furfural water mixture forms azeotrope (b.p. 97.7°) in which 34% of furfural is present by weight. After cooling to room temperature, the furfural rich layer contains 96% furfural, whereas water layer contains 8% only. Furfural rich layer is admitted as a reflux into furfural column (6).

4.4.3.4 Duo-sol (Selecto)

Hitherto only single solvent extraction has been dealt with. Duo-sol, as its name indicates two solvents are simultaneously employed, as is the case with unisol extraction for selective removal of groups. Solvents are used in this process are propane and selecto or cresylic acids (+ phenols), propane, extensively dissolves paraffinic groups, while the other solvents are good for naphthenic groups. Both the solvents are fed simultaneously to the extractor at both the ends, while the feed enters at an intermediate point. Extract and raffinate are permitted to leave from either ends. The extractor in this case differs from the conventional types, being horizontal with separate mixing and settling zones. These different zones cause a perfect counter current extraction. In one end, the outgoing cresylic phase is stripped off paraffinic material by incoming propane, and exactly the opposite thing happens at the other END.

Selecto (mixture of phenol 35% and cresol 65%) requirements are 200% to 400% with equal amount of propane, though naphthenic oils require more propane and paraffinic oils require more cresols.

Figure 4.15 illustrates the operation. Extractor (1) is charged with phenols (cresols) at one end, while liquid propane is fed at the other end. Propane being lighter, leaves from the top, while heavy naphthenic layer leaves from the bottom. Propane from naphthenic layer is first freed by evaporating in an evaporator (2). Propane from this evaporator, joins the propane vapors of raffinate evaporator (5) to begin the cycle.

Naphthenic layer is heated and sent to selecto evaporator (3), where steam stripped vapors of selecto from extract steam stripper (4) circulates as a reflux. From the bottom of the steam stripper, extract layer finally leaves to storage. Selecto vapors from selecto evaporators [both raffinate (6) and extract (3)] mix and enter the distillation unit (8), where selecto is obtained. From the steam stripper (7) finished raffinate is obtained. It is seen from this operation that both the phases have to be systematically separated from the solvents. Propane is separated from the phases by evaporation at high pressure while selecto is recovered at ordinary pressures. In general practice raffinate layer requires 3 stages, while extract phase requires 5 stages during contact. Main advantage of this process is, its ability for removal of asphalt. The operation is supposed to be in the regions of above melting point of phenol and below vaporisation temperature of propane at that pressure.

Results of extraction work carried out by Martyn Noel are presented[41] in Table 4.8.

TABLE 4.8   Duo-sol Extraction — Effects of Paraffinic and Naphthenic Stocks.

	Stock 1 (naphthenic)	Stock 1 (paraffinic)
API gravity	15.2	26.2
Viscosity SSU (99°C)	76.0	47.4
VGC	0.876	0.839
Raffinate yield %	44	73
Dewaxed raffinate API	26.5	29.9
Dewaxed raffinate viscosity (SSU at 99°C)	143	48.0
Carbon residue	0.4	0.002
Dewaxed raffinate VI.	95	95

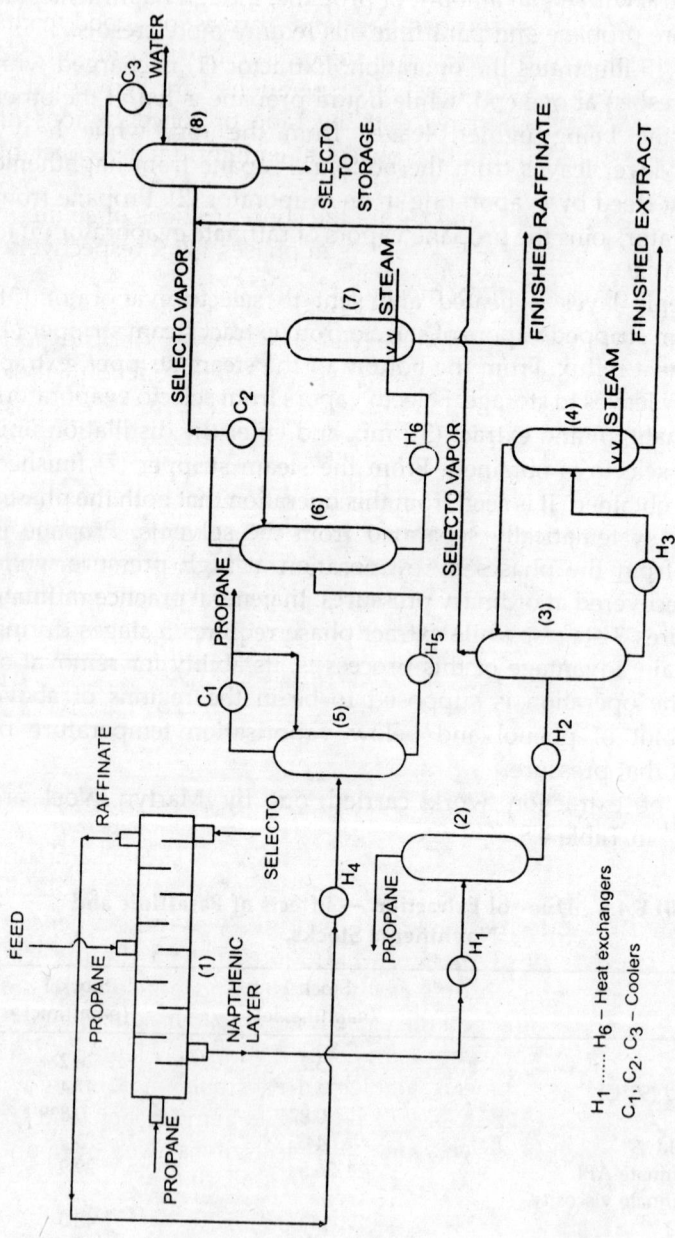

Figure 4.15   Duosol Extraction Process.

$H_1 \ldots \ldots H_6$ – Heat exchangers

C_1, C_2, C_3 – Coolers

4.4.4 About Extraction

Extraction is a physical operation and thrives on good contact between feed and solvent; and separation of the solvent from resulting phases. All extraction operations are governed by distribution law which states that the ratio of concentrations of a compound distributed into two phases at equilibrium is a constant.

i.e. $\dfrac{C_1}{C_2} = K$; where C_1 and C_2 are the concentrations of solute in phases 1 & 2 respectively

K = distribution coefficient.

Selectivity term is more often used in extraction work, it is expected that selectivity away from unity will lead to better extraction with less difficulty. Solvent power is the capacity of a solvent to dissolve a substance at a given temperature. A good solvent should possess high solvent power and good selectivity.

Conventional procedure of representing ternary system is done in triangular plots. Equilateral triangles serve the purpose most, as the apex to opposite side perpendicular distance remains same. Each apex is assigned with 100% purity of a component[33]. The concentration at any point inside the triangle can be found by drawing perpendiculars to opposite sides from that point. Percentage concentration can be found by measuring the height. Binodal curve demarks the miscible region with immiscible; thus acts as a boundary of the heterogenous area and bears the tie line points too.

With higher heterogeneous region better extraction is possible, due to the flexibility in handling feeds of quite variable concentrations. As the temperature of extraction increases, the area shrinks impairing extraction very much. Though higher temperature increases solubility and selectivity the operational region shrinks very much, hence the flexibility choosing very high concentrations of feed diminishes. A method to increase such area is by adding and antisolvent (some times water) which can open the closed hionodal curve to give liberal access to extraction. For example in DMSO, Benzene, Heptane system (Fig. 4.11a and b, heptane is in fact a better solvent for removing benzene from DMSO, but DMSO is used to remove benzene from heptane-benzene mixtures, because the selectivity is away from unity.

4.4.4.1 Tie Lines

The equilibrium compositions of extract and raffinate lie on binodal curve; line that joins these two points is known as tie line. The slopes of the tie lines vary even for the same system and distinction disappears between extract and raffinate at one point, which is known as plait point.

Slope of a tie line furnishes valuable information regarding the capacity of solvent; if the slope of the tie line is away from the solvent apex, then, the solvent is considered to be very good, otherwise not. In most of the cases the solvent may not be satisfying the needs theoretically.

Tie line concentration can be plotted on x-y diagram, as in the case of McCabe Thiele equilibrium diagrams which simplifies the evaluation of number of stages involved in operations[34].

Minimum solvent required in the operation can be obtained by extending the tie lines, the larger the distance between meeting point of highest tie line and lowest tie line from solvent apex, the higher the solvent required.

When extraction is conducted in conventional packed bed extractors the column diameter and packing size are important. Packing size to column size should be as large as possible[35] (Minimum 8) to avoid channelling. The minimum size of packing required for a system is given by critical packing size[36] (d_{FC}) defined as

Critical packing size $(d_{FC}) = 2.42 \, (\gamma/\Delta\rho)^{0.5}$

where γ :   Surface tension of the system at boundary

$\Delta\rho$:   difference in densities of phases.

Problem 4.1

In a phenol extraction operation conducted over a packed bed at 80°C the following data is obtained. Judge the necessary size of packing and column diameter.

Data : Surface tension of

Phenol at 80°C	= 37.3 dynes/cm
Lube oil at 80°C	= 26.8 dynes/cm
Extract	= 35.88 dynes/cm
Raffinate	= 29.14 dynes/cm
Density of Phenol at 80°C	= 0 .9692 gms/cm
Lube oil at 80°C	= 0 .8161 gms/cm
Extract	= 0 .8873 gms/cm
Raffinate	= 0 .8043 gms/cm

d_{FC} determined on the basis of properties of solvent and Lube oil.

$$d_{FC} = 2.42 \left(\frac{37.3 - 26.8}{(0.9692 - 0.8161)\,980} \right)^{0.5} = 0.6402$$

On the basis of Extract and raffinate

$$= 2.42 \left(\frac{35.88 - 29.14}{0.8873 - 0.8043 \times 980} \right)^{0.5} = 0.6966$$

The higher size of packing is chosen. To avoid channelling a minimum diameter of column required is eight times critical packing size.

Problem 4.2

Phenol extraction operation conducted at 100°C to extract aromatics from pressable waxy distillate; given the following data, find out the number of stages used in extractor:

Feed gms/min	160
Solvent gms/min	100
Extract gms/min	110
Solvent free extract	— 40.28 gms/min
Aromatics (As) in feed	— 21%
Aromatics (As) in extract	— 21.85%

Equilibrium data:

As in extract Solvent free extract	As in raffinate Solvent free raffinate
40.00	12.14
63.69	27.08
50.90	14.79
44.12	14.08
10.05	3.8

Solution : Total aromatics in feed : $\dfrac{160 \times 21}{100} = 33.6$

Aromatics in raffinate $= 33.6 - 21.85 = 11.75$

$$(E_1) = \frac{AS}{SFE} = \frac{21.85}{40.28} = 0.5426 \times 10^2$$

(% As in solvent free extract)

$$(R_1) = \frac{AS}{SFR} = \frac{11.75}{120} = 0.098 \times 10^2$$

(% As in solvent free raffinate)

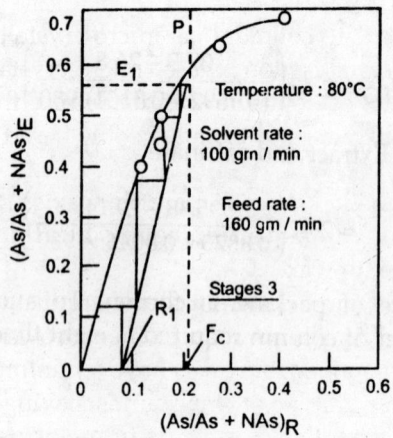

Figure 4.16   Determination of Number of Stages in Extraction.

Solvent free raffinate (SFR) = Feed ϕ - solvent free extract.

Plot the equilibrium $(x - y)$ diagram as shown in Fig. 4.16. In phenol extraction the feed is introduced at the bottom of the column, so the extract and feed are in contact (equilibrium) with each other i.e. the feed concentration (F_c = 0.21) and extract concentration are in equilibrium. So F_c and E_1 shall be fixing the point P. R_1 lies on x axis, as initial concentration of aromatics in solvent is zero. These two points are joined by a straight line, which forms the operating line. Steps are drawn from top to bottom, joining the operating line and equilibrium line.

Another method is by fixing raffinate composition and drawing a straight line with slope equal to feed to solvent ratio until it touches the concentration of extract.

Number of stages are found to be three.

4.5   WAX AND PURIFICATION

All heavy fractions of crude oils contain at least some amount of waxy material; this has been referred as paraffin wax although the composition does not reveal paraffins alone. True to a great extent, *n*-paraffins are present, however naphthenes do extend their presence. All waxy materials contain carbon atoms twenty to thirty or even more in exceptional cases. High molecular weight paraffins are prone to exist in more branched forms; this helps in contributing different structural appearances ranging from crystalline to colloidal form to wax. Severe branching fosters unsaturation, which in return stretches the crystalline state to colloidal i.e. micro-crystalline. Noted for their resistance to crystallisation these micro-crystallines invade oils. Ordinary methods of dewaxing cannot reduce the content of micro-crystalline waxes to desired extent. Zeolite based catalyst can reduce wax in the products.

Wax, commercially known as paraffin wax, is not as pure as is sold in the market; an elaborate stripping and bleaching operation gives a shiney white look to wax.

In paraffin wax three different forms are distinctly visible namely microcrystalline in one form and other two are plate and needle forms. All these commercial waxes have a minimum melting point of 50°C. These semisolid type of compounds, having a melting point less than 50°C are differentiated as cerols from waxes, although chemical constitution does not differ much. Crystalline forms of wax fall into the category of plate type. These are constituted of straight chain compounds only.

Important properties of wax

  (a) Specific gravity
  (b) Melting point
  (c) Viscosity
  (d) Colour
  (e) Penetration index

Microcrystalline waxes are produced from tank bottoms or from residuums. Needle crystals are pure form of wax and can be obtained by urea adduction. It is well established that only n-paraffins are adductable. Branched[37, 42] chains and certain ring compounds are also reactive with thio-urea. The demarcation between mal (or plate) and needle crystals is not sound although it is great with microcrystalline wax.

The specific gravity of waxes is shown to have an increasing tendency with the melting point. Thus it is a good property to characterise wax. Refractive index and melting point of wax are true characteristic properties that can be distinguished between microcrystalline and other forms. When melting point and refractive index of waxes are plotted, the congregation of points results into a straight line, and nearer to that straight line lie the paraffinic waxes[38] and away are microcrystallines. Viscosity of microcrystallines is high owing to the high melting point. Microcrystallines have higher molecular weights of the order 580-700; while mal or needle crystals have molecular weights of 325-500. Another interesting property of microcrystallines is the profound affinity for oils, often an oil content of 40 to 70% is present in these waxes also known as petrolatums. Only a highly polar solvent is adeptive for such kind of waxes to remove from oils. All other types of waxes, can be dissolved in hydrocarbons very easily. X-ray studies have indicated paraffinic wax as large well formed crystals with brittleness.

Commercial waxes are classified on the basis of penetration index and melting point. Higher the penetration index higher the melting point is and *vice versa*.

4.5.1 Dewaxing

Dewaxing operation is broadly classified into types, one with the use of solvents and other without solvents. Wax removal, based on the natural principle of decreasing solubility of wax in oil at lower temperatures, is commercially known as chilling and pressing.

4.5.1.1 *Chilling and Pressing*

This operation is simple although time consuming; starts with initial warming of oil a few degrees above its pour point to keep the whole mass of wax in a homogeneous phase. This mixture is now cooled in agitators fitted with coils for circulating cold water or refrigerant as the case may be. Cooling must be uniform and should not be more than 3°C per hour. Under such conditions the growth of crystals is significantly accelerated. When heavy oils or residuums are chilled, the wax in it does not crystallise easily due to its colloidal nature, and in such situations addition of diluent is necessary. Crystallisation of wax is a slow process, and hence it should be carried out for a long time till the final temperature of the mixture falls 10 to 15°C below the required pour point of dewaxed oil.

Separation of Wax

Any convenient filter or a centrifuge may be sufficient to separate wax. Usual practice is to detail a filter press which can operate upto 100 atmospheres pressure. Filtering pressure of this order is inevitable to force the oil out from the cake; surprisingly even with these conditions the cake contains about[39] 50% oil. This necessitates a centrifuge to decrease the oil content to roughly 20%. Sharpel centrifuges are in good demand for dewaxing operations, although they are best fit for oils of end point 400°C. High boiling stocks are predominantly rich in microcrystallines are these are not efficiently removed in this operation also. Despite these efforts the cake still contains some oil. This contamination makes wax impure and hence it is given the name crude wax or slack wax. Final removal of oil from slack wax is accomplished in sweating operations.

4.5.1.2 Sweating

Sweating consists of gradual rise and maintenance of temperature, in large shallow sweat pans. This facilitates in oozing oil from cake in the form of drips. The sweat pans are shallow and have perforated plates at the bottom and sometimes these plates are covered with convenient size of screens. Diameter of sweat pans goes up to 3 meters and the depth to half a meter. These pans are equipped with heating coils through which tepid water flows. Oil droplets released from the voids and interstices of wax collect in drips into the trays kept beneath the pans. After sweating is over in the top pan for a sufficient time (when no more oil drips are noticed) at that

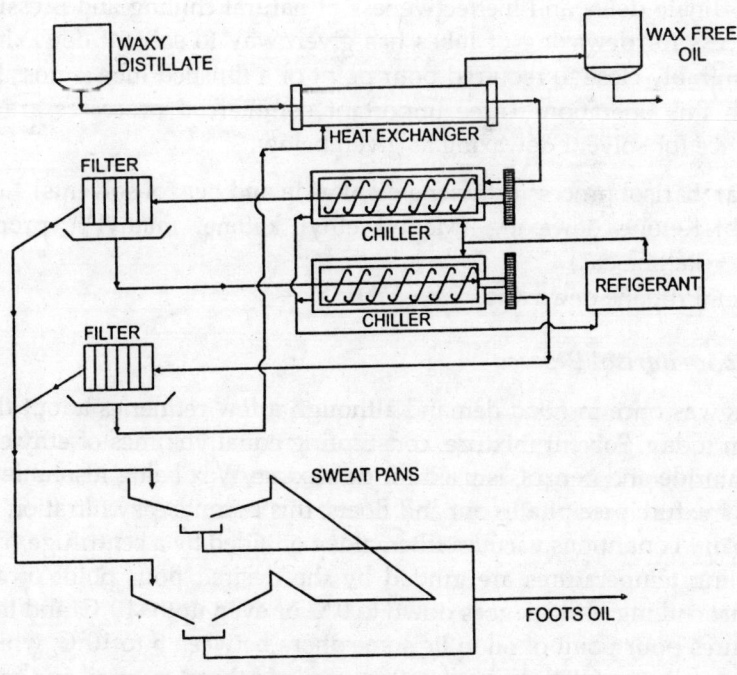

Figure 4.17 Dewaxing and Sweating Operations.

temperature, the cake is now transferred to another pan that is kept underneath the first one and maintained at slightly warmer condition. Because of slightly higher temperature more oil is sweated out from cake and process is continued in a number of sweat pans. There are eight or nine such pans imbricated by the time wax reaches the last pan when the deoiling of cake is completed. This wax is a better form of wax with well defined melting point range to suit the commercial requirements.

Figure 4.17 shows the pressing and sweating operations. Waxy distillates or lubes with high wax content cannot flow easily when chilling takes place; special type heat exchangers are required. In this heat exchanger, a provision for scraper blades rotated by external power system is created; this helps not only in scraping the wax that condensed along the wall of exchanger but also in forcing out the chilled massed. The oil collected in sweat pans is called foots oil and is fed back to act as diluent for waxy stock.

4.5.2 Dewaxing with Solvents

Inordinate delay and ineffectiveness of natural chilling and pressing process for dewaxing of lubes has given way to solvent dewaxing. Admirably close to required pour point of a finished lube is possible with this operation. Three important commercial processes are in service for solvent dewaxing as given below:

(a) Barisol process (Ethylene dichloride and benzol solvents)
(b) Ketone dewaxing (Methyl ethyl ketone, methyl[31b] propyl ketone etc.)
(c) Propane dewaxing.

4.5.2.1 Barisol Process

This was once in good demand although a few refineries adopt this even today. Solvent mixture, constituting equal volumes of ethylene dichloride and benzol, is used for dewaxing. Wax being insoluble in this mixture precipitates out and floats; this permits easy filtration by a rotary continuous vacuum filter, alone or aided by a centrifuge. The chilling temperatures are guided by the desired pour point of oil. Usual chilling of stock goes down to 0°C or even upto -10°C, and that ensures pour point of oil to lie somewhere between 5 to 10°C, which satisfies most of the consumers. Recovery of solvent from oil and cake is done by fractionation of these phases separately. Barisol is flexible in treating different feed stocks ranging from long residuums to waxy distillates.

4.5.2.2 *Ketone Dewaxing*

It is credited to be a modern process and employs methyl ethyl ketone as a solvent, hence it is known as MEK dewaxing. The development of solvent dewaxing started with commercialisation of acetone and benzol mixtures in 1927. These mixtures were gradually replaced by MEK-aromatic mixtures and which have creditably retained their position till today. Lower molecular weight ketones with low viscosity are preferred, obviously for their noted contributions towards high filter rates and effective dewaxing.

Ketones enhance the crystalisability of wax without dissolving oil. So the lyophobic tendency of ketone has to be counteracted with suitable hydrocarbons to keep the oil phase in solution. This directs the precipitation of wax at lower temperatures. Benzene and toluene mixtures are good in dissolving oil, so a suitable mixture may be made of by mixing 60-65 parts of ketone with 30 parts of benzene and 10 parts of toluene. Highly naphthenic stocks are to be encountered with rich mixtures of aromatics. MEK dewaxing enjoys advantages like complete recoverability of solvent and achievement of low pour point over acetone mixtures.

Description of the process (Figure 4.18)

Wax bearing oil and solvent (1 : 1 to 1 : 4 for paraffinic to naphthenic stocks) are proportioned and slightly warmed by out going steams (3 & 4). This mixture is now cooled by filtrate to moderate temperatures slowly, and finally chilled by refrigerant to the desired pour point. The chilled mixture then goes to an intermediate accumulator (5) and finally to continuous rotary filter (6) which operates under vacuum. The scraped cake is transferred to a storage tank (7), from where it goes to evaporator (8). Evaporation of solvent from wax cake is done in this. The filtrate from filter passes through heat exchangers and enters a flash tower (12). This contains two different sections; bottom one is low pressure flash tower and upper portion is high pressure tower. The filtrate is submitted to high pressure flashing first, later it is flashed in low pressure section of the tower. Most of the solvent is expelled in this process. After solvent evaporation is over, the remaining oil is steam stripped in a stripper (11) to free the oil from contaminants or residues of solvent. Cake is steam stripped

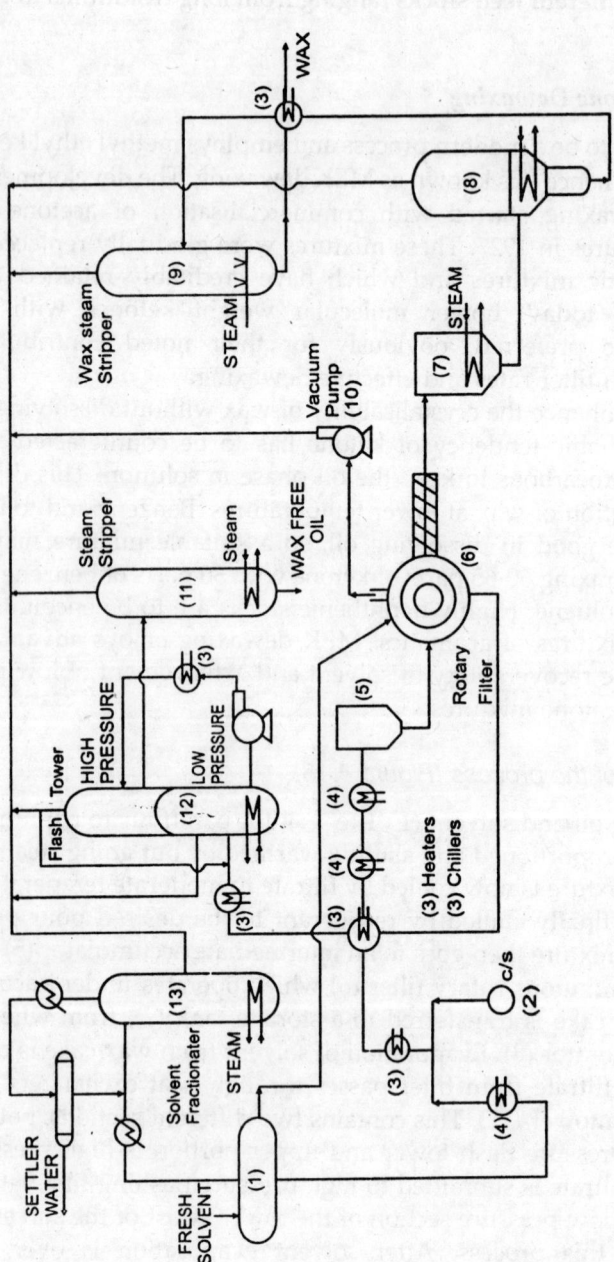

Figure 4.18 MEK Dewaxing.

separately in another stipper (9). Solvent vapours from all the strippers are combined and sent into a distillation column (13); where solvent is regenerated. It is likely that the resultant dewaxed oil may have a pour point less than the chilling temperature. However chilling is continued to reach 4 to 5°C below the desired pour point of oil. Cake from this operation does not carry oil more than 6%, this is a commendable achievement over chilling and pressing operations.

4.5.2.3 Propane Self Refrigeration Process

Propane is available in abundance in all refineries and its application for such extraction operations has aroused the interest of refiners. Its dual function in precipitation of wax and rejection of asphalt is considered to be invaluable. It has become invincible in this context as it could overcome chilling difficulties offered by feed stocks (because the stocks are viscous at that low temperature, rendering ineffective to heat transfer) by self evaporation; resulting in rapid rates of chilling, Recovery of solvent with ease and high purity cannot be substituted. For the same extent of dewaxing, it requires only half the time than that required by an effective operation like MEK. Propane is prized to handle all stocks with almost the same efficiency.

Description of Operation (Figure 4.19)

Initially oil and propane mixture is warmed to 25°C by charging through a heater and stored in a tank (1). The charge solution is now chilled in series of chillers (2) by evaporation of propane at a controlled rate till the filtering temperature is achieved. Chilling aids in crystallisation of wax, and these crystals remain suspended in propane-oil solution.

The solution is then filtered to remove wax. Filtration is conducted in a rotary vacuum filter (3), and simultaneous washing of cake with fresh solvent is also done. Wax free solution i.e. filtrate is led through process heat-exchangers to evaporators (4 & 5). In the beginning evaporation is done at low pressure and later at high pressure. Propane evaporated from these evaporators is gathered and sent back to compression and liquefaction system. The remaining oil is finally steam stripped (6). In the same way, recovery of solvent from wax is done. In strippers only traces of propane are obtained, hence by condensation water separates out (7) leaving propane as gas, which is sent to liquefaction section. Dewaxing is conducted with oil and propane ratio fixed at 1 : 3. High pressure is needed to keep propane

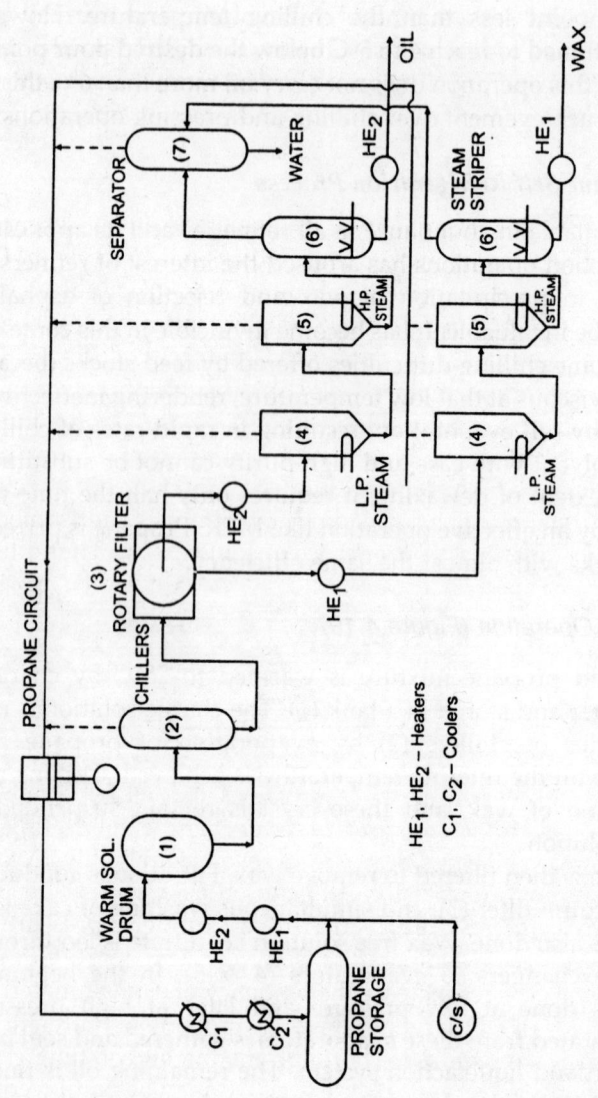

Figure 4.19 Propane Dewaxing.

HE_1, HE_2 – Heaters

C_1, C_2 – Coolers

in liquid state. Warming of solution upto 25°C is necessary before chilling, to keep all wax in solution.

If deresining is attempted, the dewaxed solution is again mixed with five to ten times propane of its own volume and warmed to a temperature of 50°C. When this solution is chilled, slowly resin settles. The sequence of filtration and other operations is same as the propane dewaxing.

4.5.2.4 *Dewaxing by Catalysts*

These are solid catalysts of zeolite type. In fact it is a chemical process whereby wax is selectively cracked and hydrogenated to yield HVT oils... by addition of noble metals simultaneous structural change (isomerisation) can be achieved, Mobil's work in this respect is highly commendable. ZSM - 5 is the proprietary catalyst of Mobil. The structure of the zeolite is more porous and can selectively handle paraffins but not ring structures. Hence decreasing the pour point i.e. like visbreaking at the same time giving high viscosity index and saturation to structure are more appreciated in this way of cracking. Further catalyst formulations by loading noble metals can isomerise the structure, thus altering the properties to some extent. Thus a feed stock having a pour point of 40°C can be converted into a free flowing liquid of 5°C pour point. Because of the presence of hydrogen in the reactor, the pressure is usually maintained around 40 to 50 atmospheres and the reactor is kept at 400-450°C temperature.

4.5.3 Propane Deasphalting

Deasphalting is another important operation and all lube oils of naphthenic base invariably require it. As propane exhibits unique capacity in separation of asphalts and resins even from highly viscous stocks like residuums industrially, it is chosen as a solvent. In this operation contacting of residuums and propane under a pressure of 15 atmospheres and at a temperature between 30°C to 70°C takes place. Higher temperatures increase the tendency of asphalt materials to separate out. Relatively large volumes of propane about eight to ten times the stock are required. Thorough mixing and settling shall separate the asphalts. RDC and other extractors may be employed.

Propane rich phase contains all the oil, while asphalts are rejected by propane and they collect as a heavy phase. The two phases are separately processed for solvent recovery as described in dewaxing operation. Virgin oils after treatment naturally give rise to light neutral or heavy neutral oils, while short residuums produce bright

TABLE 4.9 Comparison of Vacuum Distillation and Propane[31b] Deashalting.

Vacuum distillation	Medium	Cylinder stock	Residue
Viscosity at 100°C (SSU)	90.7	—	6840
Gravity °API	19.0	—	7.3
Carbon residue	2.1	—	21.4
VI. (of dewaxed oil)	90.2		
Yield % on crude base	13.0	—	14.3
Propane (deasphalting)			
Viscosity (SSU) at 100°C	94.0	239	—
Gravity °API	24.3	18.0	1.6
Carbon residue	0.66	2.6	—
VI. (dewaxed oil)	80.6	—	—
Softening point °C	—	—	55.6

TABLE 4.10 Comparison of Slack Wax Properties Obtained from Light Viscosity Index and High Viscosity Index of Aghajari Crudes; by using M-i Butyl Ketone.

	Bright[43] stock (LVI)		Slack wax Bright stock (HVI)	
Ratio of solvent to feed (vol.)	10	7.5	3	10
Deoiling temp. °C	15	15	0	15
Yield %	47.3	50.8	90.3	48.4
M. P. of wax °C	81.4	81.3	—	82.1
Penetration index at 25°C	21	25	—	23
Viscosity CS (98.9°C)	18.1	18.6	20.5	16.8

stock. The quality of asphalt obtained is very good with high softening point.

It may be true that all lighter hydrocarbons ranging from ethane to butane are as commendable as propane in operation. However, the rejection of asphalts by propane to desired extent, has in fact made its involvement indispensable.[40]

Propane deasphalting is quite comparable with vacuum distillation in production of medium to cylinder stocks as shown in Table 4.9.

Above data in Table 4.10 shows the dewaxing performance of methyl i-butyl ketone on Aghajari, bright stock of low viscosity and high viscosity index fractions. Table 4.11 shows MEK dewaxing operations on crude[44] bottoms (Ankaleswar crude). The solvent consists of a mixture of MEK, paraffin (n-Hexane) and toluene. It can be seen that lower temperatures of deoiling gives better wax yields

TABLE 4.11 MEK Dewaxing — at Different Temperatures.

Solvent composition	Wax yield % Temp. of operation			Penetration Index		
	10°C	0°C	-10°C	10°C	0°C	-10°C
40% MEK & Hexane	18.4	39.6	48.0	4	10	33
40% MEK & Toluene	19.4	33.7	47.1	4	9	—
50% MEK & Hexane	21.8	—	51.5	5	22	48
50% MEK & Toluene	24.0	36.6	48.8	4	10	40

with high penetration quality. But the lowest temperature of operation, contemplated depends upon the phase separation temperature of the system and is clearly linked to the chosen ratio of toluene/hexane to MEK with high ratio of these hydrocarbons in solvent mixtures, lower separation temperatures are attainable.

Production of paraffin wax in India was first started by Assam Oil Company. The quality of wax produced from eastern crudes is excellent and ranks high among the waxes from other sources.

Haldia Refinery has already made good strides in technical know-how for production of micro crystalline wax and is likely to go on commercial stream within one year (1983).

First unit of 50,000 tons/annum wax deoiling plant has been constructed at Barauni Refinery during 7th Plan period.

MRL uses MEK process for the production of 25,000 tons of Paraffin Wax.

4..6.1.1 Storage and Transportation

LNG (Liquefied Natural Gas) has been a viable form of energy and a raw material for petrochemicals and easily handled as in liquid form it reduces to a volume of 600 times, this is economical to store and transport. There were 15 LNG exporting countries, Qatar being highest 28 Million Tons, and 17 LNG importing countries, Japan being highest 67 Million tons(2007), followed by S. Korea 37. LNG trade volumes increased from 140 MT in 2005 to 172 Million Tons in 2008 and it is forecasted to be increased to about 300 MT in 2012. Though the industry is not without incidents and accidents, but it maintains a good safety record. LNG Carrier ship fleet in the last two decades over has delivered more than 35,000 shiploads of LNG, without any fatality LNG is mainly transported as gas/liquid through pipelines. Liquid transportation is done by bulk carriers, road and rail tankers in small quantities. The process of natural gas liquefaction, storage and vaporization is not a new technology. There are now over 120 peak-saving and LNG storage facilities worldwide, some operating since the mid-1960s. In addition, there are 58 import (regasification) terminals worldwide

4.6.1.2 StorageTanks

The LNG storage tanks at the modern facilities are constructed of a cryogenic container operating at 160°C consisting of tank the interior usually made of 9% nickel steel, stainless steel, aluminum or other cryogenic alloy. The outside wall is made of carbon steel or reinforced concrete. A thick layer of an insulating material is also used between the container and outer containment. [walls] this is known as a double containment. This is to protect the surroundings from disasters out of explosion . For land-based facilities, single containment tank designs have a secondary earthen or concrete containment surrounding the LNG tank(s). For tall tanks, concrete wall acts as a casing for the LNG tank,. When the outer tank wall and roof are reinforced concrete then the tank is considered to be a full containment type. In smaller storage volumes under ground tanks are used. In all cases, the objective is to minimize the risks and exposure of the public in case failure of the LNG primary containment based on a tank failure. Many tanks are equipped with top tank penetration only.

4.6.1.3.1 LNG-CarriersDescription

Most LNG tank ships have two hulls, so that, if a collision or due to any reason damage to the outer hull takes place, even then ,the ship will still float without spilling LNG . LNG tanks are mostly spherical although box-shaped ones are also available. A typical LNG ship is 950 feet long or more and 150 feet wide,. Today the majority of the new ships under construction can handle 120,000 m^3 to 140,000 m^3. or even more upto 260,000 m^3., There are more than 350 [2012] LNG ships engaged over deep seas. LNG ships are designed for safe operation in accordance with international marine classification standards, (U.S. Coast Guard and international regulations) The cargo tanks are located away from the ship's hull to avoid cargo tank damage. Gas detectors and safety alarms are placed between the steel hulls to continuously monitor gas leaks. In the past 40 years there have been voyages without a significant accident. Gas carrier tanks, according to International Maritime Organization (IMO) rules, must be one of three types. Tank design (Type A) and others are of pressure vessel design (Type C), and other , tanks that are neither of the two types (Type B). Again there are three general Type B tank designs for LNG. The first type of design is known as the membrane tank; is supported by the hold it occupies. The other two designs, spherical and prismatic, are self-supporting. [Gaz Transport & Technigaz (GTT)]. Membrane tanks are composed of a layer of metal (First barrier), a layer of insulation, another liquid-proof layer, over this a layer of insulation. The primary barrier, made of corrugated stainless steel of about 1.2 mm thickness is the one in direct contact with the cargo liquid. The membrane consists of stainless steel with 'waffles' to absorb the thermal contraction when the tank is cooled down. This is followed by a primary insulation which in turn is covered by a secondary barrier made of a material called "triplex" which is basically a metal foil sandwiched between glasswool sheets and compressed together. This is again covered by a secondary insulation which in turn is supported by the ship's hull structure from the outside.Those several layers are then attached to the walls of the externally framed hold. The primary and secondary barriers are made of Invar [less contraction and suitable for cryogenics] sheets. The insulation is usual wood material. . All membranes are built up from the surface of a hold using discrete units of insulation. The alternative to a membrane tank is a self-supporting tank The early sphere designs are shells of aluminum . The sphere is installed in its own hold of a double-hulled ship, so that it is supported around its equator by a steel cylinder or skirt The second type of self-supporting tank is the Self-supporting, Prismatic, Type B (SPB) tanks by Ishikawajima Heavy Industries (IHI).These tanks are reminiscent of the tanks on old single-skin oil tankers; the framing is internal to the tank. The material for tank construction can be aluminum, 9-percent nickel steel, or 304 stainless steel, The tanks are installed in the hold of a double hull ship and are insulated with covered polyurethane foam; also is able to serve as channeling for any possible tank leakage to drip trays. There is a trend towards the use of the two different membrane types instead of the self supporting storage systems. This is most likely because prismatic membrane tanks utilize the hull shape more efficiently and thus have less void space between the cargo-tanks and ballast tanks However, self-supporting tanks are more robust and have greater resistance to sloshing forces, and will possibly be considered in the future for offshore storage where bad weather will be a significant factor. While going from the inside of the tank outwards, it can be seen serially LNG | Primary barrier of 1.2 mm thick corrugated/waffled Stainless Steel | Primary Insulation (also called the interbarrier space) | Secondary barrier of triplex membrane | Secondary Insulation (TGZ Mark III)

4.6.1.4 *Safety Management*

The vessel and cargo tank arrangements, the types and thickness of the hull and cargo system materials, and the types of monitoring, safety and alarm systems, including requirements for back-up systems, Industry standards, codes, training, inspections and operating procedures as well as government regulations are in place for the safe design, construction and operation of ships. The ship's crew must be certified for LNG single significant accident or safety problem , neither in port nor at sea. LNG Tankers are equipped with a full range of safety equipment such as radar and positioning systems that continuously keep the crew informed of other shipping movements and potential hazards on board. A number of emergency systems and beacons will automatically transmit signals should the ship experience difficulties. One of the safety features of the loading system is a very comprehensive suite of instruments that automatically shuts down the system as soon as certain preset parameters are exceeded. The ships are also fitted with gas and fire detection LNG terminals have been operated near built-up areas for decades, some for more than 40 years, without a single serious safety incident. The only incident at an LNG terminal that resulted in fatalities occurred in Cleveland, Ohio (US) in 1944, when much less was known about the safe storage of LNG. The use of incorrect materials in a storage system caused LNG to be released. However, present multiple enclosure systems and the materials used nowadays, make such an accident virtually unthinkable. 0.1%-0.25% of the cargo converts to gas each day, depending on the efficiency of the insulation and the roughness of the voyage. In a typical 20-day voyage, anywhere from 2%-6% of the total volume of LNG originally loaded may be lost.. This gas is sent into the fuel system of the ship.

4.7.1.1 Pipeline Transportation

When liquid flows at -160°C into a large diameter pipe/tank which is at ambient temperature, the bottom of the line rapidly cools while the top of the pipe stays relatively warm for some time. Unequal cooling results (Contraction of S S is at a rate of 125 mm per 50°C), this leads to bowing the pipe upwards, due to lower part contraction. Thus substantial stresses placed on the system has to be taken care of. An operator must perform a periodic evaluation based on data integration and risk assessment and implement a program to continually assess the integrity of its pipelines. Operators should install new and more preventive and alleviate measures to protect surroundings and assure public safety. Such additional measures are many, such as:

Installing Automatic Shut-Off Valves or Remote Control Valves Better computerized monitoring and use improvised leak detection systems from time to time; Using sturdier pipes ,thicker wall pipe.

Some times installing of larger size pipe lines is also a wasteful on two accounts one is energy for pumping, secondly if the required gas is not available the volume of the pipe line becomes redundant. Such as HBJ line ONGC's Hazira-V(B)ijaipur-Jagdishpur (HVJ) pipeline did not receive any gas from the Vasai East Field between March 22, 2012, and April 8, 2012, due to problems at ONGC's end The Non-HVJ pipelines , the Krishna-Godavari region pipelines, the Cauvery region pipelines as well as the Gujarat and Rajasthan region, the Dahej-Vijaipur (DVPL) and Dahej-Panvel-Dabhol pipelines,. the two LPG pipelines-Jamnagar-Loni and Vizag-Secundrabad-also functioned normally.

4.7.1.2 Natural Gas Transmission

GAIL has built a network of trunk pipelines covering length of around 11,000 km. Leveraging on the core competencies, GAIL played a key role as gas market developer in India for decades catering to major industrial sectors like power, fertilizers, and city gas distribution. Currently GAIL transmits more than 160 MMSCMD of gas through its dedicated pipelines and have more than 70% market share in both gas transmission and marketing. However, at present there are regional imbalances in gas supply across the country. To bridge this gap in infrastructure, Ministry of Petroleum and Natural Gas, in the year 2007, authorised five new pipelines to GAIL covering a length of over 5,500 km. GAIL is the first company in India to own and operate pipelines for LPG transmission. It has 1,900 km LPG pipeline network 1,300 km of which connects the western and northern parts of India and 600 km of networks is in the southern part of the country connecting Eastern Coast. The LPG transmission system has a capacity to transport 3.8 MMTPA of LPG. LPG transmission through pipelines was 3337 TMT in the year 2010-11.

S. No.	Pipeline	Length km/ Capacity in MMSCMD	Commissioning
1.	Dadri Bawana Nangal*	610 km/31 MMSCMD	2011-12
2	Chainsa Jhajjar Hissar 450km	300 km/35 MMSCMD	2011-12
3.	Jagdishpur Haldia 2050km	2000 km / 32 MMSCMD	2013-14
4.	Dabhol Bangaluru 1,114km	1386 km/ 16 MMSCMD	2011-12
5.	Kochi Kanjirikkod Bangalore	860 km / 16 MMSCMD	2012-13
6.	Dadri Nangal	610km	2012=13
	TOTAL	5156 km / 130 MMSCMD	

GAIL's new pipelines - under implementation
(length and investment figures include spurlines)

7.	Dabhol - Bengaluru pipeline	1,389 km.
8.	Kochi - Kanjirkkod - Bengaluru - Mangalore pipeline	1,114 km.
9.	Capacity Augmentation Dahej - Vijaipur section	610 km.
10.	Capacity Augmentation Vijaipur - Dadri section	505 km.
11.	Jagdishpur - Haldia pipeline	2,050 km.
12	Karanpur - Moradabad - Kashipur - Rudrapur pipeline	275 km.
13	Capacity Augmentation Bajera - Agra - Ferozabad section	70 km.

4.7.1.3 GAIL & LNG

By the end of 2009-10, the gas consumption in India stood at 165 MMSCMD with LNG occupying 15% (25 MMSCMD) of the entire gas market. The proportion of imported LNG is expected to increase to anywhere between 20% and 30% by 2015. GAIL has been playing a principal role on its part in ensuring that Government's objective of achieving energy security is achieved through a judicious mix of energy portfolio. As a dominant player in the gas market, GAIL plays a major role in sourcing of LNG and creation of pipeline infrastructure to form an efficient national grid that would ensure connectivity to all demand centers. To achieve these objectives, GAIL is actively pursuing LNG sourcing from major LNG producers/sellers all across the globe and has been adopting a strategy to have a mixed portfolio of spot, short/mid-term, and long-term deals. To ensure long-term supplies in the past, GAIL has promoted Petronet LNG Ltd (PLL), along with oil majors Oil and Natural Gas Corporation (ONGC), Indian Oil Corporation (IOC) and Bharat Petroleum Coporation Limited (BPCL) for the import of LNG into India. PLL is currently importing 7.5 MMTPA of LNG from Qatar for its Dahej Terminal on long term contract basis. PLL will also be importing 1.44 MMTPA from Gorgon LNG project, Australia for its Kochi Terminal. GAIL is also the sole transporter of the entire RLNG and a major off-taker from both these contracts.

GAIL has seven LPG Plants, two at Vijaipur and one at Vaghodia, and one each in Lakwa (Assam), Auraiya (UP), Gandhar (Gujarat) and Usar (Maharashtra), producing over 1 million TPA LPG and other liquid hydrocarbons. LPG is sold in bulk to LPG retailing companies such IOCL, BPCL, HPCL and other liquid hydrocarbon products are sold to industries. GAIL has a share of about 10% of the Indian LPG market in LPG production and 7% in LPG sales.

GAIL is the first company in India to own and operate pipelines for LPG transmission. It has 1,900 km LPG pipeline network 1,300 km of which connects the western and northern parts of India and 600 km of networks is in the southern part of the country connecting Eastern Coast. The LPG transmission system has a capacity to transport 3.8 MMTPA of LPG. LPG transmission through pipelines was 3337 TMT in the year 2010-11. GAIL has a share of about 10% of the Indian LPG market in LPG production and 7% in LPG sales. GAIL is currently participating in 31 exploration blocks, in Basins such as Mahanadi, Mumbai, Cambay,Assam-Arakan, Tripura Fold Belt, Gujarat Kutch, Krishna Godavari, Cauvery and Cauvery Palar. GAIL has partnership inthese blocks with various companies such as ONGC, OIL, GSPC, Hardy Exploration & Production, Petrogas, JOGPL, Eni and Daewoo as Operators. Out of these 31 E&P blocks, 2 blocks are overseas (A-1 and A-3 blocks in Myanmar). The blocks are in various stages of exploration, appraisal and development. Hydrocarbon discoveries are placed in 7 E&P blocks where GAIL is participating. The blocks with hydrocarbon discovery are: MN-OSN-2000/2, CB-ONN-2000/1, Block A-1 and A-3 Myanmar, CY-OS/2, AA-ONN-2002/1, CB-ONN-2003/2. Production of crude oil is in progress from Cambay Onland block (CB-ONN-2000/1) @ 1250 barrels per day. Development activities are in progress in 2 blocks in Myanmar (A-1 and A-3) ard production of gas is expected from May 2013. GAIL is an active member of multi-organisation team (MOT) set-up up for assessment of shale gas potential in Indian basins. The other representative in MOT are from DGH (Directorate General of Hydrocarbons), ONGC and Oil India Limited (OIL). GAIL is also a member of National Gas Hydrate Programme being coordinated by DGH and is actively involved in activities related to Gas Hydrate exploration. Gail India, the country's biggest gas utility, has booked 2.3 million tonnes of liquefied natural gas (LNG) production capacity in the US, from where it hopes to bring the scarce resource to India Maryland "Dominion is marketing 4.6 million tonnes per annum and Gail has booked 50% of such capacity for 20 years," Presently, GAIL owns and operates a gas based integrated petrochemical plant at Pata, Uttar Pradesh with a capacity of producing 400,000 TPA of Ethylene and 410,000 TPA of Polymers i.e. HDPE and LLDPE. GAIL has also signed an MoU for participation in an integrated Refinery and Petrochemical Complex at Vizag in a joint venture with Total of France, HPCL,

Mittal Energy and Oil India Limited. The project will be completed in 60 months from the date of approval. The petrochemical complex will comprise of a cracker unit, downstream polymer and integrated off-site/utilities plants. The complex has been configured with a capacity of 220,000 tons per annum (TPA) of Ethylene and 60,000 tons per annum of propylene with Natural Gas and Naphtha as feed stock.

4.7.1.4 *Petronet LNG Limited (PLL)*

PLL has been formed for setting up of LNG import and regasification facilities. PLL has a long term LNG supply contract with RasGas, Qatar, for import of 7.5 MMTPA of LNG. PLL Dahej terminal in Gujarat has been expanded to 10 MMTPA capacity. PLL has successfully implemented a pilot project for supplying LNG through cryogenic road tankers. PLL is also coming up with a LNG terminal at Kochi, Kerala, with an initial capacity of 2.5 MMTPA, expandable up to 5 MMTPA and it is scheduled to be operational by end of 2011. GAIL has 12.5% equity stake in PLL, along with BPCL, ONGC and IOCL as equal partners.

REFERENCES

1. (a) Fitzgerald, K.J.& Richardson, J.A., *HCP*, July 1966, p. 126.
   (b) Chyuan-Chung Chen & Andrews Ng: *HCP*, April 1980, p. 122.
2. Dunstan, *Oil & Gas J.* 27, No. 142, 1929, p. 165 & 175.
3. (a) Campbell & Lawrence, *Pet Ref.* Part I & II Aug. 1952, p. 65.
   (b) Campbell & Lawrence, *Pet* Oct. 1952, p. 106.
4. Lom, W.L., Liquefied Natural Gas, Applied Science Publishers, London 1975, p. 34.
5. Kalchevesky & Stagner: Chemical Refining of Petroleum, Reinhold Publishing Co. 1942, p. 62.
6. Evans, E.V. and Stainer, H., *Prac. Roy. Soc.* (London), 1924 105 A: 626.
7. Kohl, A., Gas Purification, McGraw Hill, p. 625.
8. Maxted, E.B. & Priestly, J.J., Gas *J.* 1946, 247 : 471 : 515, 556.
9. Belov, P., Fundamentals of Petroleum Chemicals, Mir Pub. Moscow: p. 50.
10. Samaniego, J.A., Pet. Processing, 12, No. 7; July 1957, p. 128.
11. Kenneth, A., Kobe & John McKetta. Advances in Pet Refining and Chemistry, Vol. IV, p. 176.
12. Venkateswarulu, K., Rao, B.K.B. & Banerjee, T.S., Chemical Industry Developments (CID) (India), Aug. 1972, p. 21.
13. (a) Devandran, M. & Rao, B.K.B., *HCP*, Aug. 1976, p. 127, Nov. 1976, p. 237.
    (b) Rao, B.K.B., *HCP*, Dec. 1978, p. 105.
14. Choffee, B., *HCP*, May 1966, p. 188.
15. Johnson, G.C.& Francis Afred, W., *I & E.C.*, Vol. 46, No. 9, August 1954, p. 1662.
16. Broughton, D.B. & Assedin, G.F., 7 *WPC*, 1967, Vol. 4, p. 65.
17. Wilkenson, W.F., Ghublikian, J.R. & Obergfell, P: *Chem. Engg. Prog.* 49, 1953, p. 257.
18. Cinelli, E., Noe, S.& Paret, G, *HCP*, April 1972, p. 141.

340

19. O'Connor & Boyd., Standard Handbook of Lubrication Engg., McGraw Hill, 13-8, 14.
20. Mantel, C.L., Adsorption, McGraw Hill, p. 530.
21. a, b, c. Kalichevsky & Kobe, A., *Pet. Ref.* 7, 8 & 9, 1953, p. 119, p. 135, p. 215.
22. Felix Heinemann & Henz Heinemann., *Pet. Ref.* 6, 1954, p. 159.
23. Leo Garwin & Barber, C.E., *Oil & Gas, J.*, March 22, 1954, p. 137.
24. Knox, W.T., Weeks, R.L. & Hirbshuman, H.J., *IEC*, 39,1947, p. 1573.
25. Kalchichevsky, V.A., *Petrol. Engg.* 28(2), 1956,11.
26. Hournac. R., *HCP*, January 1981, p. 207.
27. Carter, R.C., *Oil & Gas J.* March p^2, 1954, p. 137.
28. (a) Rao, B.K.B.& Banerjee, T.S., Ist Law Symp. IIT, Bombay, Nov. 22, 23, 1978.
    (b) Rao, B.K.B & Banerjee, T.S., *Chemical Age of India*, Vol. 4, No. 4, January 1970, p. 159.
29. Henry Rushton, J., *IEC*, 27. 1937, p. 309.
30. Pass, F.J., Poll, H. & Schuster, J.F., Vth WPC, Section III, p. 455.
31. a, b, c. *HCP*, Sept. 1972, p. 198, 195, 190, 206.
32. Speight, J.G., The Chemistry and Technology of Petroleum (Marcel Dekker), p. 412.
33. Alders, L, Liquid-Liquid Extraction, Elsevier Pub. London, p. 51.
34. Laddha, G.S., *CEW*, Sept. 1971, p.41.
35. Baker, T, Chilton, T.H. & Vernon, H.C., *Trans. Am. Inst. Chem. Engrs.* 31, 1935, p. 296.
36. Treybal, R.E.. Liquid Extraction. McGraw Hill, 1951, p. 60.
37. Kenneth, A. Kobe & William, G. Domask., *Pet. Ref.,* March, 1952, p. 106.
38. Bennett, H., Industrial Waxes, Chemical Publishing Co., Vol. I, 1975, N.Y., p. 30-33.
39. Wrath, A.H., The Chemistry and Technology of Waxes, Reinhold Publ. 1947, p. 217.
40. Bell, H.S., American Petroleum Refining. Van Nostrand, p. 386.
41. Henry Martyn Noel. Petroleum Refinery Manual Reinhold Publishing, 1959, p. 139.
42. Karanth, P.K. & Sinha, R.K., *Chemical Age of India*, Oct. 1972, p. 216.
43. Kumar, Y., Ghosh, S.K. & Gulati, I.B., *Chemical Age of India*, Jan. 1971, Vol. 5, p. 133-140.
44. Patwardhan, S.R., Gurjar, V.C. & Mohan Rao, G., 2nd LAW Symp. IIT, Bombay, Jan., (3-4), 1980.
44. Further reading: Safety in Chemical Plants/Industry & its Management, Khanna Publishers, New Delhi.

CHAPTER 4

1. Compare and contrast liquid sulfur dioxide process with DMSO extraction for improving smoke point of kerosene.

2. (a) From the Fig. 4.11 a find the concentration of benzene in extract phase. The feed contains 10.3% benzene in benzene and heptane mixture and the solvent to feed ratio is kept 2.5.
   (b) If the solvent is mixed with 10% water repeat the above calculations.
   (c) Assuming the above extraction is conducted in a packed bed extractor, find the critical packing size.

3. (a) Describe Merox operation.
   (b) How do you remove sulfur compounds from tube stocks?
   (c) How is wax classified?
   (d) Describe in detail, production of paraffin wax.

5

Thermal and Catalytical Processes

5.1 CRACKING

Dissociation of high molecular weight hydrocarbons into smaller fragments through agency of heat alone is termed as thermal cracking or pyrolysis. Cracking can also be conducted in presence of catalysts, and this has been differentiated from the former one by adding 'catalyst' term before cracking i.e. catalytic cracking. In either of these operations, the reactions widely follow the same lines yielding more or less similar products, however the course of reactions is explained by a different mechanism. Necessity of cracking operations has become a reality in refinery operations to augment the market demands by converting less valued fractions or unwanted fractions to more commercially valued products.

Developed countries like USA, convert more than 70% of crude by thermal processes or other techniques to motor spirit. While a developing country like ours has scant demand for gasolines though imposing demands for middle distillates exist. Thus thermal cracking operations are warranted in every country though not to same extent or for the same fraction at all times.

5.1.1 Thermal Cracking Reactions

Heat sensitive, high molecular weight paraffins fragment when temperature exceeds 400°C. A molecule of C_nH_{2n+2}; where 'n' ranges from and above 25 easily splits into two, almost at the middle, resulting in one saturated molecule and other unsaturated molecule. All cracking reactions ultimatily give rise to Coke and Hydrogen.

$$C_nH_{2n+2} \rightarrow C_{n/2} H_n+2 + C_{n/2} Hn.$$

The above reaction is foremost and conclusive at moderate temperatures around 400°C, but fragmentation continues with increasing temperature, giving rise to often simple products and occasionally complex molecules. Appearance of different intrinsic reactions is common, depending upon the molecular size. For a paraffin of less molecular order, the following reactions illustrate the mode of cracking.

$$C_{12} \rightarrow C_6+C_6$$
$$C6 \rightarrow C_4+C_2/C_3+C_3 \qquad \text{(one is always unsaturated)}$$

Unsaturates (olefins) obtained in the process crack again

$$C_4H_8 \rightarrow C_2H_6+H_2/C_2H_4+CH_4+C$$
$$C_4H_8 \rightarrow CH_4+C_3H_4 \text{ (diolefin/alkyne)}$$
$$C_2H_4/C_3H_4 \rightarrow C+H_2$$

Olefin cracks or dehydrogenates to diolefin or an alkyne. Further severity in conditions results in production of methane, hydrogen and carbon. These are regarded as stable and end products of cracking operations. Unsaturates being active under congenial thermal cracking conditions form[1] dimers, trimers, etc. i.e. condensation to bigger molecules.

$$2C_2H_4 \rightarrow C_4H_8$$
$$C_4H_8 + C_2H_4 \rightarrow C_6H_{12}$$
$$2C_3H_4 \rightarrow C_6H_6 \text{ (ring) } + H_2$$
$$3C_2H_2 \rightarrow C_6H_6$$

Hydrogenation also occurs but to a less extent

$$C_3H_4+H_2 \rightarrow C_3H_6$$

Aromatics, saturated ring structures follow a different pattern of cracking.

Chain detachment is frequent followed by dehydrogenation as shown below:

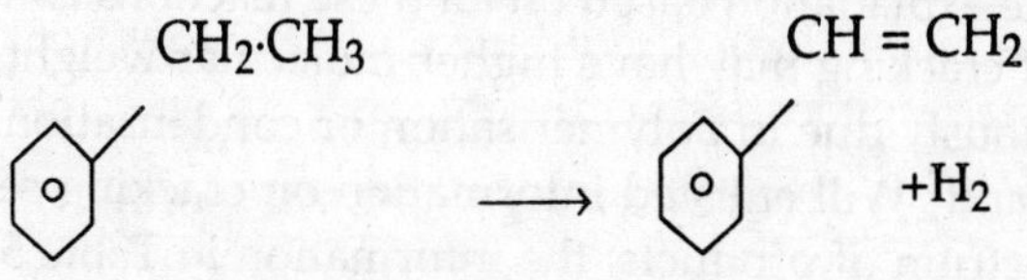

$$CH_2 \cdot CH_3 \quad\quad CH = CH_2$$

(ring) $\longrightarrow$ (ring) $+H_2$

Saturates are converted to unsaturates

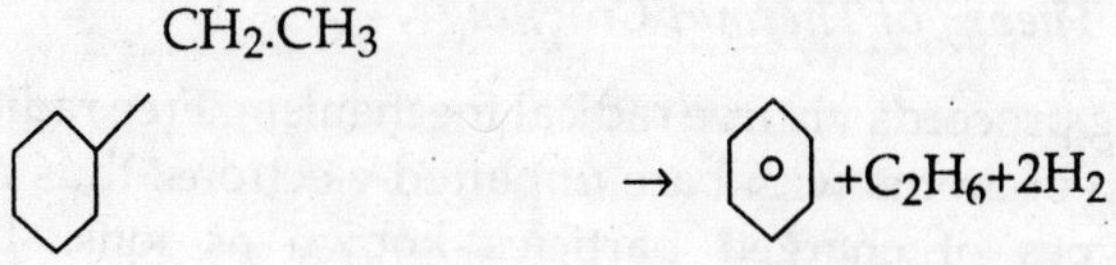

$$CH_2 . CH_3$$

(ring) $\rightarrow$ (ring) $+C_2H_6 + 2H_2$

Ring opens in the extreme conditions of cracking

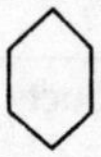

(ring) $\rightarrow CH_2 : CH_2 + CH_2 = CH.CH = CH_2 + H_2$

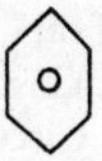

(ring) $\rightarrow 3C_2H_2./CH_2 = CH.CH = CH_2 + 2C$

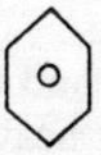

(ring) $\rightarrow C_2H_4 + 4C + H_2$

Above reactions illustrate carbon and hydrogen formation.

Light paraffins when subjected to cracking de-hydrogenation reaction shall be imminent and it dominates over other cracking reactions:

$$C_4H_{10} \rightarrow C_4H_8 + H_2$$
$$\rightarrow C_3H_6 + CH_4$$
$$\rightarrow C_2H_6 + C_2H_4$$

The tendency of dehydrogenation becomes less, when the chosen molecule is above C_4. Dehydrogenated molecule normally exhibits isomeric forms, such as

$$C_4H_{10} \rightarrow CH_3CH_2CH = CH_2 + H_2 - 131.5 \text{ kJ/mole}$$
$$i\text{-butane}$$

$$\rightarrow CH_3CH = CH.CH_3 + H_2 - 11.68 \text{ kJ/mole}$$
$$\text{Cis-2-butane}$$

$$\rightarrow CH_3CH = CH\text{-}CH_3 + H_2 - 115.8 \text{ kJ/mole}$$
$$\text{Trans 2-butane}$$

A plausible explanation based on all these reactions concedes that the product of cracking may have higher-molecular weight than the feed itself, obviously due to polymerisation or condensation.

Martin and[2] Will enlisted information on cracking severity and its innate spectrum of products: the information in Table 5.1 is based on it.

5.1.1.1 Theory of Thermal Cracking

Cracking proceeds via free radical mechanism. Free radicals are atoms or group of atoms with bare unpaired electrons, thus differing from other types of charged particles known as ions. The following reactions explain the difference.

TABLE 5.1 Thermal Cracking Operations.

Cracking temperature °C	Nature of operation	Products
425-460	Visbreaking	Fuel oil
460-520	Thermal cracking	Gas, Gasoline. Tar oils, Circulating oils.
520-600	Low temp. coking	Gas, gasoline soft coke,
600-800	Gas production	Gas and unsaturated
800-1000	High temp. coking	Gas, heavy aromatics pitch or coke.
above 1000	Decomposition	H_2, gas, carbon-black.

$$NH_4Cl \rightarrow NH_4^+ + Cl^- \quad \text{ions}$$
$$H_2 \rightarrow H^+ + H^-$$
$$CH_4 \rightarrow CH_3^+ + H^- \left.\right\} \quad \text{charged particles}$$
$$H_2 \rightarrow H\bullet + H\bullet \quad \text{radicals.}$$

In a molecule of hydrogen there are two electrons sharing the spin in opposite directions.

For hydrogen paired electrons are arranged in S-shell like $\boxed{H\downarrow H\uparrow}$. When atomic hydrogen is formed without charges unpaired electrons

result in S-shell: $H_2 \rightarrow \boxed{H\downarrow} + \boxed{H\uparrow}$. Similarly

$$R.CH_2.CH_3 \rightarrow RCH_2\bullet + CH_3\bullet$$

These unpaired electrons are very active, and try to form stable compound by acquiring unpaired electrons from sources available.

Motive power of free radicals to combine with other radicals or metallic bodies is immense; generous dissociation into small radicals is also high.

Hydrogen (charged) ions are usually referred as carbonium ions or carbanium ions depending upon the charge they carry.

$$CH_3.CH = CH_2 + H^+ \rightarrow CH_3.CH^+.CH_3 \text{ carbonium ion}$$

$$CH_3CHO + OH^- \rightarrow CH_3CHOOH \text{ carbonium ion.}$$

The reactions of free radicals are fast; indeed only one radical is sufficient to bring about a series of reactions as shown below:

$$
\left.
\begin{aligned}
H_2 &\rightarrow H^\bullet + H^\bullet \\
\text{or} \quad O_2 &\rightarrow O^\bullet + O^\bullet \\
C_5H_{12} &\rightarrow C_5H_{11}^\bullet + H^\bullet
\end{aligned}
\right\} \quad \text{radical production}
$$

$$
\left.
\begin{aligned}
C_5H_{12} + H^\bullet &\rightarrow C_5H_{11}^\bullet + H_2 \\
C_5H_{12} + C_2H_5^\bullet &\rightarrow C_7H_{16}^\bullet + H^\bullet \\
&\rightarrow C_5H_{11}^\bullet + C_2H_6 \\
C_3H_7^\bullet + C_5H_{12} &\rightarrow C_4H_{10} + C_4H_9^\bullet
\end{aligned}
\right\} \quad \text{Propagation}
$$

$$
\left.
\begin{aligned}
C_3H_7^\bullet + H^\bullet &\rightarrow C_3H_8 \\
C_5H_{11}^\bullet + C_3H_7^\bullet &\rightarrow C_8H_{18} \\
H^\bullet + H^\bullet &\rightarrow H_2
\end{aligned}
\right\} \quad \text{Termination}
$$

Different free radicals have different lives. Hydrogen free radicals have the highest stability and so a higher life. The stability or life of a radical decreases with the increasing size of molecule. Minimum bond energies invoke the scission at a convenient position during the reaction. The bond energy of C-C is 262.7 kJ/mole while C-H bond requires 358 kJ/mole to break open. Some of the bond energies given by different authors is furnished below[3]:

Bond	Energy kJ/mole		
Author	Polanyi	Stevenson	Kistiakowsky
CH_3-H	428.76	422.90	426.67
C_2H_5-H	407.84	404.91	409.93
n-C_3H_7-H	397.39	414.12	—
iC_3H_7-H	372.29	391.11	—
CH_3-CH_3	363.92	348.86	358.06
CH_3-C_2H_5	355.56	339.24	350.54
C_2H_5-C_2H_5	—	330.46	344.68

Due to the high stability of tertiary carbonium ions, the asymmetric carbon bonds are easily ruptured than the symmetrical ones. This is amply sufficient to explain why isomerisation does not result in thermal cracking operations. The free radical mechanism offered by Rice, though it cannot correctly predict the composition in kind and quantity of the products, still, is the best tool in view of the complexity of reactions.

5.1.1.2 Properties of Cracked Materials

Products of cracking have different properties because of unsaturation. These properties are entirely dependent upon the conditions of cracking. A cracked product acquires refractory nature. This calls for severe conditions for further cracking: hence they are named as recycle stocks. For economic way of cracking the heavy components from the cracked product are mixed with incoming charge and allowed to crack. Evidently some of the properties of the cracked products are very much different from the parent substance. The following properties are expected to change:

(a) Characterisation factor (Decreases)
(b) Boiling point, pour point and Viscosity (Decrease)
(c) Unsaturation and aromatisation (Increase)
(d) Octane number (increases)
(e) Sulfur in cracked products-heavier (Increase) although much may excape in the form of hydrogen sulfide.
(f) Oxidation stability (Decreases).

Cracking being endothermic, external heat is required. Cracking is preferentially conducted to obtain the desired products as closely as possible. Because of the complexity of reactions theoretical predictions may not be convincing. Best way to produce more of the desired low volatile fractions is by cracking the intermediates obtained in the operation.

Pressure, temperature and time are the main parameters which govern the cracking operations. At a given pressure and temperature the yield of light fractions is a function of time. Time of cracking, at a set conditions of temperature and pressure, increases with increase in API gravity of the feed. This explains why that heavier fractions crack easily. Yet at times the operation becomes extravagant due to high gas production. Deposition of coke checks the run lengths too. Thus the parsimony of the process is guided by gas and coke yields.

In an operation of a high crack per pass with low recycle ratio of intermediates, the resulting gasoline exhibits low and better volatility and high octane number. The effect of pressure on the production of gas at different cracking rates is given in Table 5.2.

TABLE 5.2 Effect of Pressure on Cracking.

| Gasoline conversion per pass | Gas (M^3/bbl) yield Pressure | | |
	6.6 atms	13.5 atms	27 atms
10%	3.3	1.6	1.2
20%	8.0	4.1	3.5
30%	14	8.3	6.5

TABLE 5.3 : Yield of Gasoline and Heat of Decomposition.

At a set of conditions of operation	
Yield of gasoline	Heat of decomposition; kJ/kg
5%	5585
10%	3595
15%	2492
20%	1860
25%	1777
30%	1672

With increase in yield per pass the heat of decomposition also decreases. This is due to the increase in yield of gasoline. Proportional increase in yield to heat input is much higher. So at an yield of 30% gasoline the heat of decomposition is 1600-1700 kJ per kilogram. Yield of gasoline versus heat of decomposition is shown in Table 5.3. Summing up all the facts:

(1) Pressure for all practical purposes has no direct effect on velocity of reaction. If more gases were to be produced then low pressures are desirable. With high pressures, gas to gasoline ratio becomes less.
(2) Reaction velocity is directly dependent upon temperature only.
(3) Recycling increases refractory nature of stocks, and hence, recycling should not exceed 2 to 3 times of fresh stock for economic operations.

5.1.1.3 Effect of Pressure on Cracking

Obviously pressure retards cracking reactions. But in practice a positive pressure of 10 to 15 kgs/cm^2 is enclamped to minimise coke formation. Increase of pressure, decreases the yield of light fractions but in the earlier stages it may be quite favourable for production of diesels or circulating oils. Even though the decrease in heavier fraction may be anticipated the fall may not be so sharp.

5.1.1.4 Depth of Cracking and Soaking Factor

In the initial stages of cracking, practically the concentration of feed stock remains unchanged. But with progress in cracking, mixed products result; naturally product concentration shall be different. To express the conversion in a better way, terms like severity or depth and soaking factor are introduced. Severity is mainly connected with temperature. Soaking factor on the other hand is connected with temperature and volume of feed per unit time. These two can furnish a lot of information on cracking operations; and may percept the operations too. Soaking factor is a comparative term, i.e. comparison is done by bringing the conditions of operation to some known conditions (Standard). The specified conditions for such comparison are: temperature 426.7°C and pressure 52.7 kg/cm². The soaking factor may be now defined as

$$SF = \int_{o}^{v} R \frac{K_T}{K_{426.7}} \frac{dv}{F}$$

where

R : constant, correction factor when operating[4a] pressure differs from 52.7 kg/cm².

$K_T/K_{426.7}$: Relative cracking reaction velocity at any temperature (T).

$\dfrac{dv}{F}$: Differential coil volume in liters per bbl of feed stock daily throughput.

The above equation is suitable for vapour phase cracking only.

Rate of Reaction

Cracking reactions are described as first order reactions, when cracking per pass does not exceed 25%. Rate is given by

$$K_t = \frac{1}{t} \ln \frac{a}{a-x}$$

where

t in seconds
a % of material in feed
x % of material that disappears in time t.

For a feed stock which does not contain any cracked product 'a' becomes equal to 100. Frequently x is expressed in moles or by volume. When gaseous feed stocks are handled x is presented in moles, otherwise in volume only.

Temperature and reaction rate-constant are related by Arrehenius equation given below:

$$ln\ K = -\frac{E}{RT} + C$$

where

R = 8.3143
T = °K
E = Energy of activation, kJ
C = Constant

E and C in the above equation are usually obtained from investigations. Such investigations are plenty in the literature.

Keith, Ward and Rubin, reported after extensive studies, activation energy for gas oils as –239 kJ and 'C' as 31.924. Genisse and Reuter's[5] values are slightly different and given as –224 kJ and –28.8 respectively. Approximate values given by Nelson[46] as E = –227.9 & C = –30. All these would be accurate enough for all theoretical considerations.

Heat of Decomposition

It is not descretely possible to ascertain the accompanying heat changes in the cracking reactions due to the complexities involved in bond ruptures and formations. Heat[4a] of cracking can be calculated by knowing molecular weight of charge stock and products. Such simplifications are amply available; as an example, an equation of immense help is presented as:

$$\text{Heat of cracking (kJ/gm)} = \frac{C\left(M_c - M_p\right)}{M_c M_p}$$

where

C = –1150.20
M_c/M_p mole weights of charge and products.

The best method of finding out heat of decomposition is from the principles of conservation of energy *i.e.* heat of combustion.
Heat of decomposition = (Heat of combustion of products)—
(Heat of combustion of feed stock).

Problem 5.1

A thermal cracking operation of heavy gas oil has given rise to the products as given below.

(a) Find out the heat of decomposition per kg gasoline.
Data : Feed °API-37
          Temp. of cracking-450°C
          B.P. of feed 330-380°C

Products:

Gases–12.3% average calorific value at 30°C
                          50.16×10^3 kJ/kg
Gasoline–30.6%, 68 °API, 44.22×10^3 kJ/kg
                          (Cal. Value)
Residuum–15%; 15 °API; 35.53×10^3 kJ/kg.
                          (Cal. Value)

Losses–2%
Uncracked feed–40.1%; 41.8×10^3 kJ/kg.
          (balance)   (Cal. Value)

Basis 100 kg of heavy gas oil

Constituent	°API	Amount (Kgs)	Heat of combustion at 30°C (kJ/kg) ($\times 10^3$)	Total Heat content kJ($\times 10^3$)
Gases	–	12.3	50.16	622.38
Gasoline	68	30.6	44.22	1,353.13
Residues	15	15.0	35.53	532.95
Losses	–	2.0	–	–
Remaining stock	37	40.1	41.80	–
		Total heat content of products		2,508.46

Heat content of feed (including losses)

$$(100 - 42.1) (41.8) 10^3 = 2,420.22 \times 10^3$$

Heat of dissociation $= (2,508.46 - 2,420.22)10^3$
                      $= 88.24 \times 10^3$ kJ/kg

Per kg gasoline $= \dfrac{88.24 \times 10^3}{30.6} = 2.556 \times 10^3$ kJ/kg

(b) If the cracking is conducted for 28 min. at 450°C, find the reaction constant and activation energy.
Assuming $C = 30$

$$\ln K_1 = \frac{E}{RT} + C$$

$$K_1 = \frac{1}{t}\ln\frac{100}{100-x} \qquad \text{where } x = 30.6 \text{ gasoline produced.}$$

$$= \frac{1}{1680}\ln\frac{100}{100-30.6} = 2.174 \times 10^{-4} \text{ sec.}$$

Substituting for K_1 in the equation

$$R = 8.3143$$
$$T = 723° \ K \ \& \ C = 30$$

Activation energy, $E = 230.726 \times 10^3$

Problem 5.2

In a low pressure, low temperature liquid phase thermal cracking operation a mixture of heavy oil and residuum is processed. The mixture is found to have 15.3 API (Mol. wt. 310).

The cracked products after separation show the following properties.

	Wt %	Mol. wt.	API
Wet gases	2.1	48	95
Light fraction	11.8	112	48
Furnace oil	38.8	270	22
Pitch (Residuum)	46.8	425	—

From the above information, find the API gravity of residuum and Heat of decomposition per kg furnace oil.

(assume C = -84, 166 kJ/gm mole).

Solution

API of the mixture is equal to the feed, assuming that the properties do not abnormally change.

$$100 \times 15.3 = 2.1 \times 95 + 11.8 \times 48 + 38.8 \times 22 + 46.8 \times \text{API.}$$

From this API is found to be −2.6 or

$$\rho = 1.09$$

Heat of decomposition is found by calculating no. of moles of reactant and no. of moles of products:

$$\text{Reactant}: \frac{100}{310} = 0.3225$$

$$\text{Products}: \frac{2.1}{48} + \frac{11.8}{112} + \frac{38.8}{270} + \frac{46.8}{425} = 0.4149$$

$$\Delta H = \frac{-84,166 \times (0.3225 - 0.4149)}{0.3225 \times 0.4149} = 58,121$$

	Feed	Mol. wt.	Products %		kJ/gm mole Mol. wt.	API
API	15.3	(310)	Wet gas	2.6	48	95
			light fractions	11.8	112	48
			F O	38.8	270	22
			Deposit	46.8	425	$\rho = (1.09)$
						API = –3

No. of moles of furnace oil: $\dfrac{38.8}{270} = 0.143$.

$$\Delta H = 58.121 \times 10^3 \times 0.143 = 8352 \text{ kJ/gm mole.}$$

5.1.2 Visbreaking

In the wide spectrum of thermal cracking operations (Table 5.1) visbreaking operation is in the first stage. The name suggests the reduction of viscosity. Earlier days the fuel oil demand is mostly met by cracking of heavy residuums or a process devised to utilize the bottom of the barrel. Later on the application of this technique has been adopted for upgradation of heavy crudes, lube extracts and most unwanted heavy components whereby they could be used as feed stocks for subsequent operations. Materials like residuums of not direct utility or stocks besieged with difficulties due to high pour point usually form palpable feeds for this operation. The liquid products from visbreaker may be used as cycle stocks for catalytic cracking although most of the product is used as fuel oil. Thus this unit helps very much in conservation. In India where the middle distillates have higher demand, it is the cheapest device to produce feed stocks for various units as well as fuel for heavy duties. During Seventy-Eighties the visbreaker played a major role in Indian refineries touching a capacity installation of 600,000 bpd. However with the introduction of FCCs in almost all refineries followed by hydrocrackers, the interest on visbreakers has commendably fallen. But every process has some advantages for sometime over the other. The catalytic operations always require precise operating conditions besides consistence in the quality of the catalyst. Sometimes activation and disposal of the catalyst is a major concern. In reality if parochial interests are neglected visbreaker forms the cheapest device for producing valuable intermediates for the secondary units. Thus the advantages of visbreaker are listed as:

a. Cheapest device, b. Can handle different feeds, c. Quality of the feed is not deterrent for operation, d. Operating conditions are flexible, e. A good process for upgradation of bottom of the barrel as well as heavy crudes, f. Not much technical knowledge is needed for installation or operation of the unit.

Like any unit it works upon two basic parameters namely the cracking temperature and the residence time. Based on these factors two types of units are very much popularin to the industry, one is Soaker type another one is Coil type.

In Coil type visbreaker the coil is maintained at the cracking temperature and the reaction is allowed to proceed in shortest possible time in the coil itself. At the outlet the products are quenched/cooled and separated. Thus a long coil sufficient to permit the reactions to conclude is envisaged accordingly the time of travel of the stock in the coil is fixed.

In Soaker type installations the reaction is closely approached or initiated in the furnace but the cracking reactions progresses only in a soaker drum because of the time given. Thus the reaction continues at a lower temperature than the coil type and coke deposition in this process is less.

5.1.2.1 Visbreaking Operation and Description

Figure 5.1 shows the operation. Feed stock comprising variety of materials ranging from asphalts, short residuums to medium oils is blended separately and passed through heat exchanging system (1):

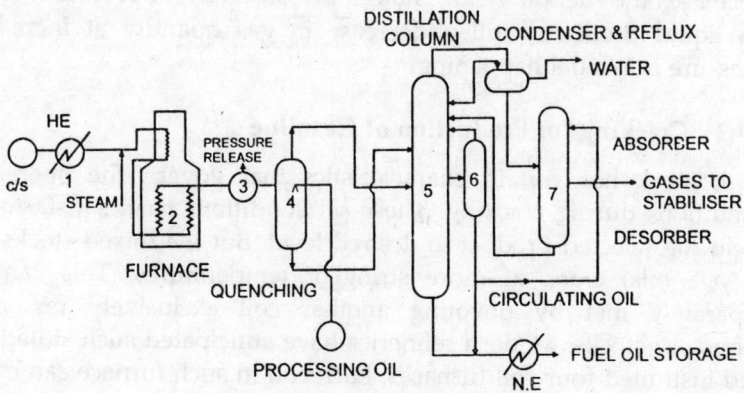

Figure 5.1 Visbreaking.

the temperature rise in such exchanger is close to 250°C. The preheated stock, is heated in a furnace (2) to attain a temperature of 470°C. A pressure of 10-15 kg/cm² is essential to keep the coke forming reactions in dormant state. Obligingly steam admission into feed stock checks coke formation, thereupon the life of tubes can be increased. The cracked products pass through a pressure releasing value (3) only to be quenched in a quencher (4) effectively with processed heavy oils. The light fractions and bottom fraction from this quencher go into a distillation column (5). At the top of column, the condenser system separates water and supplies adequate light fraction as reflux to column. These light fractions with gases are sent to an extractor (7) to absorb gases. The top portion of the extractor is meant to dissolve gasoline like fractions in circulating oils, and the bottom section of the extractor acts a stripper to free the gases. The circulating oil with dissolved fraction is again sent back to the distillation column (5). From the side stripper (6) the circulating oil is tapped as a bottom product, part of which is mixed with the bottoms of column (5) and sent to storage system through process heat exchangers. The addition of circulating oil to fuel oil maintains the required plasticity. Gases from stripper are sent to gasoline stabilizer for recovering butane and the rest are sent to gas stream circuit of refinery.

5.1.2.2 *Operating Conditions and Products*

Operating data on visbreaking of Nahorkatiya[6] long residuums are furnished in Figs. 5.2a, and b. With increase of cracking time and temperature, light fractions increase, which correspondingly decreases the fuel oil yield. Though pressure has no profound effect on liquid products, still a decrease in gas quantity at increased pressure is a usual happening.

5.1.3 Cracking for Production of Gasoline

Each stock has certain characteristics that govern the operating conditions during cracking. These set conditions shall satisfactorily yield the selected product to desired level. But the mixed stocks (or recycle oils) crack at more stringent temperatures. This can be separately met by devoting another coil exclusively for such refractory stocks. Modern refineries have anticipated such situations and instituted four coil furnaces. Each coil in such furnace can carry

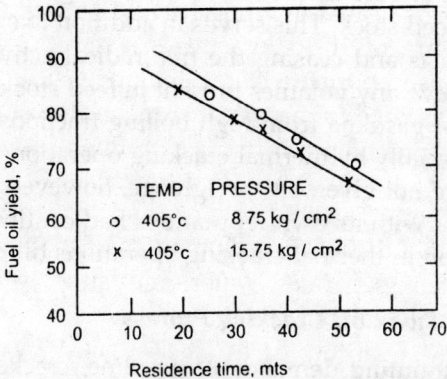

Figure 5.2a   Yield of Fuel Oil as a Function of Residence Time and Pressure.

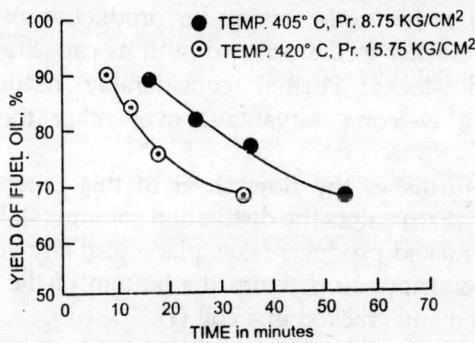

Figure 5.2b   Yield of Fuel Oil as a Function of Temperature.

out specific duties. To elaborate further, one coil may be assigned to visbreaking operations (low temperature and pressure). Second coil may be reserved for cracking heavy oils where temperature and pressure may be 520°C and 25-35 atmospheres respectively. Like this, the other two coils may be operating at different conditions (for reforming or treatment techniques etc.). This type of furnace excels the other heating methods in rendering economic benefits to the refiner. Present pipe still heaters are capable of raising the temperature of feed stock to 500°C in a short time, may be less than three minutes.

Cracked products leaving at that high temperature are still active with abundant free radicals, hence immediate quenching is indispensable. For economic reasons quenching is deliberately done with the fresh feed stock. This serves in addition to extraction of he.t from the products and ceasing the free radical activity, as a partial stripper to remove any volatiles present in feed stock.

Production of gasoline from high boiling fractions, was in fact an old process, specially by thermal cracking operations. At present the old methods are not given any weightage, however, Dubbs two coil cracking process with its own reputations had continued to occupy a valuable position in thermal cracking operations till today.

5.1.3.1 Dubbs Two Coil Cracking Process

To meet the mounting demands of gasoline, cracked gasoline was first produced by Burton in 1912. Ever since that time various processes are in service to meet the fast rising demands for gasoline. Presently, refiners are not fond of producing thermal gasoline due to the extolling performance of catalytic crackers. Growing tendency of using this Dubbs two-coil-cracking for production of fuel oils from residuums or cracker feed stock is due to its capacity to handle less valuable feed stocks. Further considerable reduction in coke formation is a welcome advantage over other thermal cracking operations.

Figure 5.3 furnishes the flow sheet of this process. Firstly, the reduced crude (F/S) enters the distillation column (4). Here extraction of heat from cracked products takes place; also any volatiles present in the feed are stripped off. From the bottom of the column heavy fraction is taken and cracked in a coil (1).

The temperature and pressure in this coil are maintained at 520°C and 30 atmospheres respectively. Light oil (circulating oil) a fraction

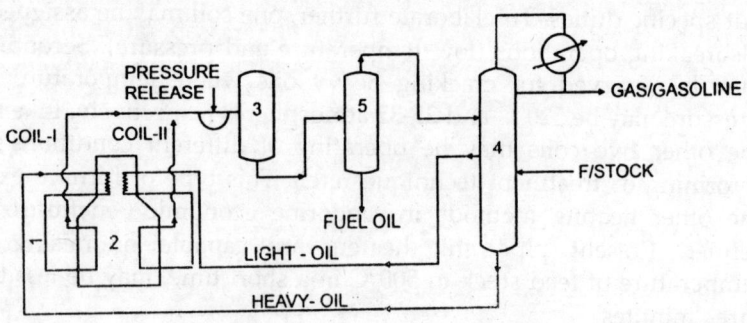

Figure 5.3   Two Coil Cracking.

obtained as a lower side cut from fractionation column (4), is routed through light oil coil (II), where the temperature shall be atleast 30°C higher than the other coil. The cracked products from both the coils are streamed into reaction chamber (3). Pressure in this chamber is maintained at 15 to 20 atmospheres. This pressure is essential to prevent excessive coke formation and increase liquid fractions. The products are now led into an evaporator (5) where heavy fraction is obtained from the bottom of the evaporator. Overhead fraction goes into fractionator (4). Heavy fraction from the evaporator is fuel oil and its pour point can be regulated to certain extent by recycle ratios of light to heavy oil. Usually, this ratio should be within 1 to 2. High severity and low recycling ratios result in better volatility and octane number of gasoline, although the quantity produced is less. With increasing severity, more gas and carbon formation are obvious. Gasoline, from low pressure operations is enriched with unsaturates and aromatics. An octane number of 70 is common, though it can be more if feed is more naphthenic. As a rule, sulfur in cracked products increase, though maximum concentration shall be in the bottom product. Gases are reported to be 50% unsaturated and form good feed stocks for polymer and alkylation purposes.

5.1.3.2 *Suspensoid Catalytic Cracking*

This is once through process, and owes its birth from thermal cracking operations. Main idea of this process is to deposit all the coke that is formed in the operation on catalyst itself.

The process consists of first mixing the catalyst and feed stock in proportion and passing through heater. Cracking takes place in presence of catalyst. At the delivery end the products of cracking are separated from the catalyst in a separator. Heavy fraction (tar) with catalyst particles is carried to a filter, where separation of catalyst takes place. Carbon on the catalyst is burnt off. In this operation high temperatures can be used.

5.2 CATALYTIC CRACKING

Catalytic cracking is the most widely used refinery process for converting heavy oils into gasoline of high octane value. This forms the heart of U.S. Refineries and is reputed to remove the ill effects of thermal cracking, such as unsaturation of products, high carbon formations and low yields of gasoline. In 1955 catalytic cracking capacity of the World was 13.3% of crude, while this is reported to be

6.9% in 1966, though in this decade demand for gasoline remarkably doubled. Alternative methods of production of quality gasoline have successfully permeated into industry.

Basically, catalytic cracking is distinguished from thermal cracking in the reaction mechanism, which is called carbonium ion mechanism. Carbonium ion mechanism leads to the formation of more stable saturated compounds with abundance of isomeric reactions accompanying. The stability of these ions is mainly guided by the nature of molecule undergoing cracking, than the environments of cracking.

5.2.1 Carbonium Ion Chemistry

Most of the reactions like alkylation, polymerisation, cat-cracking, hydrocracking, reforming etc., involve organic chemistry of carbonium[7] ion. Unlike free radicals, these are charged groups. A positively charged organic ion acting in chemical reactions, as if the charge were localised on one carbon atom, is known as carbonium ion. Electrophillic additions (Lewis acids) give such kind of charged ions. The formation of such ions is known as protonation (with Bronsted acids).

Protonation of an olefin

$$RCH : CH_2 + H^+ \rightarrow RC_2H_3^+ \rightarrow RCH^+CH_3 / RCH_2CH_2^+$$

Additions of Lewis acids

$$RCH : CH_2 + (BF_3/AlCl_3/FeCl_3) \rightarrow RCHF.CH_2^+ + BF_2^- \text{ and the}$$
$(BF_2 + F)$ reaction goes on till $(RCHFCH_2F)_3B$ results. Breaking of carbon halogen bond

$$\begin{array}{c} Cl \\ | \\ R-CH_2-C-CH_3 + AlCl_3 \rightarrow RCH_2.CH^+.CH_3 + AlCl_4^- \\ | \\ H \end{array} \quad + FeCl_4^-$$

Aromatics react with electrophiles (Cl^+)

Cl^+

The life time of a carbonium ion may vary from a fraction of second to minutes (indyl cations exist even for months in sulfuric acid). Tertiary carbonium ion is the stablest one followed by secondary and primary carbonium ions. When protonation takes place, a primary carbonium ion may result,

$$\underset{\underset{CH_3}{|}}{R.C} = CH_2 + H^+ \rightarrow \underset{\underset{CH_3}{|}}{R.CH.CH_2^+} \text{ is less favoured}$$

$$\text{than } \underset{\underset{CH_3}{|}}{R.C^+\,CH_3}$$

Now addition takes place according to Markovinkov's rule

$$\underset{\underset{CH_3}{|}}{R.C^+}.\,CH_2 + F^- \rightarrow \underset{\underset{CH_3}{|}}{R\,.\,CF}.\,CH_3$$

Number of steps are involved in these reactions like (a) proton elimination, (b) hydride shift, (c) methyl shift, (d) addition, (e) cracking, (f) hydride transfer, (g) termination of carbonium ions.

Methyl shift is similar to hydride shift

$$\underset{\underset{C}{|}\,\,\underset{C}{|}}{\overset{\overset{C}{|}}{C-C-C^+-C-C}} \rightarrow \underset{\underset{C}{|}}{\overset{\overset{C}{|}\,\,\overset{C}{|}}{C-C-C^+-C-C}}$$

Addition to an olefin

$$\underset{\underset{C}{|}}{\overset{\overset{C}{|}}{C-C^+}} + C = C - C \rightarrow \underset{\underset{C}{|}}{\overset{\overset{C}{|}}{C-C-C-C^+-C}}$$

Cracking to an olefin (β. Scission)

$$\begin{array}{ccc} & C & & C \\ & | & & | \\ C-C-C-C^+-C \rightarrow & C-C-C^+ + C = C & \text{(olefin will not} \\ & | & & | & \text{have charge)} \\ & C & & C \end{array}$$

Hydride transfer

$$\begin{array}{ccc} C & & C \\ | & & | \\ C-C-H + C-C^+-C-C-C \rightarrow & C-C^+ \\ | & | & | \\ C & C & C \end{array}$$

$$\begin{array}{c} H \\ | \\ + C-C-C-C-C. \\ | \\ C \end{array}$$

Reactions with counter ions (or termination)

$$\begin{array}{c} | \ | & | \ | \\ -C-C^+ + OH^- \rightarrow -C-C \ (OH) \\ | \ | & | \ | \end{array}$$

$$C-C^+ + HSO_4^- \rightarrow C-C \ H(SO_4) + H^+$$

$$C = C + H^+ \rightarrow CH^- \ C+ \quad \text{(initiation)}$$
$$H^+ + OH^- \rightarrow H_2O \quad \text{(termination)}$$
$$H^+ + F^- \rightarrow HF$$
$$C-C^+ + F^- \rightarrow C-CF \quad \text{(termination)}$$

Carbanion ions (negative charges) are apt when reactions proceed with aldehydes or ketones.

$$\begin{array}{ccc} O & & OH \\ || & & | \\ R.C.H + OH^- \rightarrow & R.C^- \\ & & | \\ & & OH \end{array}$$

Acid sites of the catalysts provide such charges. All catalytic reactions can be explained by this way. Even naphthenes can undergo hydrogen transfer to produce aromatics, which is a good example of such transfer.

Heats of formation of unsolvated carbonium ions. kJ/mole.

$$CH_3^+ - 1095.95$$
$$CH_3 . CH_2^+ - 936.99$$
$$CH_3 . CH_2 . CH_2^+ - 903.53$$
$$CH_3 . CH^+ CH_3 - 794.77$$
$$CH_3 CH_2 CH_2 CH_2^+ - 865.88$$
$$CH_3CH_2 C^+H CH_3 - 757.12$$
$$(CH_3)_3 C^+ - 694.38$$

The heats of formation show that methyl, ethyl, ions have very high value, this explains why catalytic cracking yields little methane, ethane or ethylene.

Reviews of Sittig[8] provide an excellent thesis on the subject of early 50's. Earlier catalysts of importance were silica alumina type; the introduction of zeolite catalysts containing Ca, Mn or rare earths have really boosted the industry with unexpected dividends. All these catalysts subscribe to acid cites (Bronsted) that cause protonation.

5.2.1.1 Feed Stock and Catalytic Cracking Conditions

Catalytic cracking is conducted relatively at less severe conditions. Temperature in all catalytic cracking operations lies in the range of 450-510°C; pressure, although specified for each type of reactor, does not exceed two atmospheres in fluidized bed reactors. In addition to these operating variables, catalyst to oil ratio is a set factor for a particular kind of operation.

Feed of catalytic crackers is gas oils or blends of thermally cracked oils. With modern tendency to utilise even residuums or solvent extracts, the feed characteristics are more decisive. Feed stocks generally contain sulfur and nitrogen; these with metals present in oil, give trouble to the refiners. Sulfur and oxygen compounds exhibit pugnacious effects on catalysts. Nitrogen though available in small quantity; 30 to 50% of it is frequently basic. This harms the catalyst by reacting with acid cites. Voltz et al.[9] discussed the effect of basic nitrogen on conversion rates and on decay coefficient of catalyst (see Table 5.4).

At least half of the known metals are present in petroleum, although in traces[10], such as vanadium, nickel, iron, copper, etc. restrain the functions of catalyst by effectively blinding the active sites. Straight run and vacuum gas oils are relatively free from such metallic bodies, but residuums definitely call for special attention. The ability to feed the heavier portion of crudes to Fluid Catalytic Crackers depends to some extent on vanadium and nickel content. Phillips Research and Development[11] Facilities devised a process using an additive containing antimony that will passivate the metal contaminants. By this way catalysts may be able to take up feeds of 12,000 ppm of nickel vanadium complexes. Conradson carbon residue of the stocks should be kept well below 4% to limit carbon formation.

TABLE 5.4 : Effect of Nitrogen on Cracking.

	Cracking Temp.	K_0	K_1	K_2	Decay-co-efficient
Midcontinent gas oil (a)	536	42.7	33.1	3.59	32.8
Fresh oil (b)	494	28.4	24.1	2.07	33.4
+0.1% wt. of basic nitrogen					
(a) +0.1% "	536	22.5	17.2	2.21	30.5
(b) +-do- "	494	13.8 ·	10.3	0.22	29.3
(a) +0.2% wt. of basic nitrogen	536	16.3	12.7	1.97	27.3

where, K_0 = rate const. for gas oil cracking, K_1 = rate const. for gasoline formation, K_2 = rate const. for gasoline cracking.

Pretreatment of such stocks is immensely regarded as benefaction to catalytic cracking operations, not only in safeguarding the life of catalyst, but also saving the time and amount in purifying each fraction. Sulfuric acid treatment furnishes effective removal of sulfur and nitrogen compounds. 85% acid strength is common for such treatments. Even some of the base metals may be dissolved, but the probability of elimination of the metals by sulfuric acid is not satisfactory. Phenol or furfural treatment is indispensable for aromatic (condensates) rich stocks. Deasphalting of residuums generally eliminate—highly condensed structures. These condensed structures contain metals too, in addition to sulfur and nitrogen bearing resins. Sulfur present in aromatics usually favours more coke formation during cracking.

5.2.1.2 Commercial Cracking Catalysts

The first cracking catalysts were synthetic amorphous silica-aluminas and silica-magnesias. Later on zeolites with rare earth exchanged ions were added to amorphous matrix to improve the catalyst selectivity and activity. Small amounts of noble metals were added to zeolites to assist the oxidation of CO and CO_2 during regeneration. Oxides of nitrogen and sulphur are primarily obtained from feed stocks. Base metals can retain sulfur in the form of sulphate during regeneration and the sulfur is belched off as H_2S during cracking which can be separated easily from products. However, nitrogen poses problem, as there is no way of elimination or suppression except hydrogenation. Heavy metal complexes, like porphyrins increase coke deposits. This can be attended by two methods, either by demetallization of feed or by adding metal passivators like bismuth and sulfur. The following relationship shows when the catalyst is to be replaced due to fouling[80].

$$4 V + 24 Ni + Fe + Cu > 1000 \text{ ppm.}$$

Cracking of hydrocarbons takes place on acid sites. Synthetic silicates are found to possess higher activity and give greater product distribution. Silica by itself has no active sites, but by adding aluminium, acidity begins to rise. Though the amorphous silica-aluminas can be expected to exhibit wide range of active site-acidity and pore sizes, ambiguous structural assemblies create problems for any scientific correlations. Thus the need for crystalline alumina-silicates known as zeolites has begun to increase. The zeolites have all the desired properties like high adsorption capacity, high surface area and acid sites. Basically in all crystalline catalysts aluminium and silicon atoms form tetrahedra, which are linked by shared oxygen atoms. Aluminium atom when forced into tetrahedral configurations negative charges appear on each atom. More than 150 synthetic zeolites have been thus formulated, though commercially important catalysts occupied small spectrum. X and Y zeolites are explored commercially for cracking operation. These are represented by the formula

$$Na_pAl_pSi_{102.p}O_{384}. \text{ g } H_2O$$

where p ranges from 96 to 74 for X

74 to 48 for Y

While 'g' goes from 250 to 270 as aluminium decreases. In zeolites, silica and alumina are the primary building blocks which are arranged at the truncated octahedron known as sodalite cage. Negative changes on tetrahydrally coordinated alumina atoms are balanced by cations which precisely makes zeolite acidic. In case of Na, X and Y zeolites sodium ions are initially exchanged by ammonium ions to reduce the structural collapse which predominates at low P_H values. The exchange takes place in a number of steps after which calcination takes place at 500°C to decompose ammonium ions leaving behind protons in place of sodium.

Thermal stability of zeolites is a major concern. Used catalyst always shows less activity because of the coke deposits, concentration of heavy metals and other acidic groups. During regeneration, steam stripping, followed by exposure to below 800°C is done. Each cycle of regeneration gives some abrasion and increasing concentration of

Bronsted site → Lewis site

after ammonia exchange

undesirable elements. Presence of steam during regeneration decreases surface area which causes loss of active sites leading eventually to a fused non-porous inactive material. In absence of steam, regeneration, if attempted, makes the hydroxyl group from the surface escapes leading to the loss of active sites.

In addition to the physical changes, chemical changes also appear on the surface of the catalyst; Bronsted and Lewis sites are developed. With progress of heating Bronsted acid sites are converted into Lewis sites. However by adding water the lost sites can be regenerated, provided the activation temperature does not exceed 700°C. Zeolites are more resistant to heat and steam than the amorphous catalysts.

5.2.1.3 *Difference between Amorphous Catalysts and Zeolites*

Zeolites have more active sites and adsorption capacity which helps in faster cracking rates. Zeolites have more selectivity for intermediate (gasoline) products. Further, hydrogen transfer is also very high, hence saturation of products at the primary step of cracking takes place. Silica-alumina amorphous catalysts do not transfer hydrogen, so ultimate cracking with unsaturation results. Dealumination of zeolites increase intermediate acid sites and simultaneously reduce the weak sites, thus acid sites fall in a narrow brand. The acid strength depends upon Si/Al ratio, zeolite type, cation type and pretreatment conditions. Commercial catalysts thus possess 100 to 400 M^2/g surface area and pore volume range from 0.2 to 0.5 ml/g with average pore dia. of 50-80 Å. In all such formulation Si/Al ratio is about 4.5, larger pore sizes such as X-Y zeolites can crack big molecules. Zeolites have sieving effect, the pore size is certainly oriented for specific materials only. As an example 5 Å catalysts can crack n-paraffins only. ZSM-5 is suitable for production of aromatics and alkylation reactions.

With the invigorous catalysts formulations FCCs are able to handle high asphaltic, bitumen fractions as shown by Asphalt Residual Treatment Process (ART)[12]. Asphaltenes are decomposed on the fluidising medium to hydrocarbons, thus increasing the yields of liquid products. Even heavy crudes can be handled to form artificial crudes.

5.2.2 Catalytic Crackers—Types—Working

Catalytic cracking is conducted over fixed beds or in moving beds. 20% of catalytic cracking[12] as it stands now is done by moving beds; while 80% of cracking is left with fluidized beds, fixed beds are now obsolete. Pellet form catalysts are suitable for fixed bed operations. While moving beds function with beads. Moving beds, initially, used bucket elevators to lift the catalyst upto disengager but now air lifts have successfully replaced such mechanical designs. Zeolite catalysts[13] in the form of beads, (of Calcium, Manganese and rare earths) are new arrivals in this field since 1965, and the commercial production of these has become immense. Dura bead type catalyst preparations[14] or even zeolite catalysts with active sites crack ten times more than low alumina catalysts or twice as much as high alumina catalysts. Cracking with less deposition of coke on surface of catalyst is an admirable thing with these zeolites. At 70% conversion, zeolites yielded only 6% coke while low alumina catalysts produced twice this amount. Yield of gasoline is also up by 10% with zeolites (see Table 5.5). Zeolites further improve the selectivity characteristics that may result in quality rather than in quantity of products. Gulf's FCC[15] catalysts were developed to suit high temperature operations that usually happens in short contact risers. These produce 7% more yields of gasoline than the long contact risers. Short contact risers are capable of producing gases rich in olefins. Moving and fluidized beds sustain the endothermic cracking reactions with the residual heat left in the catalyst during regeneration. Regenerators and reactors are placed close enough to save the heat energy of the operation. Perfect heat balance in reactor and regenerator system is left with adulation. However, fixed beds have to survive by constant circulation of molten salt from regenerator to reactor, the only method of supplying heat. Obvious advantages of fluidized and moving beds have terminated the expansion of fixed beds. The old units are still continuing, without any advancement.

TABLE 5.5   Fluid Bed Cracker Yields with Different Catalysts.

	Low alumina	High alumina	Zeolites
Conversion %	61	68	74
$C_1^+ C_2^+ H_2$ wt%	1.674	1.672	1.755
C_3% vol.	8.4	8.2	7.4
C_4% vol.	6.2	7.4	9.4
Gasoline % vol.	50.5	58.0	63.5
LGO % vol.	12	14	13
Coke % wt.	6.0	6.0	6.0

Recycling is a common phenomena with some types of feed, and fresh feed to recycle stocks[16] are kept in 1 : 2 to 5 :1.

5.2.2.1 Reaction Variables

Pressure, temperature and catalyst to oil ratio are the usual parameters that take care of cracking. Catalyst to oil ratio is a variable factor, as the activity of catalyst swiftly dimishes with usage. Even after regeneration the carbon content of catalyst does not go below 0.25%; this causes fatigue in catalysts, consequent effect of which is improper cracking. Surface area is another factor; the sharp drop in such active area in equilibrium catalysts, followed by increase in bulk density with cracking time, vastly reduces the conversions. Thus, catalyst oil ratio, catalyst activity and space velocity should be treated as dependent variables in the reactor operation. Physical composition and some important properties of catalysts used in moving beds and FCC units are shown in Tables 5.6 and 7.

TABLE 5.6 : Moving Bed Catalysts[9b]

| Chemical composition | Durabead catalysts (two types No. 1 & 5) | | | |
| | No. 1 | | No. 5 | |
	Fresh	Equili-brium	Fresh	Equili-brium
SiO_2	90		85	
Al_2O_3	9.7		12	
Cr_2O_3	0.15		0.15	
Na	0.1		0.1	
SO_4	0.1		0.1	
Rare earth oxide	–		2.5	
Density gm/cc	0.7	0.78	0.67	0.73
Particle dia. mm	3.5 to 3.6	3.1	3.5 to 3.6	3.1
Surface area sq m/gm	195	85	200	135
Average pore dia Å	72	141	92	110

Reactor temperature and pressure are truly independent variables. In short contact risers (where the time of contact between oil and catalyst lies between 1 to 4 sec)[17], a high improvement in octane[18] quality is achieved due to cracking at higher temperatures. Pressure has no direct impact on cracking reactions. Pressure is often set by the unit capacity. Regenerator pressure is always coupled to the reactor system. The increase in partial pressure of oil causes much coke formation. Catalyst being a fine powder in FCC units excessive regeneration pressure will give problems.

5.2.2.2 Impact of Catalyst to Oil Contact Time on Selectivity

60% of total conversion results in short contact risers where time of contact between oil and catalyst may not exceed 4 seconds. Catalyst separation is done immediately after effluents leave the riser. This is achieved to a maximum extent by decreasing the velocity. The dense bed environments cause back mixing and lead to secondary reactions which contribute excessive coke formations. Avoiding such contacts may be possible by direct admission of vapours into cyclones. In the riser system[19] the activity and selectivity of the catalyst is largely maintained till the cracking is over; and coke does not build up so rapidly in this period.

TABLE 5.7 Comparison of Properties[a] of Fresh and Equilibrium FCC Catalysts.

	Amorphous high alumina		Zeolite XZ-2	
	Fresh	Equili-brium	Fresh	Equili-brium
Chemical analysis (dry basis), wt.%				
Alumina[b]	28	21.5	31	25.4
Carbon	0	0.20	0	0.25
Sulfate	0.7	0.1	0.3	0.1
Sodium	0.03	0.02	0.04	0.04
Iron (ppm)	300	3900	700	3700
Metals (ppm of V+Ni+Cu)	0	162	0	259
Physical properties				
Surface area, m^2/g	415	140	335	97
Pore Volume. cm^3/g	0.88	0.43	0.60	0.45
Av. bulk density g/cm^3	0.39	0.70	0.52	0.68
Particle size, micromesh sieves, wt.%				
0 to 20 μm	2	0	2	0
0 to 40 μm	17	8	19	6
0 to 80 μm	68	69	75	75
Av. particle size, μm	65	63	62	62
Catalytic activity Microactivity, vol. %				
Conversion	60[c]	59[c]	85	73
Carbon factor	1.0	1.1	0.6	0.6
Hydrogen factor	1.0	1.1	0.2	0.7

(a) By courtesy (Marcel Dekker inc. New York) taken from "FLUID CATALYTIC CRACKING with zeolites" by Paul B. Venuto, p. 46.
(b) Equilibrium values are low due to presence of some residual low alumina of long age.
(c) After 24 hr/1050°F/60 psig, 100% steam.

5.2.3 Catalytic Cracking Processes

(1) Fixed bed, Houdry.
(2) Moving beds—Air lift—Thermofor catalytic cracking Houdri-flow.

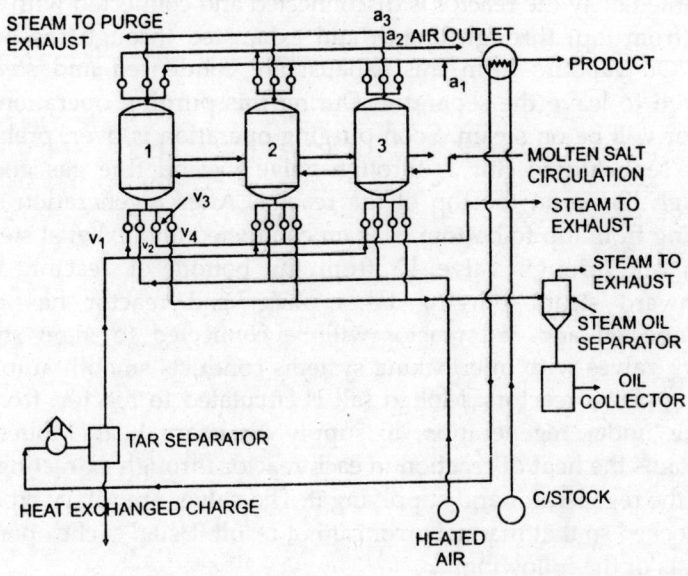

Figure 5.4 Houdry Fixed Bed Process.

(3) Fluidised Beds: (a) Gulf Research (Fluid catalytic cracking) FCC, UOP, Texaco.
   (b) Kellog (Ortho flow)
   (c) ESSO—Flexi cracking
   (d) Standard oil—Ultra cracking.

5.2.3.1 Fixed Bed Crackers

Figure 5.4 illustrates Houdry Fixed bed process. At present, though no new plants based on this technology are being set up the old ones still continuing. There are three reactors in parallel (1, 2 & 3), in this system. At the bottom of each reactor there are four connections (V_1, V_2, V_3 and V_4) provided through a rotary disc. Only one line at a time is to be connected to the reactor; when the first reactor is on process stream, 2nd reactor will be on steam purge followed by air purge, and the 3rd reactor will be in regeneration process. At the start, charge oil vapor, free from tar, will enter the reactor through valve V_1. After cracking, the vapours escape through the reactor outlet as product and go to the fractionator through line a_1. (At the top of the reactor also; same type of rotary arrangement is there, but connects only three valves; these values connect line a_1 for products; a_2 for air outlet and a_3 for steam purge alternatively.)

Immediately the reactor is disconnected and connected with steam line (from top) through line a_3 and exhausted through the bottom (V_4). Oil vapours from this exhaust are condensed and steam is allowed to leave the separator. During this purging operation, 2nd reactor will be on steam. Soon purging operation is over, preheated air is sent into reactor 1, through valve V_2 and flue gas goes off through line a_2 at the top of the reactor. After regeneration steam purging from top to bottom is again conducted and exhaust steam is taken out through valve V_3 from the bottom of reactor. When downward steam purging takes place, 2nd reactor has to be disconnected and 3rd reactor will be connected to main stream. Rotary valves with interlocking systems conducts smooth automatic cycling of the reactors. Molten salt is circulated to reactor, from the reactor under regeneration, to supply necessary heat. Molten salt maintains the heat of reaction in each reactor through extracting heat from the regenerator and supplying it. The valves are rotary type and interlocked so that manual error cannot result. Usual cyclic operation consists of the following:

1. On stream 10 minutes.
2. Downward purging of steam to recover hydrocarbons— 5 min.
3. Regenerations by blowing hot air through bottom of the reactor — 10 min.
4. Air purge followed by steam purge downward— 5 min.

It can be seen that on stream will be equal to regeneration period; which will be equal to the sum of steam purging periods, before and after regeneration.

Operating conditions

Reactor temperature — 450 to 520°C

Operating velocity (gauge) — 1 to 2 kg/cm^2

Space velocity (Liquid) — 1 to 1.75 (1/Hr.)

Regeneration temperature — upto 520°C

Air pressure, during regeneration — upto 4 kg/cm^2

Catalyst shape — Beads/pellets

Composition — Silica-alumina

Catalyst processing capacity — 20 to 40 Bbls/kg.

Vapour inlet temperature — 430°C.

5.2.3.2 Moving Bed Crackers

Thermofor catalytic cracking units fall into this category. Catalyst shape in these units may be spherical or beads. The shape and size of the catalyst permit free flow of catalyst in the reactor. Catalyst and oil move cocurrently from top to bottom in the reactor. Reactor is stacked directly over the regenerator as shown in Figure 5.5. In between reactor and regenerator there is a constriction, where steam purging is done. This helps firstly in removing hydrocarbons from the surface of catalyst, secondary steam prevents the reactor products from going downwards into regenerator; which is highly essential otherwise the situation may become unwarranted.

Spent catalyst is regenerated by air blowing. The flue gas passes through cyclone separators, leaving catalyst dust, if any, in the out going stream. At the bottom of regenerator, there is a cooling section to cool the catalyst. Such regenerators shall be in two numbers, kept below the reactor. The catalyst collected from the regenerator is lifted

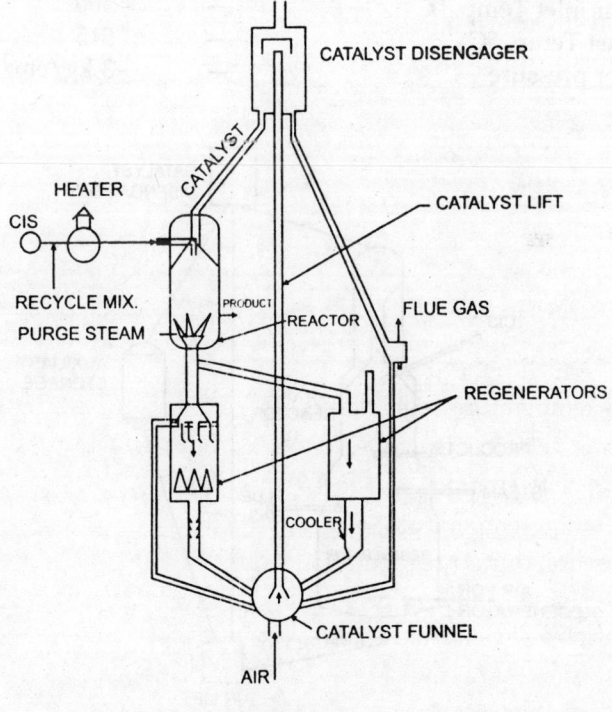

Air lift T.C.C.

Figure 5.5   Air Lift Thermofor Catalytic Cracker.

into catalyst disengager, placed at a height more than the reactor, which automatically feed the catalyst into reactor due to gravity. Old designs employed bucket elevators to lift the catalyst, while modern practice is to use air. Thus the name Air lift Thermofor catalytic cracking has been accomplished. Catalyst from both the regenerators is conveyed into a single funnel, from where the pneumatic conveyance can take place. Feed and catalyst always travel downwardly in the reactor, and products escape from the side of reactor. Proper distribution of catalyst for good performance is essential. The disadvantage of this process is in low catalyst to oil ratio which increases the coke deposition on catalyst.

Operating conditions

Regeneration Temperature	—	600°C
Recycle ratio	—	0-0.5 Vol./vol.
Catalyst to oil ratio	—	2 to 7
Space velocity	—	0.7 V/hr. V.
Catalyst shape	—	beads
Vapour inlet Temp. °C	—	500
Catalyst Temp. °C	—	515
Reactor pressure	—	3 kg/cm^2

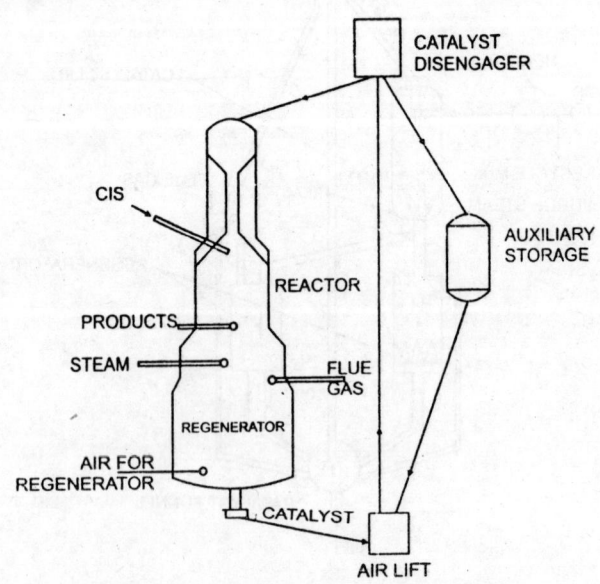

Figure 5.6 Houdri Flow Catalytic Cracking.

5.2.3.3 Houdri Flow Process

Houdri flow process is distinguished from the TCC process in its (a) ease of operation (b) low cost of maintenance (c) high catalytic activity (low catalyst make up costs) (d) flexibility in handling charge stocks and conversions attained.

In this design, reactor is placed above the regenerator, in a single shell. The regenerated catalyst (either natural or synthetic type) is carried by pneumatic lift to disengager, from where the catalyst can either go into reactor or auxiliary storage bin. The system works on self heat balance, so no additional heating or cooling of catalyst is anticipated. Usually operations[20a] are conducted in temperature range of 450-520°C, at pressure 1.5 to 2 atmos; with a space velocity of 1.5 to 4 V/hr./V. Catalyst to oil ratio of 3 to 7 is maintained with high recycle ratios of 1:1. A conversion of 90% can be obtained with recycling, while once through cracking permits a conversion of 60%. The octane value of gasoline lies in the range of 85 to 90. Yield can be improved by using zeolite catalysts. Fig. 5.6 shows the flow diagram of the above process.

Houdri flow yields[21]

Naphthenic gas oil	BPSD	API		
Charge	4200	32.6		
Recycle	4200	31.7		
Total	8400	32.0		
Products			Vol %	Wt%
C$_2$ & Lighter	–	–	–	3.9
C$_3$	306	–	7.3	4.4
C$_4$	714	–	17.0	11.5
Gasoline (FBP 215°C)	2705	59.7	64.4	55.2
LGO	521	31.0	12.4	12.5
HGO	200	–	4.8	4.9
Coke	–	–	–	7.6

5.2.4 Fluid Catalytic Cracking (FCC Complex and Hardware)

Fluid catalytic cracking enjoys an impregnable position in catalytic cracking operations. All types of fractions from naphtha to deasphalated oils can conveniently form the feed stocks. Numerous designs of FCC units have appeared in the literature[22a]. The FCC complex usually consists of three major sections:

(a) Reactor-riser

(b) Regenerator-flue gas separation

(c) Distillation and recycling.

Associated with the cracker, other units like polymerisation, alkylation are maintained for gaining economy.

Reactor riser design had been subjected to number of changes since its inception and still better types of risers are being introduced. The function of the riser is in fact to bring a thorough contact between vapor and catalyst in a short time[22b]; short contact times are less th n 5 seconds. In the reactor, the cracked oil vapours escape through a perfect separation system, that is an inbuilt cyclone system. Disengaged catalyst from cyclones falls into reactor and is steam stripped. The catalyst is then regenerated after burning off the coke in a separate unit, i.e. regenerator. Fluidisation is carried out in dense phase and dilute phase. Dense phase is prevalent in the bottom portion of the reactor, while dilute bed prevails in the upper part of reactor.

Cracking in risers

From the moment of contact between oil and catalyst at the point of introduction in the riser, the complexity of reactions go on increasing as the material moves through the riser.

With increasing cracking, deposition of coke, degradation in catalyst[23] activity and a sharp increase in velocity of vapour are some inherent complications. Though the increase in velocity of vapour is regarded as a benefit to fiuidization, the relative velocity (slip velocity) between catalyst and oil decreases, and the result is improper cracking. In the Figure 5.7 the profiles of various parameters with respect to height of riser is shown.

Slip velocity is defined as the velocity of vapor divided by velocity of catalyst[24].

$$\text{Slip velocity} : \frac{V_V}{V_C}$$

where,

V_v velocity of vapor

V_c velocity of catalyst

The terminal velocity of catalyst particle may become the difference in these two velocities hence the slip velocity can be rewritten as:

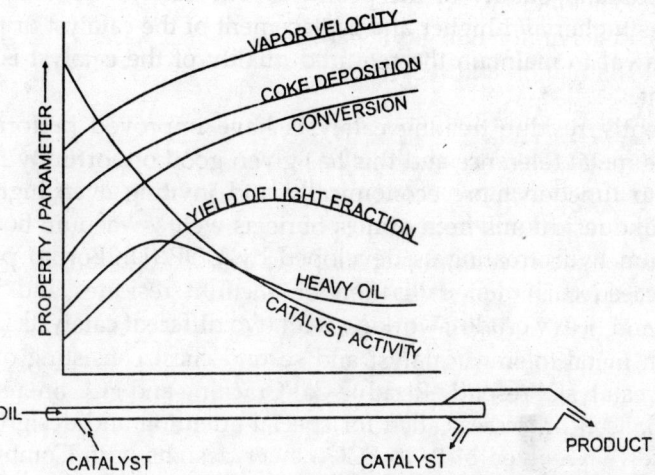

Figure 5.7   Profiles of Different Parameters in a Riser Cracker with Distance.

Slip velocity : $\dfrac{V_V}{V_V - V_f}$

where,

V_f is terminal velocity of particle

V_v velocity of vapor

Generally the slip velocity should be of the order 1.5. Each configuration has its own riser design, assembly of reactor, regenerator and fractionator.

An ideal addition and recovery system for fluid catalytic cracker would be one, which minimises additions of fresh catalyst and maintains an optimum size distribution of circulating equilibrium catalyst. In fact, loss, frequency of withdrawal of catalyst (equilibrium) are all functions of quality and quantity of fresh catalyst added. Factors affecting the FCC operations[25] are usually, shrinkage, attrition, entrainment in cyclone systems and regeneration systems.

Efficient feed injection system to take care of pressure drop and erosion of the riser with low pressure drop in atomization, (drop size of 50 microns) is very essential. Finally quick separation of products from catalyst to prevent over cracking is augmented by efficient steam stripping.

FCC feed is conventionally is of high boiling range thus the metallic as well as other harmful contaminants such as asphaltenes, porphyrines are high. These contaminants obviously adversely affect the yields and quality of the products. The catalyst contamination becomes higher and higher and replacement of the catalyst or partial withdrawal to maintain the required quality of the catalyst is more frequent.

Presently residue treating catalysts have improved performance towards metal tolerance and this has given good opportunity for the FCC's to function more economically and inviting even high feed stocks like residuums from atmospheric as well as vacuum bottoms. Residuum hydrotreating as developed by UOP (UniBoron) process has succeeded in demetallisation of vacuum residues and heavy blends and heavy crudes; working with two different catalysts i.e. one for high metal tolerant catalyst and second chest consisting of high activity catalyst. Presently Residue Cat Cracking and FCC are referred as specialised processes called for special attention and design. RCC Processes have given birth to FCC newer designs with Combustion Style designs and alternatively RCC style designs. These technical improvements have become very versatile in industry since 1983. In both the designs the cracking is facilitated in open risers while catalyst regeneration is done in combustion style or in two stage activation. In both the cases the regeneration of catalyst is done at low temperatures along with cooling of catalyst. These two have improved the performance of FCC reactors.

Figure 5.8 shows the outlines of FCC unit designed by Texaco Development Corporation. In this design separate riser crackers are

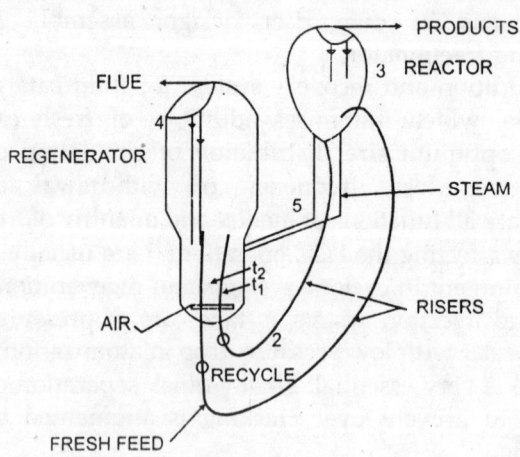

Figure 5.8   Fluid Catalytic Cracking.

detailed for cracking fresh feed and recycled feed. In the reactor the variation in dense phase and dilute phase conditions are supposed to be added advantages. Efficient reactor and the cyclone separator system furnish the necessary implements to process feed stocks of different types.

Preheated stock is mixed with catalyst at the bottom bend of the fresh feed riser (1). The products and catalyst enter the reactor space (3) in dilute phase, the catalyst disengages from vapours. The vapours are freed from catalyst dust in cyclone system. Recycle stock is permitted to enter through the other riser (2). It mixes here with regenerated catalyst and goes into reactor space. The fresh feed riser directs the products downwardly in the reactor, while recycled stock riser delivers vertically into the reactor space. Vapours from both the risers join together and go for fractionation through cyclones. Cyclones are situated in the reactor and regenerator, so that the catalyst dust can be caught.

The spent catalyst from the reactor after steam stripping passes into regenerator (4) through a pipe (5). A blast of air burns off the coke on the catalyst particles and flue gas formed escapes through the cyclones into atmosphere. Regenerator in this design is provided with two channel pipes (p_1 & p_2). These pipes are joined to risers. Through one riser, while fresh feed enters through the other one, recycled stocks enter, as mentioned already.

FCC catalysts crack 2% outside the catalyst pores remaining inside the pores. This shows the importance of pore architecture. FCCs are revamped for short contact times. Optimising catalyst accessibility is needed to allow pre cracking of heavy molecules. AKZO NOBLE[85] is launching an innovative break through FCC catalysts which is focused on cat-particle arrangement namely JADE and TOPAZ. These new assembly of building blocks are suitable for high metallic heavy hydrocarbons which encourage coke formation.

FCCs and riser crackers extensively use catalysts of variable pore size and active acid sites. Zeolities have been used extensively for this purpose. Too much activity of catalysts may results in over cracking leading to excessive coke deposition. Even continuous regeneration of the catalysts may not always attend the problem of efficiency of the catlyst. Usually catalyst can inform conversion, product distribution or selectivity and regeneration efficiency. The bigger pore sizes decrease the surface area of the catalyst, however increase the regeneration efficiency of the catalyst by successully buring of the coke. The mesoporosity is good characterization factor for FCC catalysts.

5.2.4.1 Flexi Cracking

Flexi cracking, developed by ESSO, offers two reactor configura-tions: and gives operating flexibility and expansion capacity.

Figure 5.9a shows riser cracker model. In normal operations, the riser cracker operates with no dense phase bed in the reactor. In this model high space velocities of 50^{26} (weight/hr/weight) are maintained to enhance the advantages of zeolite catalysts. Maximum liquid products and least problems in handling different feed stocks are considered to be the major benefits of this assembly. The riser cracker can work on variable space velocities ranging from 50 to 5 (V/

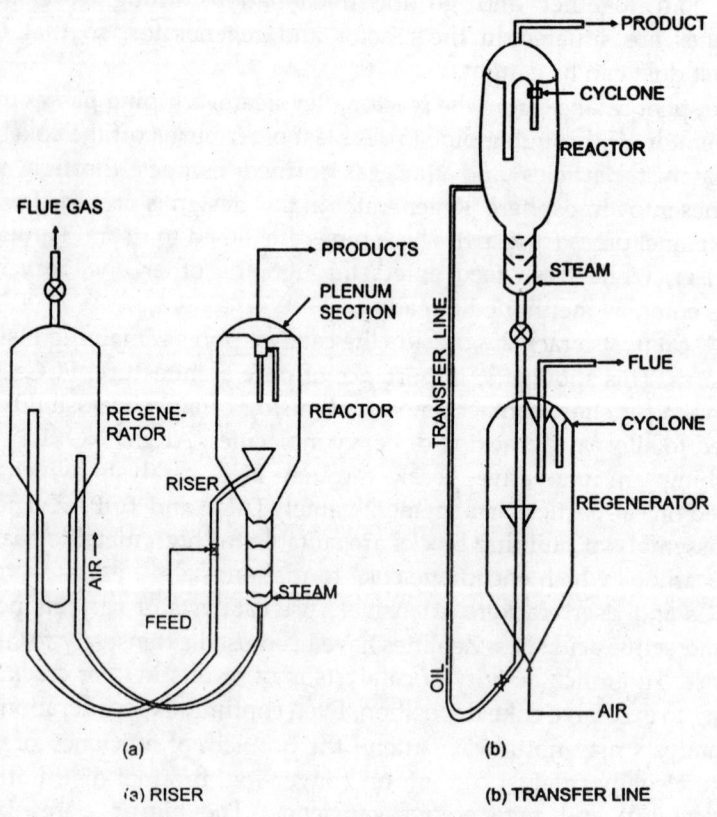

(a) RISER

(b) TRANSFER LINE

Figure 5.9   Flexi Cracking Models.

hr/V). The cyclones in the reactor are designed to provide plug flow of catalyst without back mixing the oil with catalyst. The latest modification of U bend design differs from the previous one, which abruptly changes the direction of catalyst and flow lines. Side by side reactor and regenerator placement furnishes smooth control in catalyst circulation. Model IV removes the disadvantages of the previous ones by increased recovery in catalysts and free switching of naphtha to reduced crudes. The overall yield of gasoline may be of the order of 70%, with low gas yields. The percentage of olefin content in the gases can be increased to as high as 55% by proper adjustment of the variables.

The second configuration Fig. 5.9*b* is pure transfer line configuration. In this design all the cracking takes place in transfer line, and the reaction is rapidly terminated by disengaging oil from catalyst in a cyclone separator. High to low space velocities can be worked out in this model depending upon the quality of feed. These two configurations can cover adequately the needs of a refinery either for expansion or switching to process different feed stocks. These two configurations over about 2,000,000 bpsd or 25% of the free world capacity.

Silica-manganese type catalyst or natural catalysts with activated earths are also tried in the above configurations for better conversions.

5.2.4.2 Ortho Flow Reactor

Ortho flow riser reactor developed by Kellogg is highly selective in converting heavy petroleum fractions to valuable light olefins and furnace oils in addition to gasoline.

Figure 5.10 shows the ortho flow reactor. It covers the specific needs of the refiner in converting low value feed stocks to high value products, with less investment and operating costs. Variety yields can be obtained by sacrifying efficiency. The cracking takes place in the riser[27] part of the reactor.

The fresh feed enters into external riser at the point where it becomes vertical, as shown in figure. Catalyst fluidised by steam from the stripper enters the riser. Oil and catalyst mix in this riser, where cracking takes place. Cracked products now enter the disengager, where the catalyst is separated. The products escape through a cyclone. The catalyst is stripped off with steam. Hydrocarbon free

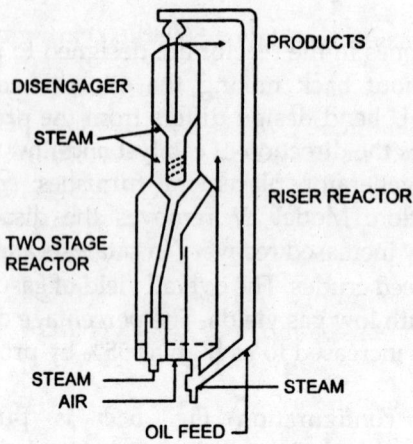

Figure 5.10 Orthoflow Reactor.

catalyst then enters the regenerator, where it is regenerated by blowing air either in one or two stages. After regeneration the catalyst is steam stripped and fluidised to enter the riser cracker to repeat the cycle. The riser part may be kept in the reactor-regeneration combination itself when it is called internal riser model.

Operating[20b] conditions

Temperature	470-500°C
Pressure	1.5 to 2.2 atm
Regenerator temperature	580-610°C
Pressure	1.3 kg/cm^2
Cat/feed	2.5 to 5 kgs/bbl.

In India catalytic cracking practice is not very much. Approximately 3% of the crude refined is used for this purpose.

There were two units of FCC, situated one in Bombay and other in Vizag Refineries (Hindustan Petroleum). One of these two is designed to crack heavy distillates of Iranian crude (350 to 550°C).

The following data belongs to the same unit.

Cat. cracker operating pressure 1.43 Kg/cm^2

Feed Temperature	315.5°C
Cat. hold up	15 tons in cracker
	19 tons in regenerators

Particle size 40-80 — 70 to 80%

(Microns) 0-40 — 5 to 7%

Surface area of the catalyst 500 to 600 m^2/gm (fresh)

	200 m^2/gm (used)
Cat. circulation	8 tons per hour
Capacity	7,500 BPSD
Reactor temperature	465—510°C
Oil feed temperature	315—455°C
Regenerator temp.	593—675°C
Dense phase temp.	660°C
Plenum chamber temp.	680°C
Spent riser temperature	470°C
Cat loss 1.4 tons/day	
Coke yield 3.8%	

Steam rate at 9 Kg/cm^2 with 20° superheat.

Injection	370	⎱ Kg/hr
Stripping	1260	⎰
Process	150	

Yields

Gas	1.9%
LPG	13.0%
Gasoline	18.8
Kerosene	16.0
Light gas oil	11.6
Heavy gas oil	26.4
Tar	12.3

There were three FCCs during 1950-60 ESSO Model IV two units and Shell model one unit. But by 82-85, 6 FCCs of UOP design were inducted into our refinery scene. 10th FCC unit at Panipat is based upon Residue cracking unit supplied by Stone Webster. The residue cracker generally can operate on high carbon residue content upto 7% and can still work on with NI + V up to 40 ppm.

5.3   CATALYTIC REFORMING *INTRODUCTION AND THEORY*

With the advent of I.C. engines of high compression ratio, the quality of fuel had to undergo changes. Prevention of knocking under such compression ratios, is achieved by increase in the octane value of the fuel. Upgrading low octane gasolines catalytically is known as catalytic reforming. Thermal reforming can also be practised, but not desirable due to the inherent disadvantages. The octane rating improvement is accomplished chiefly by reorienting or re-forming the low octane components into high octane components. Much desired reformate is influenced by the characteristics of feed stock and catalyst. The reactions, reckoned of general occurrence are given below:

(a) Dehydrogenation (Saturates to unsaturates)

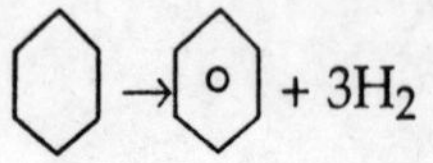

Cyclo hexane $\longrightarrow$ Benzene

$$RCH_2.CH_2.CH_2.CH_3 \longrightarrow RCH_2.CH_2.CH = CH_2 + H_2$$

n-paraffin   Olefin

(b) isomerisation

n-Hexane (32) $\longrightarrow$ 2 methyl pentane (66)

n-pentane (63) $\longrightarrow$ isomerisation (90)

(c) Paraffin cracking (chain cracking)

$$C_8H_{18} \longrightarrow 2C_4H_8 + H_2$$
$$\longrightarrow C_4H_8 + C_4H_{10}$$

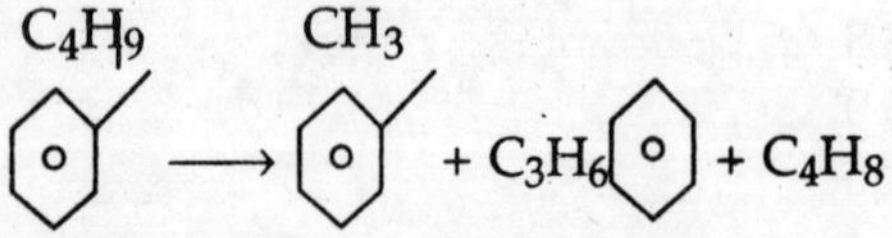

Hydrogenation of Unsaturates

$$C_4H_8 + H_2 \longrightarrow C_4H_{10}$$

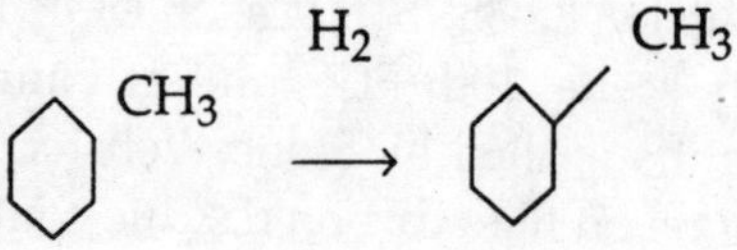

Methyl cyclohexene $\longrightarrow$ Methyl cyclohexane

Paraffin dehydrocyclisation

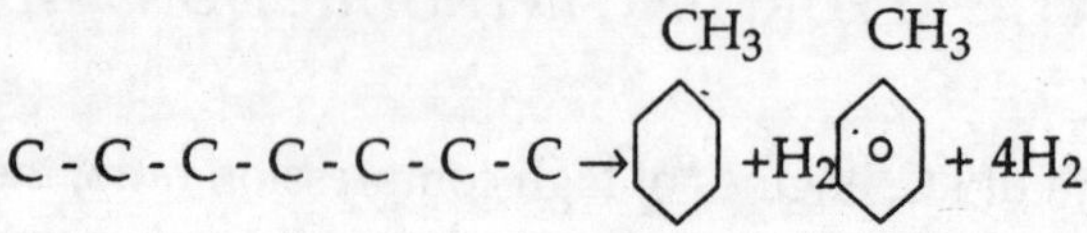

$$C - C - C - C - C - C - C \rightarrow \text{(ring)} + H_2 \text{(ring)} + 4H_2$$

Naphthene isomerisation and dehydrogenation

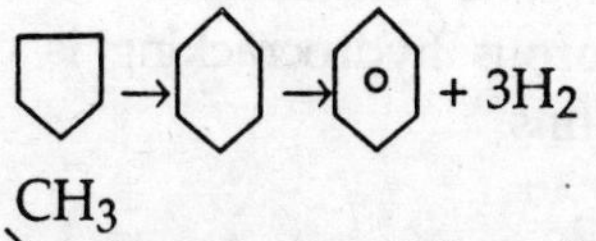

$$\square \rightarrow \bighexagon \rightarrow \hexagon + 3H_2$$

$\underset{\text{Methylpentane (91)}}{\overset{CH_3}{|}}$ (103)

(Figures. in brackets show octane no.)

Hydrodesulfurisation

$$\text{[thiophene]}_S + 4H_2 \longrightarrow C_4H_{10} + H_2S$$

$$RSH + H_2 \longrightarrow RH + H_2S.$$

Denitrogenation

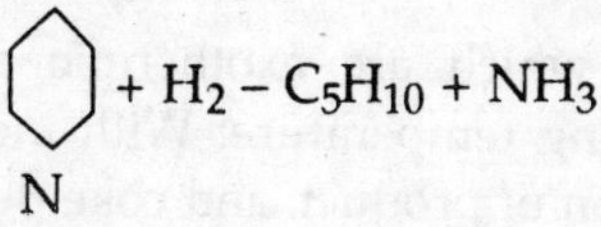

$$\underset{N}{\bigcirc} + H_2 - C_5H_{10} + NH_3$$

De oxidation

$$\overset{OH}{\hexagon} + H_2 \longrightarrow \hexagon + H_2O$$

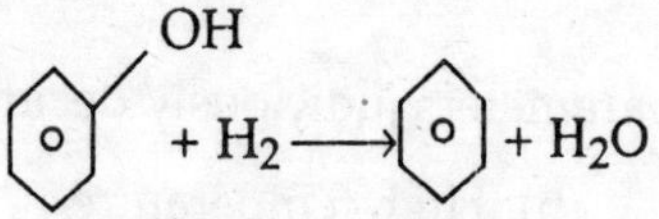

Combination reactions.

$$RSO_2 + H_2 \longrightarrow RH + SO_2 + H_2O$$

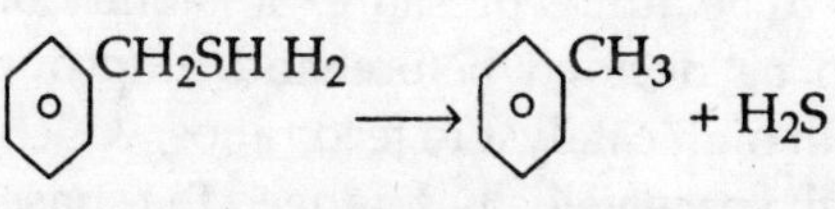

$$\hexagon^{CH_2SH\ H_2} \longrightarrow \hexagon^{CH_3} + H_2S$$

5.3.1 Reactions-Conditions

All the above reactions are likely to occur to varying degrees. Isomerisation and dehydrogenation reactions are fast.[28] Cyclisation reactions are very slow and require severe conditions. Hydrocracking and other elimination reactions may be treated as slow reactions. However isomerisation and cyclisation reactions are the best contributors of octane number. The proper and worthy reaction is dehydrogenation and cyclisation of n-paraffins; that results in a tangible improvement in octane rating (n-heptane of zero octane rating-to more than 100 octane number).

Cracking reactions produce more lighter components which have better octane quality than the heavier ones; yet the reactions are not opted due to the fear of excessive coke formation on catalyst and unsaturation in the products. Vigorous hydrocracking is desirable when the feed stock is rich in paraffins.

5.3.1.1 Effect of Pressure

Low pressure encourages dehydrogenation reactions, while no noticeable effect of pressure on isomerisation may be expected. Coke formation is more at low pressures. Increase in pressure causes dealkylation very much. Hydrocracking and elimination reactions are directly related to the partial pressure of hydrogen.

5.3.1.2 Effect of Temperature

Except hydrogenation reactions, which are exothermic all other reactions are favoured by increasing temperature. With increase of temperature, chances of degradation of product, and coke deposition are likely. Hence for smooth operations, the amicable parameters instituted in a catalytic reformer must embrace all the above considerations.

For economic operations the parameters judiciously decided are

(a) Low pressure   (b) High temperature
(c) High hydrogen circulation.

With increased coke deposition at low pressures, the reactors are preferably operated at 10 to 50 atmos pressure. Suitability of catalyst for such operations is to be decided beforehand. Improvement of octane quality is rapid with high catalyst to feed ratios, which restricts the operation to go for high space velocities. Indeed the space velocity does not exceed more than $3(V/hr/V)$.

The temperature above 500°C promptly spoils the activity of catalyst. A temperature range of 450-500°C is frequently advisable and can bring about all reactions without much severe coking problems. All dehydrogenation reactions produce good amount of hydrogen; hydrogen is consumed to some extent in hydrogenation reactions, the remaining is available as by product. High circulation of hydrogen suppresses the coking tendency too.

In conclusion it may be stated that faster reactions will tend to predominate in short residence time operation, which results in maximum aromatics and hydrogen production. High temperatures give rise to high aromatic content. Naphthenic stocks with high hydrogen to hydrocarbon ratio are liable to affect adversely the octane number of the product.

5.3.2 Catalysts

In reforming many reactions are going together; hence a catalyst to maintain its activity towards all reactions equally is rare. Haensel[29] developed platinum catalysts and these are still being used with slight or no modifications. Acid sites on catalyst are responsible for isomerisation and cyclisation reactions; while metallic bonds favour hydrogenation reactions. These two functions have to be met by a catalyst. For this reason acid sites incorporation on platinum catalyst is unavoidable; eventually halogenation (upto 0.1%) has come into service. Platinum Catalysts are graded by the amount of platinum deposited on the base. Low platinum catalyst are non regenerative type. These contain 0.3% of platinum on silica or silica-alumina. High platinum catalysts contain 0.3 to 0.5% metal and form regenerative catalysts. The cost of platinum prohibits its trials in fludised reactors. Moving bed units are however in operation but their incessant use is doubtful.

Isomerisation is one of the simplest methods followed for different purposes; however increasing octane number of gasoline is considered to be the main function, n-paraffins by isomerisation and even in iso-paraffins by changing the side chain or group placement position the octane number can be increased significantly. However a common pool of gasoline fraction when subjected to increase the octane number, it is aimed for a gain of 15 to 20. New developments are based upon Zeolites and SAPO Catalyst. Infact the energy requirement is not much in the process.

Suitable pore size and acid sites are required for the operation. In the reforming operation where platinum formulations are used, not only aromatisation but significant changes in molecules take place. Recent advances in catalysts for reforming reactions as well as FCCs modified ZSM-5 catalysts are cited. The modification with respect to acid sites and structural modification with metals has brough much interest in these formulations for valorization of light napthas or NGLS. n-paraffins available in these fractions are cyclised by these catalysts.

ZSM-5 catalysts are shape selective. These with certain modifications in both pore size and acid sites, can bring about reactions of isomerisation (Xylene), disporportionation (Toluene), and alkylation reactions or aromatics SAPO series are some kind of modified zeolites which are extensively used in alkylation reactions of alcohols with aromatics.

Other catalysts suitable for such operations are chromium oxide and molybdenum oxide on alumina. These are very good catalysts for isomerisation and hydrogenation reactions. Chromium oxides and molybdenum oxides on silica or alumina or silica-alumina are quite abrasive resistant and can be used in moving beds and fluidised beds.

Approximate operating conditions with different catalysts is presented in Table 5.8a.

High platinum catalysts are frail to poisons and occasionally contribute adverse reactions like ring opening. Houdri forming process employs platinum catalyst in fixed bed reactors, while platforming may be conducted in semi-regenerator or cyclic regenerator reactors. Cyclic regeneration is periodical and is required at an interval of 24 to 24 hrs. During this regeneration time, a spare reactor is switched into stream; thus three reactors will be on stream while the fourth one will be in regeneration. Continuous Houdri forming[30] reactions are conducted on bimetallic catalysts, platinum and rhenium compositions. These catalysts are able to resist coke formation. Semi-regeneration requires a long time before regeneration is attempted, usually a time of 3 to 6 months operation renders the catalyst ineffective.

5.3.2.1 Feed Stock Selection

Reforming helps in boosting octane value by converting most of the n-paraffins to aromatics. Hence this process can be applied to produce aromatics. Stocks that are suitable for gasoline engines are chosen as the feed material for reformer and straight run gasolines of low octane number mainly constitute the bulk of feed. Light products obtained in various cracking operations in gasoline range are also suitable. The feed stocks in gasoline range have the boiling point spread from 35 to 180°C. Upto C_6 carbon atom (i.e. 35 to 78°C) the fraction by its nature possesses good octane value and further attempt to improve octane number is not worthy. In fact this fraction enjoys a free ride in the reactor at the prevailing conditions. A minimum of six carbon atoms is essential for production of simplest aromatic ring. So, depending upon the highest aromatic compound (usually xylene), the carbon atoms range in the feed may be fixed. In other words, for production of benzene, toluene, xylene, six to eight carbon atoms must be present in the feed. Virgin or cracked gasolines, invariably contain impurities

TABLE 5.8a Approximate Operating Conditions with Different Catalysts.

Process	Thermal Reforming	Molybdenum Catalysts		Moving bed	Platinum Catalysts		Other Catalysts	
		Fixed bed	Fluidised bed		Non-regenerative	regenerative	Thermo-for	Cyclo-verson
Catalyst composition	None	10%MoO$_3$ on Al$_2$O$_3$	10%MoO$_3$ on alumina	Cobalt molybdate	Pt and halogen on Al$_2$O$_3$ max. 0.3% pt.	0.5 to 0.8% Pt on Al$_2$O$_3$	Cr$_2$O$_3$ on Al$_2$O$_3$	Bauxite
Capacity of catalyst (life)		about 10 months	10 bbls per Kg		300 bbls per Kg	300 bbls per Kg	5 bbls per Kg	–
Reaction conditions (T°C)	500-600	500-580	480-560	475	450-500	450-510	500-530	500-580
Pressure (atmos)	20-75	10-20	10-25	25-30	20-50	20-30	10-20	2-10
Spece velocity (1/m)		0-5	0.3-1.0	1.0	1.0	1-5	0.7	1-7
Contact time sec.	10-20	15		0-5				
Gas circulation H$_2$: oil mole ratio.								

like sulfur, nitrogen and oxygen compounds, and their presence is more if crude is naphthenic. Feed stocks for reformers must be free from these compounds, preferably sulfur[31a] and nitrogen compounds should be kept as low as 5 ppm. Hydrodesulfurisation of feed stock assures the required purity. Sulfur removal not only helps in reducing corrosion, but also coke deposition. Specially for platinum catalysts, feed stocks should not contain more than 5 ppm of nitrogen and 10 ppm of sulfur.

UOP Refining Catalyst R-132 is useful for:

1. Maximum hydrogen production
2. Reduction of reformate benzene
3. Highest gasoline yield
4. Lower RVP
5. Less operating costs.

5.3.2.2 *Operation with Different Catalysts*

About 90% reforming operations are conducted in fixed beds using platinum catalysts. Moving beds and fluidised beds mainly use cheap catalysts of molybdenum and chromium compositions. Fluid hydroformers[32] employ catalyst molybdena on alumina. Thermofor Cat. forming[33] is done in moving beds using chromia on alumina as catalyst. These catalysts greatly help in removing sulfur to almost zero presence, followed by low yields of butane. The temperature can go up to 525°C. Bauxite and clays have also been utilised as catalysts, but at present much significance is not attached to these catalysts. One difference between these clay catalysts and platinum catalysts is in the production of hydrogen. Not much hydrogen is produced with clay[34] catalysts hence none of the products is recycled. Usually in cat forming operations[35] 70 to 350 M^3 of hydrogen per cubic meter of feed stock is produced. Platinum combinations have been found to yield less light ends, (propane, butane and pentane) compared to conventional platinum catalysts. It is interesting to note that the yields of these components is linear and inversely dependent upon the yields of reformate. Usually the total amount of such light ends produced in reforming operations (platinum catalysts) falls from 15 to 5% while reformate yields increase from 75 to 95%.

A correlation presented, by R.R. Jacob[31b] is very interesting on the yields of potential aromatics (Ap) in reforming operations, which is given as

$$Ap = \left(\frac{0.875}{\rho}\right)(0.85N + A)$$

where,

N = % volume of naphthenes in feed

A = % volume of aromatics in feed

ρ = specific gravity of feed at 15.6°C

At lower pressures increased yields of aromatics and hydrogen are experienced.

5.3.2.3 Catalytic Reforming (Platforming)

Catalytic reforming process employs platinum and other combinations as catalysts. Hydrotreated naphtha (68-180°C) is mixed with recycle hydrogen and sent through a furnace. The outlet temperature of the vapours shall be around 500°C. Hot vapour is sent into three catalyst cases (reactors) in series. Old designs may enjoy four reactors is series: the last one is known as swing reactor. Swing reactor comes into main stream when regeneration is attempted. Pressure in reactors is kept in the range of 15 to 75 atmos. Except hydrogenation reactions, which are less prevailing, all other reactions require heat, this invites a heater to substantiate heat of reactions and is placed in between two reactors. Cyclisation reactions are pronounced at severe conditions, hence the last reactor operates generally at higher temperature by 20 to 30°C from the preceding one. Catalysts are made into pellets of 3 mm size. Halogenation is prescribed whenever regeneration is carried out. The operation is shown in Fig. 5.11a. After the reforming is over the effluents are cooled and gas is separated. High pressure separator removes recycle gases mainly hydrogen, while low pressure separators relieve most of the light fractions. Hydrogen is usually produced to an extent of 2.5% by weight of feed. High recycling ratios prevent coke deposition and the space velocity is kept below 2 in all these operations. In regenerative processes, the space velocity goes upto 5. Continuous operation as developed by UOP is classified as platforming operation, and is presented in Fig. 5.11b. As the name suggests, continuous regeneration of catalyst is achieved here. Further, continuous rates and constant quality production of reformate of high octane value and production of LPG from naphthas are some remarkable advantages in this system.

Arrangement of three stacked reactors one over the other forms the main difference in reactor arrangement. Feed and catalyst travel cocurrently in all these reactors from top to bottom. Heating the reactants in between the reactors remains unchanged. The catalyst from the bottom of the third reactor is transferred to separate regenerator. After regeneration the catalyst is transferred to the first reactor continuously.

Englehard Magnoforming employs platinum on bimetallic catalysts. The rugged nature of the bimetallic catalysts permit the operation to be conducted at much severe conditions and also tolerate sulfur up to 20 ppm.[36] Institut Francais due Petrole has formulated number of catalysts of different platinum contents, to reduce the cost of catalyst and at the same time permitting the operation at much severe conditions. Also these catalysts can work actively under pressures less then 20 atmos. This leaves an advantage to the operators in choosing low pressure or high pressure or combination of these two depending upon the necessity.

5.3.2.4 *Iso Plus Houdri Forming*

This is a combination process, laces cat. reforming with thermal reforming. In another combination reformate extraction plant is

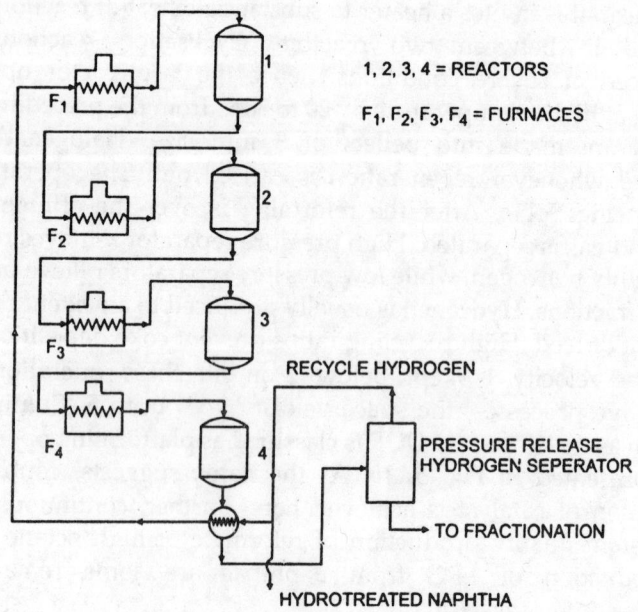

Fig.ure5.11a   Catalytic Reforming.

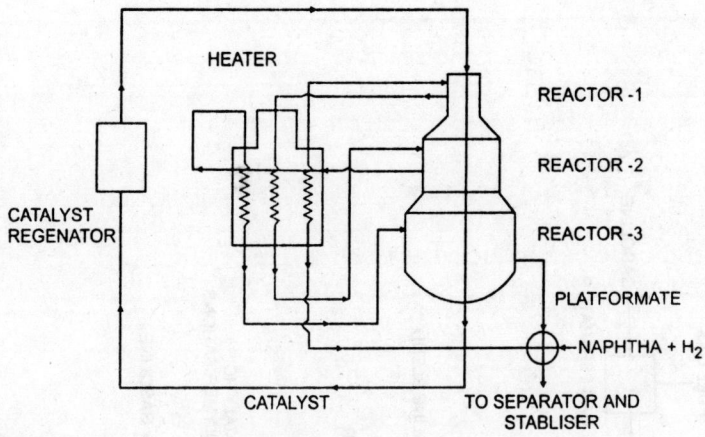

Figure 5.11b Platforming.

provided to remove aromatics, as this helps in recycling paraffins rich raffinate. This process instils the polymerisation unit for C_3 and C_4 off gases. No drastic pretreatment of feed is essential as the feed first goes through a guard case that helps in removal of sulfur and nitrogen impurities. After guard case, the feed passes through an array of reactors and Houdri reformate is obtained as an effluent from the last one and directed to other combination units. Different types of combinations are shown in Figure 5.11c.

The system lends and adjusts itself for regeneration when operating at high severities. Polymerisation of the off gases helps in obtaining quantity and quality gasoline. Conversions up to 80% of naphtha are possible yielding an octane value of 108.

Operating conditions:

Temperature	470 – 540°C
Pressure	10 to 50 atmos
Sp. velocity	1.5 to 5.0 V/hr.V
Hydrogen to oil ratio	3 to 10 (vol. basis)

5.3.2.5 Rhein Forming

This is a fixed bed regenerative process. Rhenium-platinum combinations are quite good in resisting fouling. These can be operated at pressures below 20 atmos. High space velocity with lower recycle ratios of hydrogen (3 to 3.5) are permissible. At low pressure

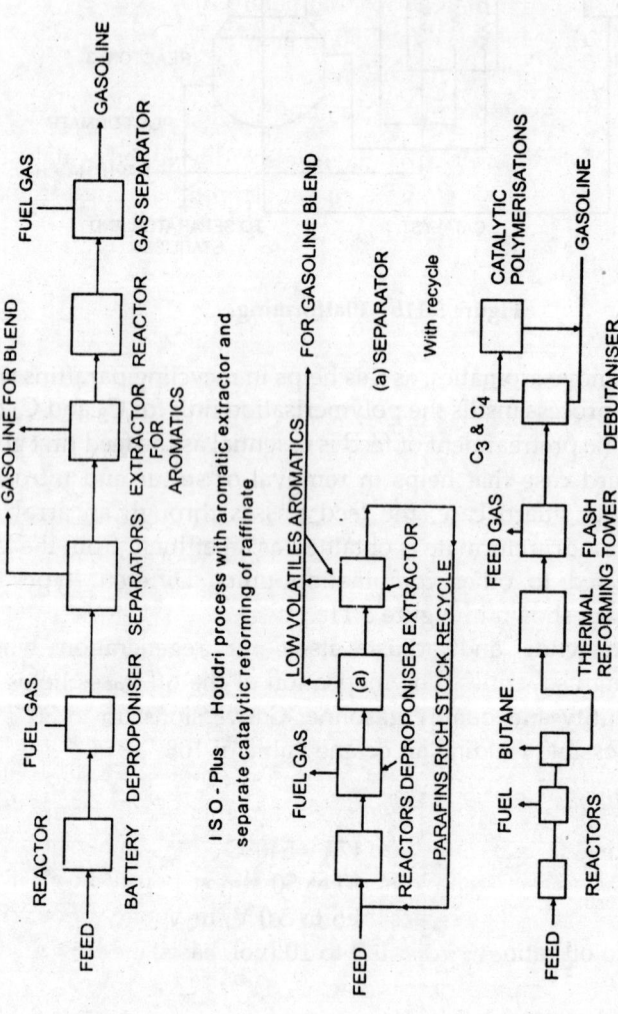

Figure 5.11c Iso Plus Houdri Combinations.

operations, increase in quantities of aromatics and hydrogen is experienced. Rheinforming was first used by Standard Oil Co. of California in 1967 and is getting popular.

Yields are approximately 75% to 84%.
Octane number of reformate is approximately 100.
Hydrogen yield per barrel of feed 50 to 60 M^3
C_1 to C_3 fractions per bbl.: 10 to 12 M^3
Outlet pressure of reactors : 10 to 15 atmos.

Universal Oil Products[37] Co. have also developed catalysts consisting of Platinum—Rhenium and Selenium for reforming operations.

5.3.2.6 Power Forming

Power forming process may be designed to operate in cyclic or semi regenerative unit. Reformates of octane value 85 to 102$^+$ are common. In cyclic regenerative process, any reactor can be isolated from the main system for regeneration. Thus one swing reactor, three process reactors are always in service. Platinum and multimetal catalysts are used in cyclic operations. A novel metallic catalyst KX-130 is found to be superior in semi-regenerative operation because it maintains the activity to a high degree compared to the previous platinum combinations. Three percent yield of hydrogen and 11 to 17% yield of C_1-C_4 fraction is a notable thing.

5.3.2.7 Selectoforming

Selectoforming[38] is a mild hydrogeneration process. Non-noble metals act as catalysts; activators like potassium enhances the capacity of catalysts. These are zeolite based dual functional catalysts and distinguish themselves from other catalysts in permitting only straight chain compounds to react. Achievement in quality of octane number is significant. Low volatility gasoline is obtained as main product. The life of catalyst is very long.

Operating conditions:

Temp.	350 to 500°C
Pressure	5 to 40 atmos.
Sp. velocity	1-10 V/hr. V.
H_2: oil	1-12 (Vol.)

The salient observations made available are

Increase in C_5^+ product with increase in octane number

Increase in propane yields (7 to 9% at 95 to 100 octane number range)

Aromatic content of reformate is reduced, if octane no. increases. 50% to 60% aromatic production is reduced if octane value is to be in the range of 95 to 100.

Front end volatility is reduced.

5.3.2.8 Ultraforming

This process is specially designed for low pressure operations with regenerative catalysts. As the composition of noble metal in these catalysts is very small, they are very rugged and active too. The catalysts are exceptionally good for high temperature operations (600°C) and possess characteristically long life.

TABLE 5.8b : Effect of Pressure.

	Pressure and yields in reforming			
	High 30 atmos.	Medium 24 atmos.	Medium 17atm.	Low 10 atm.
H_2 yield wt%	2.2	2.5	2.8	3.0
C_5^+	81.6	83.4	84.9	86
First cycle lengths months	14	12	9.5	5
Hydrogen purity %	80	83	86	87

The effect of pressure[39] can be revealed by observing Table 5.8b.

Reduction in operating pressure immensely decreases the operation time, significantly light fractions are produced more in quantity and quality.

5.3.2.9 Rex Forming

This employs a catalyst of platinum irridium and palladium deposited on inert base. Developed by Exxon Research and Engg. Co., the catalyst is credited with outstanding activity and high yields of C_5^+ products. This process can be used in combination with platforming operations and simultaneous extract ion of aromatics.

Approximate heats of reactions in different processes are shown in Table 5.8c.

TABLE 5.8c : Heats of Reactions.

	Approximate Heats of reactions	
	Process description	Heat of reaction ($\times 10^3$) kJ/Kg
1.	Low pressure vapour phase thermal cracking	169.95
2.	Vapour phase thermal cracking (gas oil)	156.69
3.	Mixed phase thermal cracking	124.44
4.	Mixed phase thermal cracking of the gas oil	113.27
5.	Thermal reforming naphtha	101.44
6.	Catalytic cracking heavy gas oil	88.68
7.	Catalytic cracking light gas oil	78.24
8.	Catalytic reforming of naphtha	69.89

Problem 5.3

A virgin naphtha (800-100°C) has been reformed over platinum catalyst. Three reactors are in series to achieve a clear RON 97 reformate. The reformer capacity is 10,000 BPSD. From the data given below:

(a) Estimate losses
(b) Find net Hydrogen
(c) Present complete material balance
(d) Find the amount of catalyst in each reactor

Data

Properties of feed		Properties of reformate	
BP-80 to 110°C		50 to 110°C	
Average mol. weight	90	85	
Characterisation factor	11.59	11.52	
API	50.2		
Sp. gr	0.7781	0.782	
S%	0.01	0.0004	
Space velocity (V/V. hr) Liquid	3		
Reactor conditions			
Pressure	20 atmos.		
Temperature	45°C		
Bulk density of catalyst	0.82		

Solution

From Fig. 5.12a Vol. of gasoline (> C^{5+} at K   = 11.59 and
                97 Octane no   = 87%

     5.12b Vol. of C_4 fraction at 87%
                gasoline   = 4.4%

     5.12c Weight of dry gas at 87%
                gasoline   = 2.2%

     5.12d Weight of H_2 = 1.3%

Feed in Kgs/day = 10,000 × 200 × 0.7786

$$= 15,57,200$$

$$\text{Reformate} = 15,57,200 \times \frac{0.87 \times 0.782}{0.7786}$$

$$= 13,60,680 \text{ Kgs/day}$$

Sulfur balance

$$\text{Sulfur in feed} = 15,57,200 \times \frac{0.01}{100} = 155.72 \text{ Kgs/day}$$

$$\text{Sulfur in product} = \frac{13,60,680 \times 0.004}{100} = 5.44 \text{ Kgs/day}$$

Sulfur in H_2S (balance) = 150.28 Kgs/day

Net Hydrogen = Total hydrogen produced – Hydrogen in H_2S

$$\text{Hydrogen in } H_2S = \frac{2}{32} \times 150.28 = 9.40 \text{ Kgs/day}$$

$$H_2 \text{ produced} = \frac{15,57,200 \times 1.3}{100} = 20243.6 \text{ Kgs/day}$$

Net Hydrogen = 20243.6 – 9.40 = 20239.2 Kgs/day

$$\text{Dry gas} = \frac{2.2 \times 15,57,200}{100} = 34258.4 \text{ Kg/day}$$

$$C_4 = \frac{10,000 \times 200}{22.4 \times 100} \times \frac{4.4 \times 58}{1000} = 227.85 \text{ Kgs/day}$$

(where 58 is the mole weight of C_4 fraction)

Losses: Reformate   =   13,60,680
           H_2S =   150.28
           Net H_2 =   20,239.2
           Dry Gas   =   34,258.4

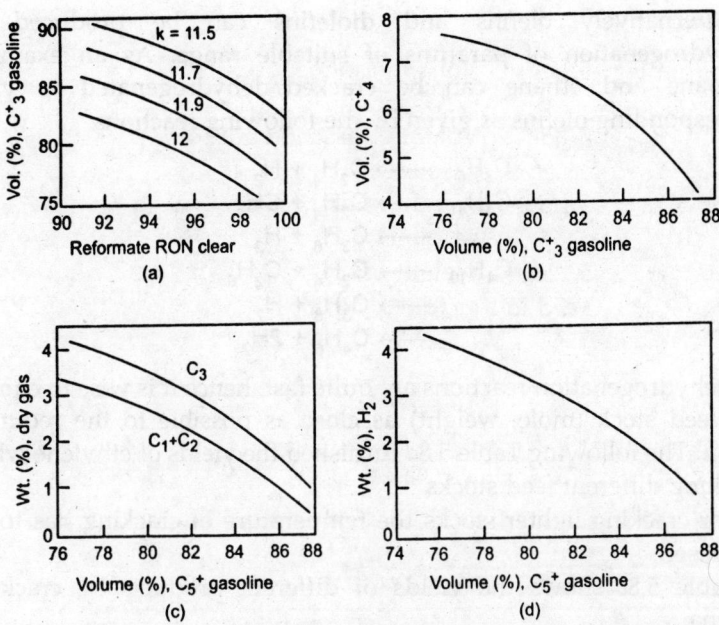

Figure 5.12 Approximate Yields of H_2, Gases and Gasoline in Reforming Unit.

    (a) Octane number Vs. C_5 yield

    (b) Vol. % C_4 fraction yield Vs. Vol. of gasoline

    (c) Dry gas Vs. gasoline yield

    (d) H_2 yield Vs. gasoline yield

C_4	=	227.85
Total	=	1,41,645 Kg/day.
Losses: Feed-Total	=	1,41,645 Kg/day (High because of graphical errors).

Catalyst volume: (at '3' hourly liquid space velocity)

$$\frac{10,000 \times 200}{24 \times 3 \times 1000} = 27.8 \text{ cu. Meteres.}$$

5.4 NAPHTHA CRACKING

In almost all parts of the world, naphtha cracking is appropriated to produce olefins, mainly ethylene and propylene. In fact any petroleum stock is suitable and is in a position to replace naphtha, which is preferentially reserved for steam reforming process.

Alternatively olefins and diolefins can be produced by dehydrogenation of paraffins of suitable range. As an example, propane and ethane can be cracked/dehydrogenated to yield corresponding olefins as given by the following reactions:

$$C_2H_6 \longrightarrow C_2H_4 + H_2$$
$$C_3H_8 \longrightarrow C_2H_4 + CH_4$$
$$\longrightarrow C_3H_6 + H_2$$
$$C_4H_{10} \longrightarrow C_2H_4 + C_2H_6$$
$$\longrightarrow C_3H_8 + H_2$$
$$\longrightarrow C_4H_6 + 2H_2$$

Dehydrogenation reactions are quite fast, hence it is wise to choose the feed stock (mole, weight) as close as possible to the required olefin. The following Table 5.8d furnished the yields of ethylene while cracking different feed stocks[40].

For cracking lighter stocks the temperature of cracking has to be increased.

Table 5.8e shows the yields of different products by cracking naptha.

When ethane and propane are chosen as feed stocks, the following points must be pondered.

Methane : It enjoys a free ride in the reactor hence it increases the burden in the system.

Propane and Butanes : Both produce olefins of same carbon atom range in the beginning; for alkylation and polyiner gasoline these can be used comfortably.

Carbondioxide : This is always present up to 20 ppm. in all cracked stocks. It should be removed before the feed is cracked.

Hydrogen Sulfide : Mostly present in all feed stocks; one way it is beneficial, as it inhibits coke formation.

Oxygen : It is a dangerous component as it leads to explosion. Nitrogen and other inerts: No problem.

Naphtha cracking for olefins is schematically presented in Figure 5.13. The hydrocarbon feed stock is preheated and cracked in presence of steam in a tubular furnace (1) (Lumus employs a short residence time (SRT) cracking furnace, while Kellog Co. achieves a high flexibility in a furnace containing entirely separate pyrolysis radiant section). The furnace is maintained at a high temperature, such that the effluents record a temperature of 800 to 850°C as soon as the furnace effluents emerge out of furnace, they are cooled in a

TABLE 5.8d   Ethylene Yield vs Feed stocks.

Charge Stock	Ethylene Yield Cracking Temp. °C	Yield %
Ethane	810	40-70
Ethane + Propane	800	40-70
Naphtha (EP 200°C)	745	37
Gas Oil (EP 420°C)	680	20-28

TABLE 5.8e   Products of Naphtha Cracking.

	Naphtha Cracking & Yields		
	Low severity yield%	High severity (%)	Naphtha cracking + Conversion of % C_3 & C_4 fractions
CH_4 & lighter	13.6	20.8%	16-22%
C_2H_4	26	36.4	34.9-42.7
C_3H_6	17.5	12.6	–
C_3H_8	1.2	1.0	–
C_4H_8	12.6	9.0	–
C_5	29.1	21.1	32 to 23

transfer line heat exchanger (2), which is set apart for production of high pressure steam (100-120 Kg/cm^2). Still hot, the products are rapidly quenched by showering with feed stock or with circulating oils. Direct quenching with water is also practised in some cases (Braun & Co). Quenching with oils is done in a pyrolysis fractionator (3), where fractionation is also carried out simultaneously. The bottom product of this fractionator shall be heavy oils or fuel oils; mostly polymer products. The lighter fractions are cooled in a reflux condenser, at the top of the fractionator. In the reflux condenser, separation of gaseous products from light distillates takes place. The gases are always infested with acidic constituents like CO_2, CO, H_2S, SO_2, oxides of nitrogen etc. and these are stripped off by suitable solvents. The acid free gases are now cooled and compressed to a pressure of 30 to 40 atmospheres whereby propane and higher components are liquefied. The mixture is separated into propane and gases in depropaniser unit. The top gases of deproponiser mainly contain uncondensables such as H_2, CH_4 and C_2 fractions. Further

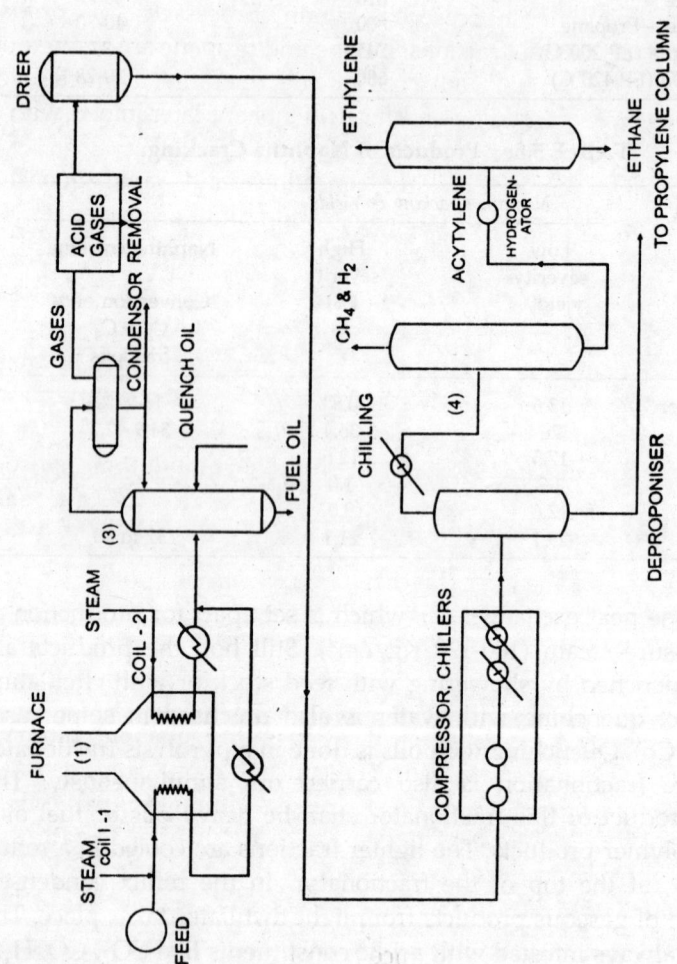

Figure 5.13 Naphtha Cracking.

chilling will liquefy C_2 fractions, from where CH_4 and H_2 steams can be easily driven off; this is done in demethaniser (4). The demethaniser bottoms are given mild hydrogen treatment to convert acetylene, which is always present in cracking operations (max. 1%) into ethylene. Ethane-ethylene fractionation is done in a column (5). Propane, from deproponiser is again processed for propane, propylene and butane fractions. Ethane and propane are again routed through furnace, but through a second coil, operating at much severer conditions than the other coil. The effluents are later joined with the main stream products.

Proper selection of cracking conditions and rapid quenching achieve high yields of desired olefins, and failure of proper selection results in undesirable coproducts like methane and polymers.

5.4.1 Feed Stock Selection

.The lower boiling range naphthas usually contain more hydrogen, hence the products after cracking also follow the same pattern; with increasing partial pressure of cracking stock, the amount of hydrogen produced falls down. Most frequently the amount of hydrogen available in cracked products vary from 10 to 16% for heavy residuums to light naphtha. Significantly for the same feed stock when operating at high partial pressure ($1.2 \, kg/cm^2$) then the amount of hydrogen drops down to 9%, while the same feed operating at $0.3 \, kg/cm^2$ contributes as high as 15% hydrogen.

Aromatic rich stocks offer more resistance to cracking, hence their presence always detracts the efficiency.

Naphthenes pyrolysis proceed by dehydrogenation, ultimately lead to more aromatic and hydrogen concentration. Paraffins the left out series, are naturally preferred sources, although n-paraffins are better than i-paraffins.

5.4.2 Effect of Steam

The function of steam is to reduce hydrocarbon partial pressure. The effect of partial pressure is shown in the Table 5.8g. In coking systems, cracking pressure will be kept always below $3 \, kg/cm^2$. The addition of steam by 0.4 to 1.0 times the weight of oil, shall effectively provide the partial pressure required for the system.

The function of steam is to reduce the partial pressure of hydrocarbon. In fact most of the naphtha crackers are operated at normal pressure, however it is closely dependent upon the boiling range of feed stock.

By increasing partial pressure of hydrocarbon, an increased coke deposition is obvious. Steiner emphasised the importance of steam in such cracking reactions, the quantity of steam is exceptionally governed by the molecular weight of feed stock operating pressure as shown in the Table 5.8g.[41]

TABLE 5.8g Approximate Yields of Ethylene from Naphtha Cracking as a Function of Partial Pressure of Hydrocarbon in Cracking Zone.

	Wt. % yield ethylene		
0:1 partial pressure Kg/cm^2	Light naphtha < 90°C	Medium naphtha 50-160°C	Heavy naphtha 90-210°C
0.1-0.13	27	25	23
0.4-0.62	23	19.0	17.5
1.2	16	–	–

Operating Conditions

Naphtha temperature	800-850°C
Ethane-Propane temp.	820-870°C
Steam: HC ratio	0.4 to 1. (wt)
Velocity of naphtha vapors	30 M/sec.

5.5 COKING

Petroleum coke is obtained in petroleum industry as an ultimate product of prolonged thermal cracking. It is a peerless raw material for electrodes. In fact demand for coke in electro-chemical and electrometallurgical industries is immense. Furnace linings in ferrous and non-ferrous industries are yet another application of petroleum coke. Indian petroleum[42] coke is the best in the world, due to less sulphur content in the parent material. Major benefactor of this coke is undoubtedly the aluminium industry.

At present petroleum[43] coke production in India stands at 152,000 tons per year. Coke producing plants and the annual turn over is presented in Table 5.9.

Coking is a thermal cracking operation falling in a temperature range of 500 to 650°C. Feed stocks otherwise not suitable to operations like thermal or catalytic cracking, are usually fed to coking units. Coking is influenced by the gravity and molecular structure of the

TABLE 5.9 : Coking Plants in India[49]

Raw coke:	
Assam Oil Company	16,000 tons/year
Barauni Refinery	78,000 "
Gauhati Refinery	36,000 "
Coke Calcining Plants Approved by Govt.	
Manali	50,000 tons/year
Visakhapatnam	50,000 "
Marogo	50,000 "
Haldia	50,000 "
Bongaigaon	30,000 "
India-Carbon Ltd. at Noonmati (1961)	65,000 (installed capacity)
(Raw coke obtained from Digboi, Gauhati (Anthracite coal also)	
Anthracite coal calcining plant Budge-Budge (1970)	12,000 tons/year
Actual Production (1972-73)	58,701 "
IOC-Plant, Barauni, Capacity	60,000 "
(1971) Production	40,000 tons/year
Goa Carbon Ltd., Panjim (1967)	50,000 (Capacity)

feed. Propensity of cracking is an outright function of Conradson carbon residue (CCR). Industrially a concise analogical term Conrandson Decarbonising Efficiency (CDE) is more familiar; it is given as :

Carbon residue in feed stock-carbon residue in

$$CDE = \frac{Liquid\ products}{Carbon\ residue\ in\ feed\ stock}$$

Estimation of coke yields can be done by the correlations available in literature. Such correlations are linked with CCR of feed stock.

	Delayed coker	Fluid coker
44, 45		
Coke yield wt%	2.0 +1.66 K	1.15 K
Gas yield during coking (wt%)	5.5 + 1.76 K	5 + 1.3 K

Where K is Conradson carbon residue of feed stock Fig. 5.14 shows the above relations.

Aromatics, asphaltenes are desirable feed stocks for a good yield of coke. Sulfur in the feed stock though it increases coke yields, it forms

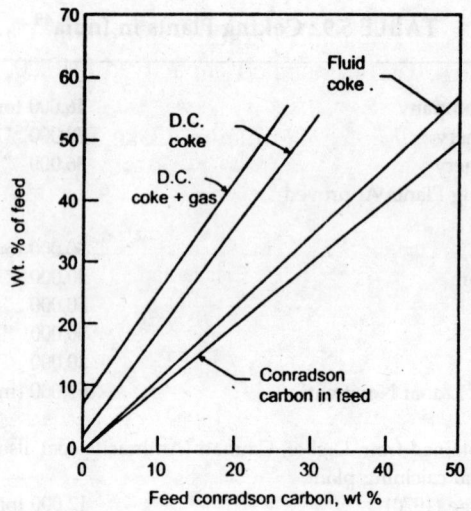

Figure 5.14 Coke Yield Vs. Conradson Carbon Residue.

an inseparable complex with coke. Once coking is complete, the removal of sulfur from such product is rather impossible; calcination of coke again strengthens the bonds between sulfur and carbon.

Sulfur crude bottoms[46] and coal tar pitches are successfully blended for getting electrode coke of comparable quality obtained from asphaltic feeds.

The recent trend of a coker units is with view on residue upgradation with maximization of middle distillates. There are two grades of Raw Petroleum Coke (RPK) calcination and fuel grade. Calcination grade RPC is produced at Barauni, BRPL, Digboi and Guwahati refineries of IOC. Fuel Grade Petcoke is product at Panipat refinery of IOC. Calcined Petroleum Coke (CPC), a pure form of carbon, is used for making anodes for aluminium smelting, apart from its uses as a source of carbon in the steel industry and in speciality consumer segments such as titanium dioxide and other chemicals. Petcoke can be used as a replacement of coal / lignite. It can be used in blend form along with coal / lignite or 100% in AFBC / CFBC boilers. AFBC Boilers are capable of absorbing SO2, and restrict GHG emissions to the desired levels by limestone injection system. Ash content in Petcoke is much lower than coal / lignite. Hence particulate emissions are always lower as compared to coal.

Goa Carbon is the second largest manufacturer of Calcined Petroleum Coke in the country with a manufacturing facility each on both of India's coasts (Goa and Paradip) and one in Central India at Bilaspur, involving a total installed capacity of 240,000 Metric Tons per annum.

The BIS specification for Petroleum Coke is IS:8502-1994. Raw Petroleum Coke (RPC)

Major parameters are as below:

Parameter	Calcination	Grade Fuel Grade
Moisture as received, % mass.,	max. 10	8–12
Total Sulphur, % mass.,	max. 1.25 (Grade A)	4.5–7.5
	2.5 (gradeB)	
Ash, % mass., max	0.45	0.2–0.3
Volatile Matter, % mass, max.	12	8–11
Fixed Carbon, %	mass, min. 85	87.73–91.01
Calorific Value, kcal/kg	–	7800–8000

RIL Petcoke consists of three grades , Grade A, Grade B and Grade C. The specifications of those are given in table 5.12

Property As Received	Unit	Petcoke Grade A	Petcoke Grade B	Petcoke Grade C
Total Moisture	%	8.0, max.	8.0, max.	8.0, max.
Air Dried -Ash	%	1.0, max.	1.0, max.	1.0, max.
Volatile Matter	%	8 min.	8 min.	8 min.
Fixed Carbon	%	87, min.	87, min.	87, min.
Gross Calorific	Kcal/kg	8200, min.	8200, min.	8200, min
Value Sulphur		7.0,% max.	7.5, max	
HGI		35, min.	35, min.	8.5, m

Air dried coke after calcination has the following proximate analysis (Foundary coke:1-5 Size mm)

Moisture content	<2%	0.5% max
FC	84-89%	
VM	8-10%	0.5% max
Ash	1-1.5%	
Ultimate analysis		
Sulfur	3-10% (varying)	0.5% max
Nitrogen	<1.5% ppm	
FC	87-89%	98.5 -99%
Hydrogen	<4%	
Oxygen	0.5-1, 5%	
Other elements Na,Ca,Ni,Fe	500-5000 ppm	
Ash	<1.4%	0.5%
Hard Grove grindability	40-80	

406

General methods of petroleum coke production are listed below:

1. Hot oven method (Koppers)
2. Thermal cracking (two coil-Dubb's)
3. Delayed coking
4. Fluid coking
5. Contact coking.

Hot Oven Method

Presently, this method has no reputation in petroleum industry. With some modification the by-product coke oven batteries were employed for coking petroleum stocks. High molecular weight stocks, asphaltenes, tars, pitches serve as raw materials to these ovens. The conditions of coking are similar to metallurgical coke production. Temperature in the range of 1000-1200°C is required and coking time of 18 hours is most desirable. With pitch of CCR value 50%, coke yield shall be approximately 70%. With ample and frugal modern technology in hand, this has become an unwanted process.

Thermal Cracking

Thermal cracking operations and coking operations can be conveniently and simultaneously carried out by Dubb's two coil cracking technique. In cracking practice heavy oil is separately cracked and the products are mixed and chilled with fresh feed. Instead of this, the cracked products are allowed to complete cracking in evaporators. After evaporators there are flash chambers where the separation of volatiles and recycle oils take place. These flash chambers in Dubb's cracking unit are exchanged for coke chambers, where coking takes place. The small difference can simultaneously yield cracking oils and coke too. Lighter products evaporate and escape into the fractionator. The coking operation is conducted at the same conditions as Dubb's cracking. Coke from the evaporator is later removed by any convenient technique. The off gases from the evaporator are opulent in unsaturated components.

5.5.1 Delayed Coking

This is the most important process of all the coking techniques. In India, all plants operate on this technology only. Delayed coking is a technical[47] outcome of two coil cracking process. Capability for cracking all types of feed materials including solvent extracts and simplicity of operation, have made this one as most adoptive in all refineries.

The principle behind delayed coking operation, is: heating is done in a furnace to initiate cracking and the reactions are complemented in huge and tall coke drums; hence the name delayed coking is used. As a rule a series of such coke drums will be pressed into service. While one drum is engaged in getting feed, the other drums shall be in the process of coking and decoking. This way by orderly rotation of the drums the process may be contrived to work continuously. Minimum two drums are essential even for small capacity plants, in fact sets of drums in series are desirable. Usual operating conditions in delayed coker are given below:

Operating conditions

Heavy oil, discharge temperature	470-520°C
Coking temperature	450-470°C
Pressure in coke drums	5 to 6 atmos
Drums diameter	4 to 5 M
Height	14 to 20 M
Thickness of drum	about 4 cms.

Recycle ratios of fresh (heavy) and light oils are adjusted to maximise yields of either liquid or solid products. An yield of 30% coke for reduced crudes or 80% for tars and pitches may be expected. Coke from these units contains volatile matter upto 8-15% and the bulk density may be around 9 kgs per liter. CDE of the plant may be reaching upto 99.8%.

Conventional delayed coking process is shown in the flow diagram (Fig. 5.15a). The process begins with the fresh feed entering the fractionator (1). The fractionator complys with all the assigned duties as in two coil cracking plant. Heavy oil from the bottom of fractionator passes through a heater (2) at a high velocity. Introduction of steam into heating coil prevents coke deposition in coil. The hot and partially vaporised mixture enters the coke drum (3). The coke drum is charged with hot mixture to half to two-third of the height of the drum; or to a convenient marked level.

The level of hot mix in drums is modulated by cathode ray monitoring. Steam and volatiles escape from the coke drums and enter the fractionator. This fractionator is also known as dephlegmator. These vapours preferentially relinquish heat to incoming feed; and in this act stripping of volatiles from feed takes place.

Changing of coke drums may require a time of 4 to 5 hours. Immediately after changing is over the drum shall be isolated from

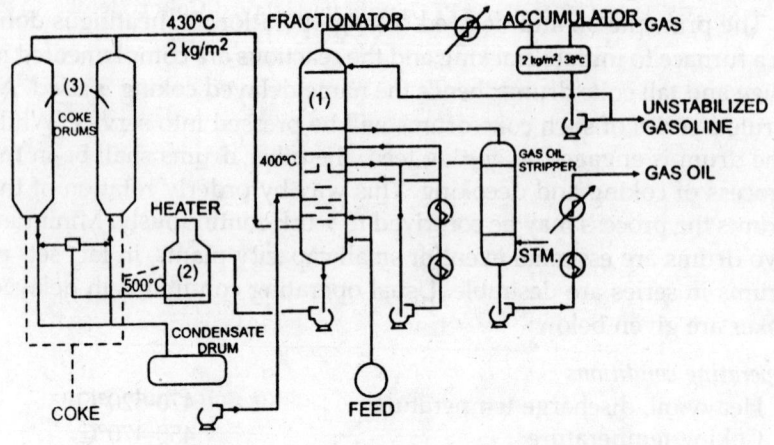

Figure 5.15a Delayed Coking Operation.

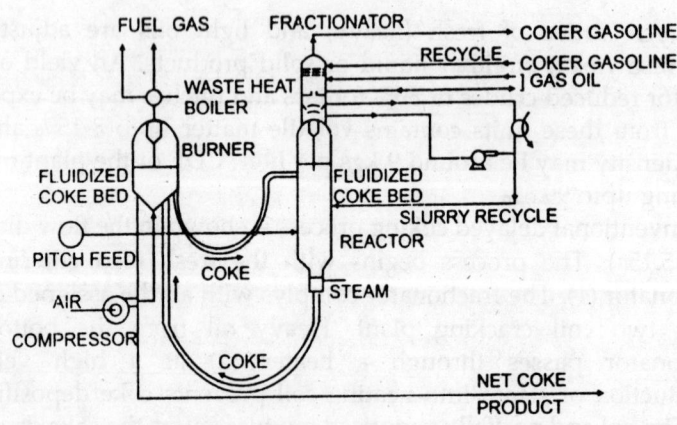

Figure 5.15b Fluid Coking.

the stream. Effluents of the heater shall now be switched in to second drum. Coking being slow usually takes a time of 10 to 16 hours. Time of charging coke drums must balance the time of coking and decoking operations.

Rough estimates of coke formed in drums is possible if calibration of drums (tons per centimeter height) is done.

TABLE 5.10a   Effect of API Gravity on Delayed Coker Yields.

CARBON RESIDUE	16%	16%	16%
Coking conditions			
Drum outlet temp. °C	435	435	435
Drum pressure kg/cm^2	1	1	1
Charge API	12.3	13.0	14.0
Product yields wt %			
Gas (Vol %)	4.2	4.0	3.6
Coke	16.6	15.5	13.6
Gasoline	8.9	8.8	8.6
Gas Oil	19.0	17.3	19.1
Heavy Gas Oil	53.3	54.4	55.1

TABLE 5.10b   Comparison of Delayed Coking and Fluid Coking.

Charge—	Delayed	Fluid
Gravity °API	15	15
Conradson Carbon		
Residue	9	9
Sulphur	1.2	1.2
C$_3$ and Lighter	6.0	5.5
Coke	22.0	11.0
CDE%	99.8	91.2
Charge—	Contact Coking	Delayed Coking
Gravity	18.9	18.9
Conradson Carbon	11.7	11.7
Sulphur	0.6	0.6
Products		
Lighter fractions %	14.9	7.5
Coke%	13.0	20.0
CDE%	99.3	99.3

TABLE 5.10c   Effect of Recycle Ratio.

Recycle ratio vol. %	0	20	40
Product yields			
Coke	15.5	18.0	20.0
Gas	4.0	5.5	6.0
Gasoline	13.0	15.0	15.5
Gas oil	21.5	26.0	28.5
Heavy Gas Oil	46.0	36.0	24.5

5.5.1.1 Decoking

Coke being set and hard, much difficulty is encountered in removing.
Decoking is a time consuming and arduous task. Lot of developments
took place in the direction of decoking operations. Presently hydraulic

jets at 150 to 200 atmospheres pressure are directed to break the coke, in place of once used mechanical breaking with hammers; an orthodoxic way still prevailing in some countries that offer cheap labour. Drilling and mild dynamiting is also allowed. Of late the ingenious chain pulling technique has been acclaimed and welcomed by the industry. The technique consists of suspending strong chains in coke drums from the hooks fixed to the thick shell of drum at the top. Other ends of the chains are free to fold and lie submerged under charge. After coking is over, usually steaming is done to drive off hydrocarbon vapours. The top flange is disconnected and the chains are pulled by cranes; this upward thrust shatters the coke to pieces, making easy for coke removal. Influence of feed stocks properties on coke yields are shown in Table 5.10a, b, c.

5.5.2 Fluid Coking

Search for continuous coking has inducted the concepts of fluidisation with necessary modifications. Fluid cat crackers use catalysts to bring the cracking and here cracking and coking are catalysed by coke particles. Explicitly, it is nothing but formation of excessive coke on surface of particles; as a special case these happen to be coke particles. Coke particles produced in the same unit assume more or less spherical shape and act as heat carriers while travelling from burner (regenerator) to reactor, and coke carriers in reverse travel. Some portion of steam stripped coke is burnt and the remaining coke is taken out. The hot coke particles are in a state of fluidisation caused by incoming vapours. Thus the effective continuous circulation of coke seeds namely coke particles is unvoidable. A close look at Fig. 5.15b reveals the process sequence. About 20% of coke produced is consumed to maintain coking reaction.

Operating conditions

Reactor temperature °C	480-560
Pressure	–Normal
Burner temperature °C	590-650
Pressure (Burner)	about 2 atm.

5.5.3 Flexi Coking

Petroleum industry, largely lured by gasification, the hub of modern petrochemical industry has given to a matrix of coking and gasification operations in the form of flexi coking. This integration

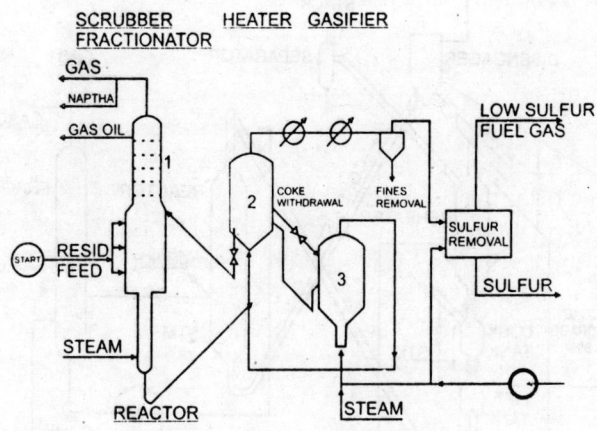

Figure 5.15c   Flexi Coking.

enables refineries to convert vacuum residuums, other heavy feed stocks into desulfurised liquids and gases. As a result 99% vacuum residuums end up as liquids or gases with less than one percent sulfur.

Figure 5.15c shows the outlines of the process. Residuum feed is injected into reactor where thermal cracking takes place.

Lighter products leave the reactor and enter the scrubber cum (1) fractionator. The purpose of fractionator is not only to rectify the volatiles, but also to entrap the coke fines leaving the reactor. Steam (admitted at the bottom reactor) stripped coke circulates into heater (2), where devolatilisation takes place; hydrocarbon gases obtained in this process are subsequently treated for purification. Coke formed in the heater is sent into gasifier (3), where coke is encountered with a stream of air and steam. Gasified products again join the main gas stream for purification. Part of coke may be withdrawn if necessary, at any point between heater and gasifier.

5.5.4   Contact Coker

This is also a continuous process. The essential steps in this process are

(a) Oil wetted circulating coke particles are allowed to flow downward in reactor in a dense bed. Sufficient time is provided, in this way to complete cracking coking and drying operations.

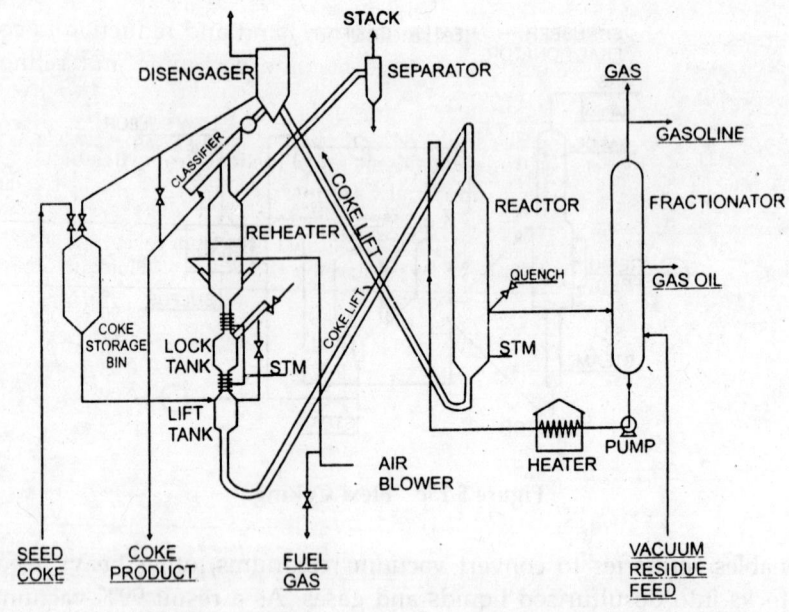

Figure 5.15d   Continuous Contact Coking.

(b) Reheater, heats the circulating coke particles to the requisite temperature levels.

(c) Coke circulation system elevates the coke particles as dense unagitated column.

(d) Fractionation system for separation of volatiles.

The operation is depicted[48] in Fig. 5.15d. Coke circulates always between reactor and heater. A part of coke is always essential for supplying thermal energy and remaining portion is separated from the disengager as shown in figure.

Advantages accrue in the process, mainly in

(a) Handling heavier residuums.

(b) Great flexibility in operation and control.

(c) Coke drums—cleaning is eliminated.

5.6   HYDROGEN PROCESSES

Hydrogen processes in a refinery play a vital role. The significance of these processes is increasing day by day. In 1955, these were inducted into refinery, but, due to high cost of operation, the growth was not as expected. Now with better technology in hand and reduction in cost of operation, these processes have become economic and refiners deem them as necessary.

TABLE 5.11 Ultimate Analysis and Physical Properties of Industrial Carbons[50].

| | Calcined petroleum coke | |
	Delayed	Fluid
Carbon wt %	98.40	93.90
Hydrogen wt %	0.14	1.27
Nitrogen wt %	0.22	0.33
Oxygen wt %	0.02	0.09
Sulfur wt %	1.2	4.90
Ash wt %	0.35	0.20
Bulk density kgs/lit.	0.89	0.99
Hardness Index HW	38	31
Real density gm/cc.	2.05	1.94
Specific Resistivity ohm	0.089	0.115

The trends in hydro processing operations can be judged from following[51] information.

	Years	
Capacity of process	1955	1966
Cat. cracking	13.3%	6.9%
Reforming	0.7%	12.9%
Hydrogen	1.2%	13.5%

Present day statistics as furnished by Carter[52] indicates

Thermal cracking	3.66 Million barrels per day
Cat. Cracking	8.20 ″ ″
Hydrocracking	1.39 ″ ″
Coking	1.80 ″ ″

Mathias Pier[53], the father of hydrogenation processes, might not have realised the importance of these. The largest hydrocracking unit at present takes up 80,000 barrels per day. This shows desirability and significance of a hydrocracker in a modern refinery.

Hydrogen processes are divided into two classes as hydro-cracking and hydrotreatment. These are dealt vividly in this chapter.

5.6.1 Hydrocracking

The name implies cracking in presence of hydrogen. In recent years, the tendency of hydrocracking has gained significance. While cracking (catalytic or thermal) is required to produce more light fractions from feed stocks prone to cracking, hydrocrackers take feeds of relatively refractory nature like cycle oils, coker distillates etc. Being refractory, these stocks resist cracking, but with high pressure of hydrogen they crack quite easily. In fact in a modern refinery both catalytic and hydrocracking work as a team. With improved technology it has become possible to handle even reduced crudes. Naphthenic stocks are preferred in catalytic cracking while paraffinic stocks require hydrocracking. Thus hydrocracking has become a powerful tool to the refiner in the production of quality gasoline or mid barrel products from heavy distillates or fuel oils.

5.6.1.1 Reactions in Hydrocracking

Thermal or catalytic cracking produce an olefin during cracking from a big molecule of paraffin. Hydrogen takes the charge of saturating the olefin produced as shown

$$C_nH_{2n} + H_2 \rightarrow C_nH_{2n+2}$$

Unsaturation in straight chains as well as in ring structures is gradually eliminated.

Detachment of side chains and isomerisation reaction are very frequent with ring structures.

Isomerisation takes place in all types of components

$$n\text{-}C_4H_8 + H_2 \rightarrow i \text{ butane}$$

$$n\text{-}C_4H_{10} \rightarrow i \text{ butane or } cis. \text{ butane}$$

Cracking of naphthalene gives the following products[54]

$$\text{(benzene oxide)} + C_3H_6/C_4H_8$$

$$\text{(epoxide)} + \text{(cyclopentene oxide)} + H_2 \longrightarrow \text{(cyclohexene oxide)} \quad C_4H_9$$

Supplementary reactions are some elimination reactions. Compounds of N_2, S, O_2 are converted to corresponding hydrides as listed below.

$$RSH + H_2 \rightarrow RH + H_2S$$

$$\underset{N}{\text{(pyridine)}} + H_2 \rightarrow C_5H_{10}\ (C_5H_{12}) + NH_3$$

$$\overset{OH}{\underset{}{\text{(phenol)}}} + H_2 \rightarrow C_6H_6 + H_2O$$

$$\overset{O\text{-}C_2H_5}{\underset{}{\text{(phenetole)}}} + H_2 \rightarrow \text{(benzene)} + C_2H_6 + H_2O$$

$$RCI + H_2 \rightarrow RH + HCI$$

Hydrogenation reactions are exothermic, while cat. cracking reactions are endothermic. Thus hydrocracking and cat. cracking are complementary to each other; in the sense cat. cracking provides olefins for hydrocracking; while hydrocracking produces necessary heat for cracking. If hydrocracking is more, then the reaction becomes wayward due to the huge amount of heat produced; cat. cracking even though it is endothermic cannot consume all the heat. A better method of heat removal is to quench the reactants with cold hydrogen during reaction. Practically all amount of feed is consumed in hydrocracking, hence the yields are likely to exceed 100%. Interestingly, small amounts of light gases are produced compared to normal cat. cracking operations. Hydrocracking produces more saturated compounds, hence for quality gasoline, cat. cracking or reforming of gasolines may be necessary.

5.6.1.2 Catalysts

Reactions of hydrocracking require a dual function catalyst that can give cracking and hydrogenation activities. Cracking is due to acid sites supplied by silica-alumina or alumina, hydrogenation function is due to metals like nickel, tungsten, platinum, palladium, molybdenum. Acid sites are sensitive to basic nitrogen. Nickel (5%) wt. on silica-alumina acts as good catalyst for treated stocks, i.e. stocks possessing not more than 50 ppm of nitrogen. These catalysts can work at a temperature of 350 to 370°C and a pressure of 100 atmospheres. The effect of nitrogen can be decreased by increasing the reaction temperature; but increasing temperature produce more gases and carbon. Also platinum or palladium (0.5%) on zeolite base can abate the nitrogen attack. By using, platinum or palladium catalysts more isomeric structures can be produced than nickel or tungsten.

Molecular sieves[55] possessing large pores are currently used for hydrocracking operations. They exhibit high activity even in presence of hydrogen sulfide and ammonia due to large pores; further they are hydrothermally stable.

Sodium contamination[56] may deactivate the catalyst to a small extent but its presence is reflected in decreasing the octane number of gasoline.

Mordenite catalysts are also available but their activity and stability rapidly decrease. They are suitable for producing gases from naphthas.

Zeolite catalysts, loaded with noble or non noble metals form excellent catalysts, as they can operate in substantial concentrations of ammonia. Unicracking JHC processes operating on heavy feeds have successfully demonstrated the activity of these catalysts up to seven years[57].

As the cracking operations go on, to counteract the loss of activity of catalyst, the reactor temperature is increased day by day by 0.5 to 1.0°C.

5.6.1.3 Hydrocracking Reaction Conditions

Usual temperatures of cracking fall in the range of 300-400°C, with high pressure of hydrogen between 60 to 150 atms. Relatively large amounts of hydrogen circulation is necessary to prevent fouling of catalyst which permits long runs of catalyst. To get high purity products, pretreatment of feed is essential. This pretreatment not only removes sulfur, nitrogen, arsenic etc. but ascertains products free from these impurities.

Hydrocracking catalyst works on Start of Run (SOR) and End of Run (EOR). When the catalyst is new it is SOR, after coke deposition and the catalyst becomes deactive just before this stage the EOR commences. Hydrocracker in Panipat refinery works at 170 Atm Hydrogen pressure and at a temperature of 365-441°C so that the products will have end boiling point of 370°C. Hydrocrackers in combination with RFCCs will maximize the middle distillates.

5.6.1.4 Processes and Description

Most of the hydrocracking operations are carried out in fixed bed reactors only. Catalysts are very similar to cat. cracking catalysts. Present trend of using molecular sieves attired-with rare earth metals in hydrocracking is in no way different from cat. cracking reactors. The process may be conducted in a single stage or in two stages. First stage reactor is bequeathed to give 40 to 50% conversion to gasoline while the heavy portion (from the effluent of the reactor) forms the feed stock for 2nd stage reactor where a 60% conversion is anticipated.

The catalyst in reactor is spread in a number of beds; this helps in locating the hot spots where cold hydrogen injection is planned. Hydrocracking processes of common interest are described below:

5.6.1.5 Isomax Process

This process employs a fixed bed catalyst system with a single stage operation. High selectivity and activity of catalyst brings all features in the design to handle stocks from naphtha to deasphalted residuums. Alternatively, Isomax offers a new route to hydro-treatment of lube oils. Important aspect of this process is low consumption of hydrogen; and with moderate temperatures of operation, a fresh charge of catalyst can maintain its activity from one to two years before any regeneration is attempted.

5.6.1.6 Isomax Hydrocracking (Figure 5.16)

Charge stock, recycle hydrogen and make up hydrogen are mixed and passed through a heater. The mixture enters the reactor (1) from the top while cold hydrogen is admitted into the reactor at different points (hot spots). The effluents from the reactor are immediately heat exchanged with the c/s, chilled and fed into a high pressure separator (2); where hydrogen is separated. This hydrogen is contaminated with H_2S and NH_3; hence a proper treatment for the removal of these constituents is required and is practised in the treater (5). The recovered hydrogen is compressed to the required pressure and

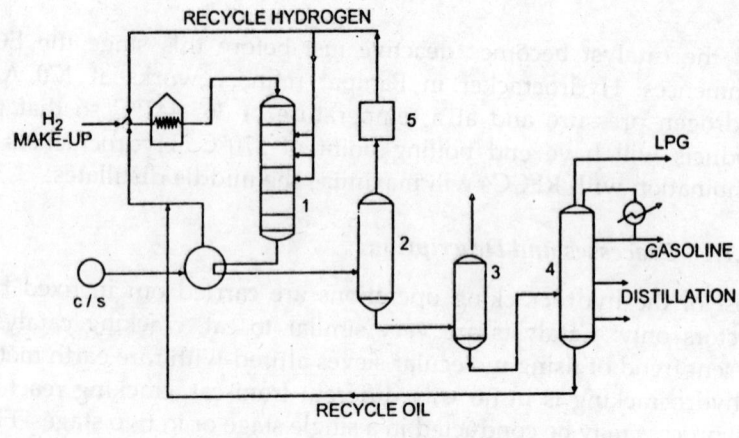

1. REACTOR
2. HIGH-PRESSURE
RELEASER
3. LOW PRESSURE SEPARATOR
4. FRACTIONATOR
5. AMINE TREATMENT

Figure 5.16 Hydrocracking (Isomax).

recycled back. Treated stock from high pressure releaser goes to low pressure separator (3), where fuel gas (up to C_3 fractions) H_2S, are obtained. Liquid fractions from the bottom of this separator enter the fractionator (4), where distillates are separated. The heavy oil from the fractionator is recycled back to the reactor. Light gases, after congenial treatment can go into the fuel system.

Operating conditions

Pressure of hydrogen	30 to 100 atmos.
Temp. of the reactor °C	250 to 370°C
Consumption of hydrogen	35 to 50 M^3/per barrel

5.6.1.7 H.G. Hydrocracking

Gulf Research and Development Co. and Houdry Divn. had jointly developed this process. In this process, conversion of light and heavy gas oils into more valuable lower boiling products is accomplished. The process is designed for a single stage or two stage reactor system.

Fig. 5.17a shows single stage operation and

Fig. 5.17b shows two stage operation.

When two stage system is employed, mild, hydrotreating operations take place in the first reactor; second stage sees through the hydrocracking reactions. Effluent from first stage reactor is cooled

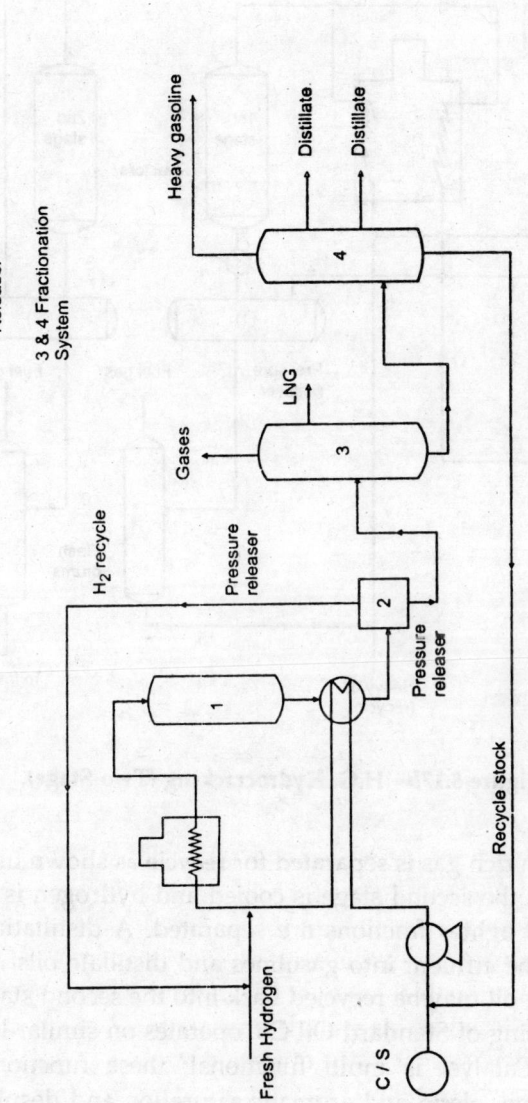

Figure 5.17a Single Stage Hydrocracking System (H.G.).

1. Reactor
2. Pressure Releaser

3 & 4 Fractionation System

420

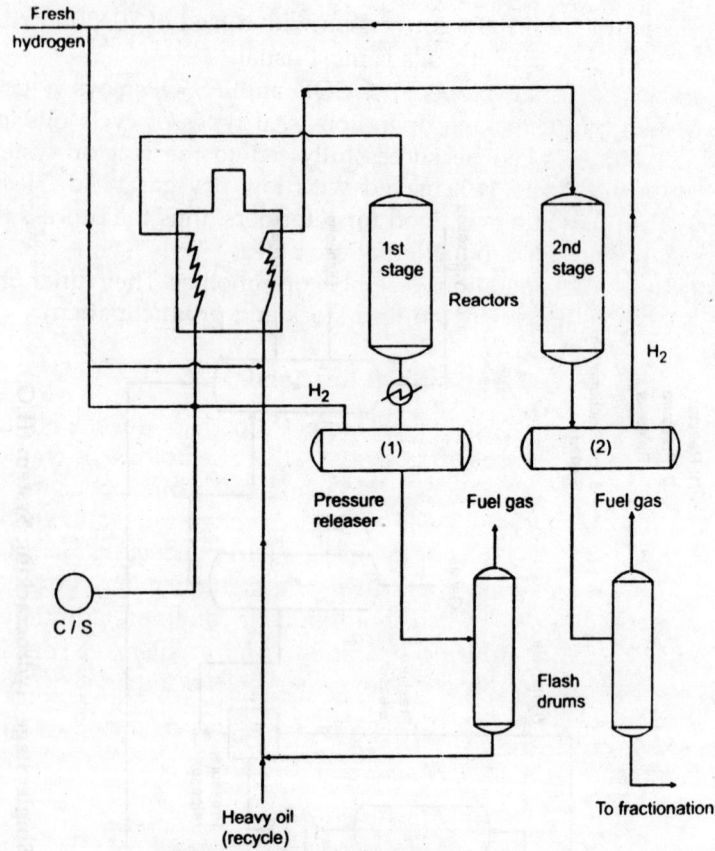

Figure 5.17b   H.G. Hydrocracking (Two Stage).

and hydrogen rich gas is separated for recycle as shown in Fig. 5.17b. Effluent from the second stage is cooled and hydrogen is separated. Fuel gas and lighter fractions are separated. A distillation column fractionates the effluent into gasolines and distillate oils. The heavy portion of the oil may be recycled back into the second stage reactor.

Ultra cracking of Standard Oil Co. operates on similar lines as the above one. Catalyst is multi functional; these functions include denitrogenation, olefin and aromatic saturation and desulfurisation. Longevity of the catalyst has made the process more economical. A difference exists from the previous operations in introducing water directly into cold reactor effluent, which absorbs ammonia and partly hydrogen sulfide too. 100% conversion with a feed of 70/30 mixture of coker and heavy catalytic oils is most usual.

Unicracking-JHC developed by UOP and ESSO enjoys a good reputation in hydrocracking operations. All types of cycle oils and paraffinic raffinates can be successfully fed to the reactor system. More than 100% yield is achieved with low dry gas yield. Heavy gasoline from these units is good for reformers, thus the process has become important for production of aromatics.

All the fixed bed operations resemble one another. They differ only in the catalyst composition and feed stock and product pattern.

5.6.2 Moving Beds (Ebulient Bed Reactors)

Cities Service Research & Development Co. had developed this process. The highlights of the process is that it employs an ebulient bed instead of fixed bed: which brings in extremely efficient contact of catalyst, oil and hydrogen. Residues are converted to low sulfur fuel oils. Heavy gas oils and cat. cracker oils are hydrogenated. Its capacity to handle short residuums or stocks of high metallic content is striking. Since the catalyst bed is in ebulient condition, addition or withdrawal of catalyst during operation is no problem. A constant quality and rate of products can be remarkably maintained without much difficulty.

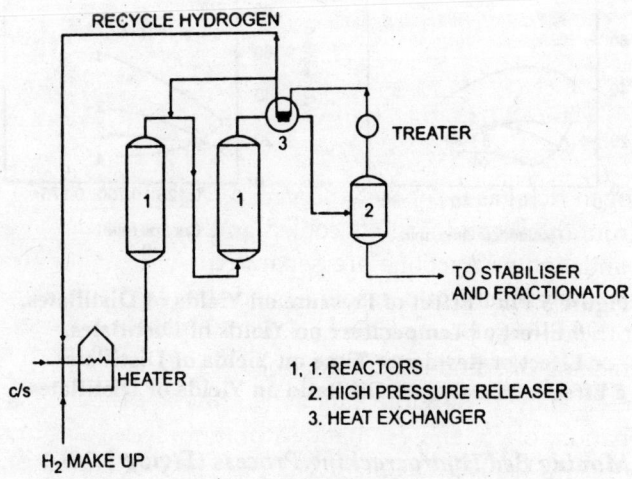

Figure 5.18 Moving Bed Hydrocracker.

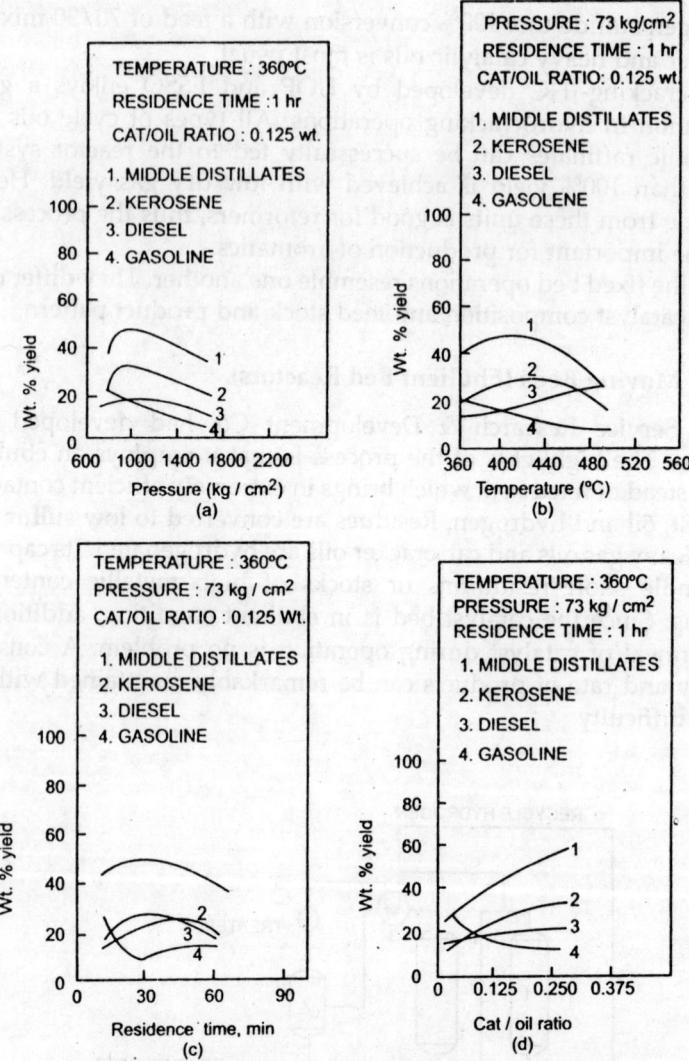

Figure 5.19a Effect of Pressure on Yields of Distillates.
  b Effect of Temperature on Yields of Distillates.
  c Effect of Residence Time on Yields of Distillates.
  d Effect of Catalyst to Oil Ratio on Yields of Distillates.

5.6.2.1 Moving Bed Hydrocracking Process (Figure 5.18)

H-oil process is designed to process residuums or other heavy feed charges and usually employs a hydrogenating catalyst. Hydrocracking process, was also developed by Cities Service Research and Development Company and Hydrocarbon Research. This process uses a dual function catalyst and can take up feeds upto a boiling point of 650°C. The major difference in these processes is that the catalyst is kept in moving condition. To achieve this moving bed conditions, charge is fed into the reactors from the bottom. The reactors two in numbers (1,1) work serially. The effluents go from the top of the reactor. The separation, recycling and fractionation follow the conventional procedure as is done in Isomax Process.

Operating conditions

Pressure	200 atmos.
Temperature	400-460°C
Consumption of hydrogen	40 M³/bbl

Some of the laboratory results on hydrocracking of vacuum distillates for middle distillates are shown[70] in Fig. 5.19*a*, *b*, *c* and *d*.

5.6.3 Hydrodesulfurization and Hydrotreatment

Next to hydrocracking operation hydrodesulfurization and treatment from major hydrogen processing operations. The difference between hydrodesulfurisation and hydrotreatment is marginal and work on the same technology including catalyst. Difference occurs mainly in operating conditions. Hydrotreatment in most of the cases may be treated as a finishing operation, while hydrodesulfurisation is mainly a technique of impurities elimination.

Though the name hydrodesulfurisation suggest the removal of sulfur, yet other impurities which can form hydrides are also simultaneously taken off.

The importance of hydrogen processes, either for desulfurisation technique or for treatment technique or production of lighter fractions by hydrocracking is ever enchanting and increasing. The refiners are thinking of introducing "All hydrogen"[58] refinery operations due to many advantages available by hydrogen processes.

5.6.3.1 *Hydrodesulfurisation (HDS)*

Hydrogen over a catalyst is capable of reacting with sulfur, nitrogen, halogens and oxygen, from the respective compounds available in the fractions. Sulfur, nitrogen and halogens escape as the respective hydrides, while oxygen escapes as water vapour. Metals found in the oil are generally adsorbed on the catalyst surface, hence the product shall always contain less metals. However increase in metal concentration on the surface of catalyst naturally rescinds the activity of catalyst.

Increase of boiling point of feed also brings alarmingly increasing difficulties to the process. Thiophenic compounds are major sulfur compounds in the high boiling fractions, and these being lethargic towards hydrogen render desulfurisation more difficult. Mercaptans do offer such difficulties but to less extent, hence may be placed in intermediate region. Nitrogen is quite adamant and requires very severe conditions.

Bauxite with impregnation of molybdenum is capable of removing 70 to 80% of metals.[59] Demetallisation (Demex) processes are being tried extensively. In this method the removal of metal concentrates is done by solvent extraction resembling deasphalting operations by solvents. Manganese nodules[60,] are found to be catalytically active in demetalation of topped crudes in presence of hydrogen. These nodules are cheaply available in marine and fresh water sediments. Manganese nodules deposited on CO-Mo on alumina base serve the best.[61] Hung and Wei[62] studied demetalation of porphyrins using Co O_3-Mo O_3 on alumina. The report shows that 90% of nickel and vanadium could be eliminated.

5.6.3.2 Catalysts

Two types of catalyst compositions are popularly used namely Nickel-molybdenum and Cobalt-molybdenum on alumina. Although both are active catalysts, still some reservations are observed with reference to Ni-Co compositions.[63] These compositions seem to be more active than Co-Mo catalysts, especially while removing nitrogen. Basically, the structure, particle size and porosity of the catalyst are irrefutable parameters in judging the activity and life of catalyst.[64] Sulfur in paraffins is easily removed, while the same in aromatics is quite resistant. Even Co-Mo catalysts at 350°C under a pressure of 100 atmos can remove 90% nitrogen.[65]

Comprehensive studies on Ni-Mo (Shell 324) and Co-Mo (Shell 344) by T.F. Kellet et al. show[66].

Ni-Mo catalysts are effective for removal of nitrogen; hence these are suitable for treating cracked stocks.

H_2 consumption with these catalysts is more.

In general, increase in space velocity increases sulfur in products and hydrogen consumption.

As reaction temperature increases hydrogen consumption too increases.

With Ni-Co catalysts, pre sulfurisation with carbon disulfide or any sulfide is essential.[67] The fullest activity of these oxides occurs in fact after interaction with sulfur and a part of catalyst is convected into sulfides.

Widely used catalyst composition is given below :

MoO_3	12.9%-13%	by wt.
CoO	4.1%-5%	do
NaOH	0.6%	do
SiO_2	1.0%	do
Sulfates	1.0%	do
Al_2O_3	(by difference)	
Surface area	m^2/gm	200-320
Pore volume	c.c/gm	0.4-0.6
Bulk density	gm/cm^3	0.65 to 0.68

Shape : Cylindrical of 1.5 mm dia × 4 mm dia.

Apparent activation energy: 7.83 to 12.58 K cals/gm.

5.6.3.3 *Process Variables*

Partial pressure of hydrogen, temperature and space velocity are the principal operating variables for a chosen catalyst composition. Increase of partial pressure of hydrogen, suppresses the coke formation and enhances the reaction rate. High temperatures favour removal of nitrogen and thiophenic sulfur; but excess of temperature retards the reaction rate due to coke formation.

Reaction variables-conditions:

Hydrogen consumption	2.9 M^3 per barrel per 1% sulfur reduction; 8 M^3 per barrel per 1% nitrogen reduction
Pressure	10-200 atmos.
Recycle hydrogen:	8 to 10 times consumption
Temperature	350-450°C

For olefin and aromatic hydrogenation extra quantity of hydrogen is required:

Yields for distillates	95 to 99%
for residues	85 to 90%

426

Sulfur reduction with
  naphthas and distillates   up to 95%
  for fuel oils and residues   up to 50%

$$\text{Bromine no. of olefins} = \frac{\% \text{ olefins} \times 160}{\text{Mol. wt. of olefins}}$$

Bromine no. reduction means increase in saturation. During hydrogen processes the saturation is attained to different degrees depending upon the boiling point of the stock for light naphtha 90 to 100% reduction is possible while for heavier cuts 10 to 15% reduction in Bromine no. is possible. For each No. reduction about 6 to 8 scf of hydrogen per barrel is required.

Figure 5.20a shows the hydrodesulfurisation operations. C/S and recycled hydrogen (fresh hydrogen) are heated in a furnace (6) and the vapor is fed into reactors (1); one or two stages may be employed. Sometimes reactors are arranged in parallel to increase the capacity of plant. The outgoing vapors are admitted into a high pressure separator (2) where H_2S, NH_3 etc. are freed along with unreacted hydrogen from the treated product. The product goes into a low pressure separator (3), where gases (hydrocarbon) are separated and the bottom product is transferred to a distillation column. Hydrogen from high pressure separator is associated with impurities formed in hydrotreatment so the gas is purified by treating with amine solution (4). Pure hydrogen recycles back.

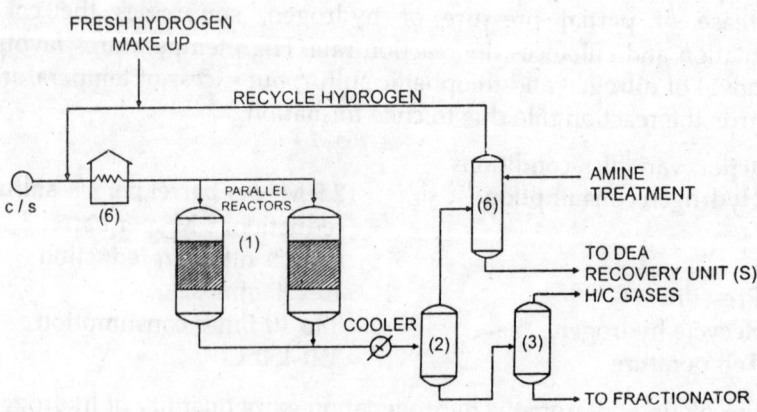

Figure 5.20a   Flow Sheet of Hydrodesulfurisation Process.

Catalyst composition	Co-Mo Metallic compounds on alumina pellets.
Temperature	350-450°C
Pressure	High, up to 100 atmos.
Space velocity	0.2 to 2 V/hr. V.
Sulfur reduction	90%
Ni+V ppm reduction	Up to 90% in light fractions Up to 60% in heavy fractions

Gulf HDS process is a fixed bed regenerative process. By varying the reaction conditions light distillates can be processed. Regeneration is carried out *in situ* by blowing air steam at 400 to 500°C. A safe period of 4 to 5 months at desulfurization levels of 65 to 75% is quite attractive, before any regeneration is attempted. Low cost catalysts with high activity are proprietary of Union Oil Comp. These catalysts enable the operation at reduced pressures (40-60 atmos.) for feed stocks of diverse nature. Gulf (HDS) process on Kuwait residual oils shows a removal of sulfur from 3.8% to 0.1 %, with improved quality of gasoline; yet requiring 33% less reactor volume.

HDS product inspection[68]

	Reduced crude —	Desulfurised gas oils
API	24.9	34.4
Sulfur % (3.8)	0.11	0.07
SUS 98.8	61	1.49 cs
38.6	456	40
P. pt. °C	–1	–1
Carbon residue	2.32	.009
N_2 wt. %	0.08	0.08
Vanadium (60 ppm)	0.3	0.1
Ni	0.3	0.1

(Figures in brackets show the amounts in material)

The catalysts are manufactured in form of extrudates of two modifications: alumo-nickel-molybdenum (ANM) and alumo-cobalt-molybdenum (ACM) ones. The both are characterized by high desulfating and denitrogenating activity, high mechanical strength and long term of service.

specifications	КГу-941		КГу-950	
	grade ANM	grade ACM	grade ANM	grade ACM
↑ MoO3 content, %wt ↑ NiO content, %wt ↑ CoO content, %wt	14,0 - 16,0 2,9 - 3,5 –	14,0 - 16,0 – 2,6 - 3,2	9,0 - 11,0 2,9 - 3,5 –	9,0 - 11,0 – 2,6 - 3,2
PABD, g/ml	0,7 - 0,9	0,7 - 0,9	0,75-0,95	0,75 - 0,95
Diameters of extrudates, mm	2 - 3 2-2,5	2 - 3	2 - 2,5	2 - 2,5
Crushing strength coefficient, average, kg/mm, not less than	1,7	1,7	1,9	1,9
Packing	Metal 200-L drums with polyethylene bags inside			

Process parameters

Feed	Gasoline fraction	Diesel fraction
Space velocity, h^{-1} (LHSV)	5 -10	2,5 - 6,0
Temperature, °C	280 - 340	320 - 400
Pressure, MPa	2 -3	3 -5
Typical inlet sulfur content % wt.	0,1	1,5
Outlet sulfur content, % wt, (guaranteed)	< 0,00002	< 0,05
Whole service life duration, years (guaranteed)	7	5

Commercial experience

The catalysts of the КГу/ series are introduced by "OLKAT" Ltd. at five plants of the refining branch (Pavlodar, Khabarovsk, Angarsk, Uhta and Komsomolsk plants). While in operation at these plants the catalysts ensure residual sulfur of less than 0.5 ppm (usually about 0,1-0,2 ppm) value. The КГу-950 catalyst is being commissioned on Surgut ZSK at 3 hydrotreatment units (gasoline, diesel and kerosene fractions) and at the hydrotreatment assembly of reforming unit at Mahachkala refinery.

It has been reported by KIRISHI OIL JOINT STOCK company by blowing hydrogen directly into hydrotreatment unit of diesel fuel the quality of the diesel fuel has been improved due to the removal of hydrocarbon gases.

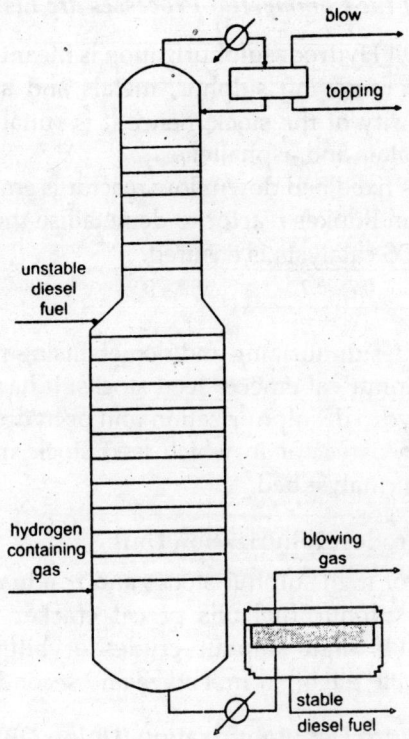

Figure 5.20b   Diesel Stabiliser.

In accordance with standard design, hydrogen-containing gas with hydrogen concentration not below 90% is used as blowing gas. Hydrogen concentration below the normative one results in the temperature decrease of the diesel fuel flash.

To remove dissolved hydrocarbon gases from stable diesel fuel at flow rate 300 m³/hr; the "PETON" Company has installed the packing in the low-pressure separator (Fig . 5.20 b).

The result:

- the necessary diesel flash point has been achieved at wide fluctuation in the content of hydrocarbon gases in hydrogen-containing gas
- the packing has been working without repair and replacement for more than 10 years.

5.6.3.4 Some of the Commercial Processes are Listed Below

Shell Residual Oil Hydrodesulphurization is meant for improving the residual oils by removing sulphur, metals and asphaltenes. It can reduce the viscosity of the stock, hence it is suitable for wide range stocks rich in metals and asphaltenes.

In this process fixed bed down flow reactor is employed. A suitable catalyst is used in bunker reactor to demetallise the feed stocks, thus protection of HDS catalysts is ensured.

H-Oil Process

This is used for desulphurizing and demetallising residua as well the produce low sulphur cat cracker feed stocks. It has the advantage of functioning as hydro desulphurization unit or hydrocracker unit. This used ebullated bed reactor in which feed stock and hydrogen pass upward through catalyst bed.

Gulf Resid Hydrodesulphurization Unit

This is suitable for high sulphur stocks and residua and is capable of producing low sulphur fuel oils or cat cracker stocks. It can be modified to tackle high sulphur crudes or bitumens. It has the provision to recycle the oil at first stage and second stage.

Uni-cracking Hydro Desulphurization (Union Oil)

This is fixed bed catalytic process mainly suitable for atmospheric and vacuum residua. Here, the formation of light fractions is not allowed, and the process is conducted at relatively low operating conditions. The high efficiency of the process is believed to have obtained from the excellent distribution of feed and hydrogen in the reactor. The specially designed Co-Mo on alumina catalyst with specific pore sizes is responsible for high degree desulphurization with coke deposition tendency minimised.

EXXON Residifining Process

This is fixed bed down flow reactor, and coking and fouling tendencies are protected by guard chamber. It is a low conversion process operating at less pressure.

Chevron RDS Isomax

The process is designed to eliminate besides sulphur, nitrogen, metal and asphaltene contents. It can accept whole crude or topped crudes. Maximum yields is low sulphur fuel oil. Moderate temperature and pressures are used in down flow reactor complex. Many combinations of fuel stocks are also handled.

5.6.4 Hydrotreatment/Hydrofining

These operations are treated as finishing operations. Different fractions require characteristic finished properties, hence the operation is selective to each fraction. Colour, stability and odour are the selective parameters and improvement in these makes the product more valuable.

For naphthas and gasolines

Treatment improves colour, odour and stability. Removal of sulfur is essential before feeding to reformers. Susceptibility to lead is increased if the sulfur is removed.

For Kerosene and Jet Fuels

Besides improvement in colour, odour and stability; smoke point improvement is added advantage. Conversion of aromatics to naphthenes results in quantity and quality of these fuels. Diesels also fall into this category. Though much saturation is not anticipated, still removal of polluents is highly commended.

For Lubes

Viscosity index improvement is a great achievement.[69] This treatment thus saves the valuable material in going as loss. Eliminates to certain extent refining and acid treatment.

Some of the trade names of the processes instituted for different purposes:

5.6.4.1 *Autofining/Objectives*

Desulfurization of distillate stocks. Preparation of feed stocks for steam reformers and SNG plants.

Arofining/Objectives

Reduction of aromatics helps in producing quality jet fuels. Aromatics free solvents and less aromatic feed stocks for some cracking operations.

Gulfining

Production of low sulfur fuel oils or cat. cracking stocks.

HPN

Selective hydrogenation and desulfurisation. Hydrogenation of C_3-C_4 olefins and diolefins.

Hydrogenation of cracker naphthas.

Hydrobon

Objectionable materials are removed from petroleum distillates. Sulfur in naphtha may be reduced to 0.5 ppm.

5.6.4.2 Hydrofining

Removal of sulfur from wide range of distillate feed stocks. Improvement in colour, stability and odour.

Lube hydrofinishing

Improves colour and stability of lubricating oils. Reduction in impurities and improvement in viscosity index.

Ferrofining

Solvent refined lubes are treated. This eliminates acid-clay treatment of lube stocks, employs rugged three component catalyst.

Ultrafining

Desulfurization, denitrogenation and saturation of olefins. Naphthas, Kerosenes and diesels decanted oils are treated. Employs CO, Ni & Mo catalysts. Low pressure and low temperatures with minimum consumption of hydrogen.

Unionfining

Desulfurization and denitrogenation of wide variety of petroleum stocks. Mainly used for cat. cracking feed stocks and kerosene desulfurization.

5.6.4.3 Hydrofining operation

This process is versatile in industry, with proper choice of catalyst and conditions, impurities are removed and burning characteristics of virgin and cracked materials are improved. Naphtha to reformers (steam/catalytic) must have as low as 5 ppm of sulfur and nitrogen; otherwise the bimetallic catalysts are easily poisoned. It may be treated as a sweatening technique for such feeds. High quality diesels and kerosene can be made from even sour crudes.

Figure 5.21 shows the outline of the process.

Feed stock and hydrogen (recycle + make up) are mixed and heated in a furnace (1). The mixture is passed through reactor (2) containing Co-Mo oxide catalyst. Reactor conditions range from 250 to 450°C and 10 to 80 atmos. depending upon the degree of treatment and boiling point of feed. The reactor effluents are heat exchanged and purged in

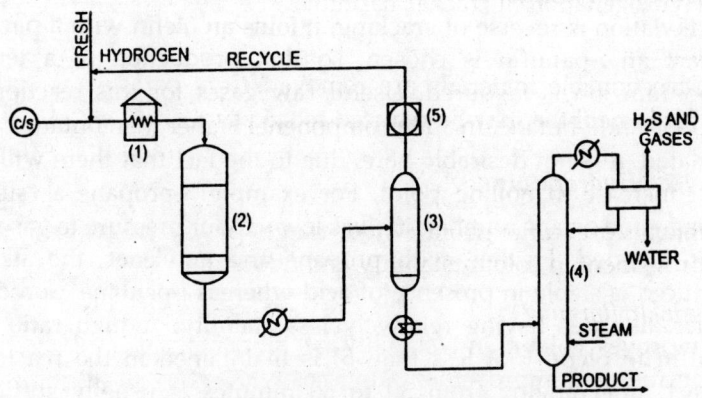

Figure 5.21 Hydrofining Operation.

a pressure releasing drum (3). Recycle hydrogen after treatment with different solutions (5) to absorb H_2S and NH_3, compressed and mingled with the make up hydrogen. The products are stripped (4) to remove light gases and remaining hydrogen sulfide; the bottoms of the stripper constitute the treated product. Catalyst is not generally regenerated but replaced after 12 to 14 months.

5.7 ALKYLATION

At present alkylation processes are gaining importance; high quality blends of motor fuels can be obtained by this method.

Alkylation processes are conducted on a large scale. Equivalent to 2% world crude refining is earmarked for this operation alone signifies this statement.[71]

Alkylation processes are conducted by Lewis acids like sulfuric, hydrofluoric acids and aluminium chloride. 60% of alkylation is done by sulfuric acid, while the rest is done by hydrofluoric acid; this shows sulfuric acid has got certain advantages over HF. Aluminium chloride is not used very much for alkylation purposes. Sulfuric acid process requires a very low temperature, usually less than 15°C to prevent oxidation of the products; HF can operate fairly well upto a temperature of 35°C. During alkylation, sulfuric acid undergoes marked changes. Fresh acids give low yields of alkylation, while circulating acids give improved yields.[72] Some additives are also added to decrease the consumption of acid and increase the yield of alkylate.[73] Alkylation reactions are exothermic hence heat removal is essential. This is done by evaporating propane in the reactor.

434

Feed Stock and Reactions

Alkylation is reverse of cracking, it joins an olefin with a paraffin. In fact an *i*-paraffin is chosen, so that production of a tertiary carbonium ion is favoured. Useful raw gases for this reaction are propone and *i*-butane. In spite components higher than butane can be alkylated, it is not desirable here, due to the fact that there will be a great increase in boiling point. For example, *i*-propane as such is having good octane number, it gives low vapour pressure to gasolines when blended. Further, with propane and amylenes, the alkylate produced is stable in presence of acid whereas *i*-pentane[74] produces higher alkylates having relatively less stability. A high ratio of *i*-paraffin to olefin (not less than 5) is maintained in the reactor. A contact time ranging from 30 to 60 minutes is usually sufficient. Unconsumed *i*-paraffin is always recycled back to increase its concentration in the reactor. Reactions proceed via carbonium ion mechanism as shown below:

$$C - C - C + C = C - C - C \rightarrow C - C - C - C - C$$

i-butane   *i*-butane   2, 2, 4 Trimethyl pentane

or

$$C - C - C + C = C - C - C \rightarrow C - C - C - C - C$$

i-butane   propene   2, 2, 4 dimethyl pentane

In sulfuric acid alkylation acid of 93 to 95% strength is used (higher strengths of acid can cause the reactions at lower temperatures). Sulfuric acid alkylation is done by two well known processes, namely Cascade and Effluent Refrigeration. The differences between the two processes are in reactor design and how propane and butane evaporate in the system to induce refrigeration.

5.7.1 Cascade Sulfuric Acid Alkylation

In this a multistage cascade reactor system (fitted in each stage with a mixer) is employed. Hydrocarbon and acid pass from one stage to other cascading serially; Olefin is split and introduced into each

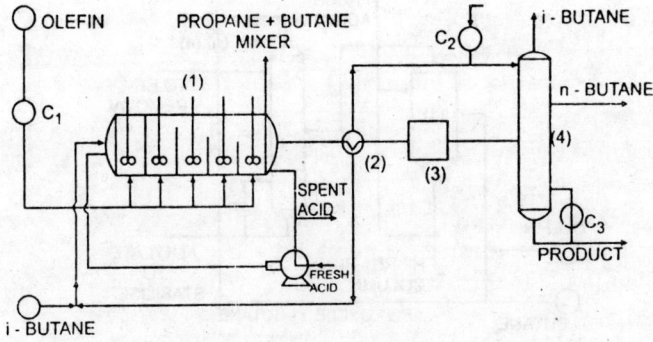

Figure 5.22 Sulfuric Acid Alkylation.
(1) Cascade Reactor (2) Heat Exchanger
(3) Caustic Wash Section (4) Fractionator
C_1 & C_2 Coolers, C_3 Reboiler

cascade. The formed alkylate is taken out from the top of cascade reactor, cooled and fractionated. i-Butane from the fractionator is mixed with the incoming feed and sent into reactor. Acid from the bottom of the reactors is taken and kept in circulation, necessary fresh addition of acid is also continued. Propane evaporation causes self refrigeration and maintains the temperature of alkylation at required low level. Propane leaving reactor with butane stream, is compressed cooled and fractionated to recover back the refrigerant for circulation. Fig. 5.22 shows the process flow sheet.

5.7.2 H.F. Alkylation

There are two commercial processes employing HF as catalyst. These are licenced by Phillips Petroleum and UOP Companies.

The essential features of the process are shown in Fig. 5.23. Olefin and i-butane feed (1 : 5) is dried and mixed with recycling acid. The mixture is passed into reactor; the reactor acts as settler also. After reaction is completed mixture is allowed to settle into two liquid layers. Acid layer is removed from the bottom of the setter, cooled and recycled. Fresh acid is supplemented whenever required. A part of acid is stripped to remove water and polymer products. Acid, free from these, is again mixed with the circulating acid.

Hydrocarbon layer removed from the top of the acid settler contains unreacted constituents and alkylate along with small amounts of acid. The components are separated by fractionation and i-butane is circulated into the reactor. The flow sheet does not show

436

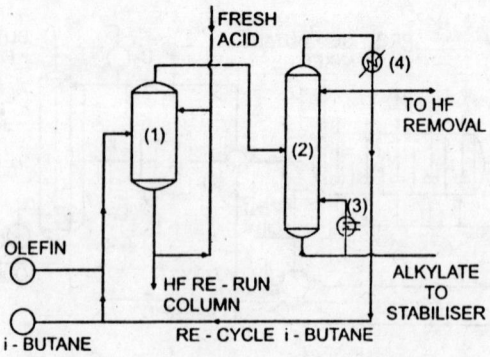

Figure 5.23   Hydrofluoric Acid Alkylation.
(1) Reactor (2) Fractionator
(3) Reboiler (4) Cooler

complete rectification system. Alkalate from the bottom is again stabilized to separate *i*-butane. *i*-Butane leaving from these columns is fed back to the reactor. Acid re-run plant is also not shown in the flow sheet.

A minimum pressure in both the alkylation systems is essential to increase the partial pressure of reactants and decrease the volatility. The properties of both these acids is shown below. It is explicit, that HF is soluble in hydrocarbon layer up to 0.5%. That is the reason, why a separate distillation plant is required to recover the HF from alkylates; while this trouble may not be there with sulfuric acid.

Properties of acids :

	Sulfuric acid	Hydrofluoric acid
B.P. °C	290	19.4
Freezing point °C	+3	–83
Sp. gravity	1.84	0.99
Viscosity (CP)	3.3 (15°C)	0.26 (0.C)
i. C_4H_{10}; in 100% acid Solubility	–	2.7 at 27°C
-do- in 99.5% acid	0.1 (13°C)	–
Solubility of acid in hydrocarbon layer:	–	0.5%

$AlCl_3$ alkylation proceeds in presence of HCl gas,[75] (8 to 15 moles per 100 moles of feed gas). The catalyst can be in complex form with

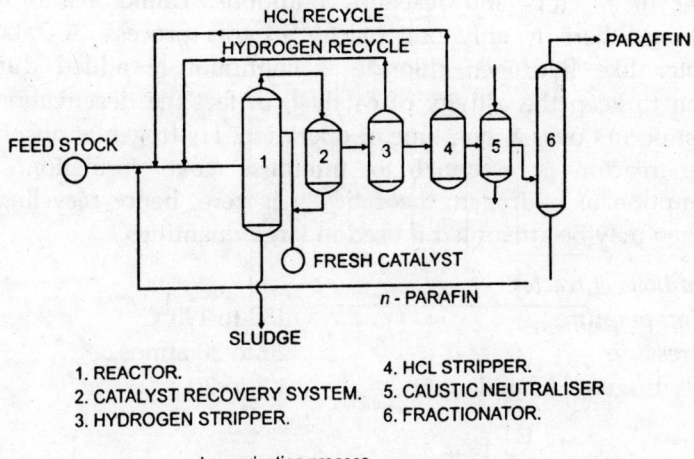

Isomerisation process.

Figure 5.24   Aluminium Chloride Isomerisation Process.

hydrocarbon diluents or can be used in dry state. About 170 gallons of alkylate is obtained per kg. of salt.

5.8   ISOMERISATION PROCESSES

Conversion of normal paraffins into isoparaffins help in two ways. Good quality gasoline due to improvement in octane number is the essential result of an isomerisation process. For alkylation purposes, isoparaffins are required because the straight run products naturally cannot meet such demands; hence isomerisation of straight chain compounds is inevitable.

Feed stocks suitable for such operations should mainly consist of C_4, C_5 and C_6 fractions. Isomerisation may take place in vapour phase or in liquid phase with a dissolved catalyst.

Process Description

All types of catalysts used in cat. cracking and reforming operations are certain to bring about this reorientation of molecules[76]. At present aluminium chloride catalyst is vastly employed. This catalyst can be used in fixed beds or in liquid contactors. Noble metal catalysts are generally employed in fixed beds only.

5.8.1   Isomerisations with Platinum Catalyst

The catalyst used in this process is similar to platinum base catalyst used in catalytic reforming operations. The main difference is in number of reactors and reaction conditions. Unlike platforming operations, there is only one reactor in this process. A catalyst promoter like hydrogen chloride is continuously added during reaction to keep the activity of catalyst; in fact the deactivation of catalyst occurs over a long time of operation. Hydrogen atmosphere in the reactor is essential to minimise coke formation; the consumption of hydrogen, theoretically is zero, hence recycling of hydrogen may be attempted if used in large quantities.

Conditions of reactor:

Temperature	150 to 190°C
Pressure	20 to 30 atmos.
Hydrogen/HC mole ratio	1.8 to 2:1

Space velocity $\left(\dfrac{V}{hr.V}\right)$ 2 to 3

The improvement of octane number for whole naphtha (C_5 to C_7) fraction may be about 8 units. For individual components like butane, pentane and hexane, it may be more.

5.8.1.1 Aluminium Chloride Process

The process can be conducted in fixed beds where catalyst is usually deposited on various carriers. The catalyst is non-regenerated. Hydrogen Chloride and hydrogen are continuously added into reactor. Fig. 5.24 shows the sequence of operations in isomerisation process using fixed bed aluminium chloride catalyst. Desulfurised and moisture free feed stock (C_5/C_6) with recycled n-paraffin is mixed with hydrogen and heated (not shown in figure). This mixture with recycled HCl gas enters the reactor (1); where the catalyst is spread on alumina or bauxite in a fixed bed.

The effluents are passed through a catalyst recovery system (2) and then through a pressure releaser (3) where hydrogen separates out. Later HCl gas is recovered in a stripper (4). The effluent is given a caustic wash (5) and finally sent into a fractionator (6) to separate i-paraffin at the top and n-paraffin at the bottom, n-paraffin is recycled back.

Reaction conditions:

Temperature °C	:	180 to 250
Pressure	:	20 to 50 atmos.

Space velocity $\left(\dfrac{V}{V-\text{hr.}}\right)$: 1.0

Hydrogen requirements : 2 to 6 M^3 per bbl.

HCl wt % : 4 to 5

Aluminium chloride, when used in liquid contractors is generally dissolved or made into a slurry in hydrocarbon liquids. The tendency for aluminium chloride to form complex with these solvents increases; hence a feed stock higher than butane is likely to carry more aluminium chloride Fig. 5.25 shows the operation in liquid phase.

Paraffin feed (1) is first dried in a dryer (2) and heated to 60 to 100°C in a heat exchanger (3). The feed now passes into a catalyst scrubber (4), where it can remove the catalyst entrapped in heavy sludge obtained in the process. The feed is then mixed with hydrogen, and hydrogen chloride vapours and allowed to enter the agitator reactor (5), where fresh/recycle catalyst is always present.

The effluents are then washed with water/alcohol to remove the catalyst in a wash column (6). The light hydrocarbon layer is led into an accumulator where hydrogen and hydrogen-chloride are separated. Heavy, settled hydrocarbon layer is circulated back to wash column while some product is taken to a stripper where remaining HCl vapours are driven off. Light paraffins recycled back, while i-paraffin is obtained at the bottom.

The conversions are likely to exceed 60% with all types of feeds.

Some processes of industrial significance are given below:

Hysomer Process: (Union Carbide/Royal Dutch/Shell).

It is aimed to improve the octane number of gasolines containing rich amounts of pentanes and hexanes. Hydro cracked stocks are generally rich with such components. In this process noble metal deposited on zeolites, capable of offering dual functions, is employed as catalyst. These catalysts are capable of taking feeds containing sulfur and water upto 35 and 50 ppm respectively. Hydrogen consumption is very low, 0.1 to 0.3% by wt. The catalyst is regenerable.

Operating conditions:

Temperature 250 to 300°C

Pressure 15 to 30 atmos.

Space velocity $\dfrac{V}{hr.\,V}$ 1 to 3

Octane number improvement is generally 6 to 10 units.

440

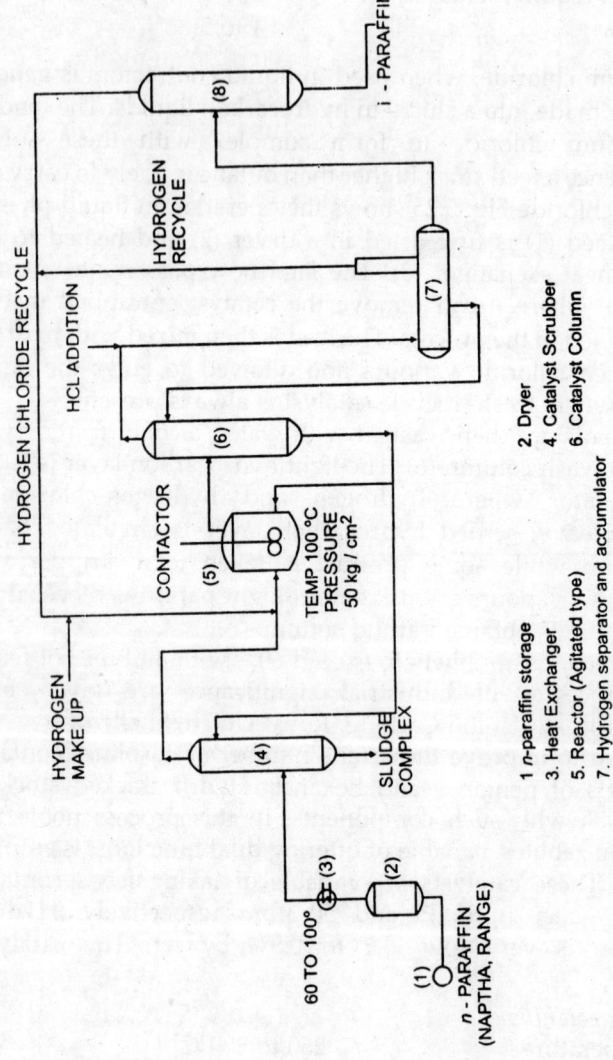

Figure 5.25 Aluminium Chloride (liquid phase) Isomerisation Process.

n - PARAFFIN (NAPTHA, RANGE)

60 TO 100°

HYDROGEN MAKE UP

SLUDGE COMPLEX

CONTACTOR

TEMP 100 °C PRESSURE 50 kg / cm²

HCL ADDITION

HYDROGEN CHLORIDE RECYCLE

HYDROGEN RECYCLE

i - PARAFFIN

1 n-paraffin storage

2. Dryer

3. Heat Exchanger

4. Catalyst Scrubber

5. Reactor (Agitated type)

6. Catalyst Column

7. Hydrogen separator and accumulator

8. Fractionator

5.8.1.2 Penex (UOP)

Pentanes or hexanes are isomerised over platinum containing catalysts.

Operating conditions:

Temperature	—	125 to 180°C
Pressure	—	20 to 70 atmos.
Octane improvement	—	6 to 10 units
Conversion	—	upto 100%

5.8.1.3 Isomate Process

Aluminium chloride complex as catalyst with anhydrous hydrogenchloride promoter offers non regenerative isomerisation process.

Operating conditions:

Temperature	110 to 125°C
Pressure	up to 30 atmos.

5.9 POLYMER GASOLINES

Polymer gasolines actually represent dimers and trimers of olefins. High molecular weight polymers form a separate chapter.

Polymerisation is intended to convert off-gases of thermal crackers. This can be accomplished in thermal units or catalytic units. In fact purification of the stream is not essential as the catalysts are capable of converting olefins even in presence of other diluents. Thermal polymerisation is not very effective, hence only catalytic polymerisation is industrially opted.

Olefin polymerisation is brought about by certain catalysts,[77] like copper pyrophosphates on charcol, sulfuric acid, phosphoric acid; at low temperatures not exceeding 200°C and at pressure of 10 to 80 atmospheres. The reaction is exothermic liberating 225 to 300 kcal/kg. of polymer formed.

Process Description

The feed stock consists of C_3 & C_4 olefins (40 to 60%) obtained from crackers.

Catalyst is phosphoric acid deposited on carriers of inert nature (Quartz, Kiesulguhr etc.). The reaction is exothermic and is controlled by admitting cold feed stock into the reactor.

If Tubular reactors are used they can be controlled by water or oil circulation. The process is shown in Fig. 5.26 and is almost similar to

442

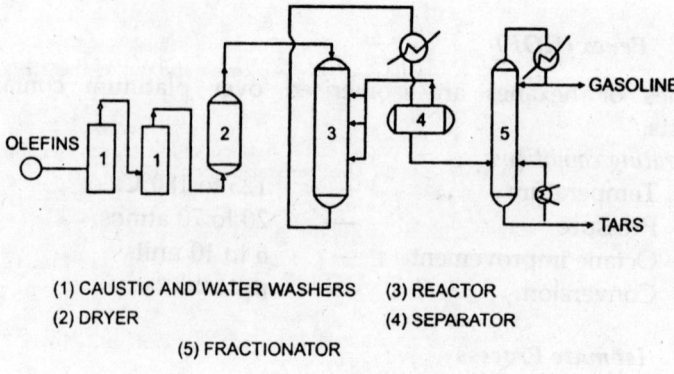

(1) CAUSTIC AND WATER WASHERS   (3) REACTOR
(2) DRYER   (4) SEPARATOR
(5) FRACTIONATOR

Figure 5.26   Flow Diagram for Polymer Gasoline.

alkylation process. The off gases from the crackers are first washed with caustic and water (1). The gases are then passed through a dryer (2), enter the reactor (3). The effluent from the reactor enters through a cooler into pressure releaser (4) where condensed gasoline is collected and recycled gas is separated. The liquid fractions are now taken into fractionator (5) where polymer gasoline is obtained from the bottom of the column. Unreacted gas escapes from the top of the tower.

The reactions can be conducted at low pressure or high pressure. The life of the catalyst is less when low pressure is used,[78] about 70 gallons of polymer can be produced by a kg. of catalyst; high pressure reactors can yield 140 gallons of polymer per kg. catalyst. The polymer can give an octane number of 94 to 97.

Sulfuric acid of 65% strength at an operating temperature of 30°C is sufficient.[79]

REFERENCES

1. Hengstebeck, R.J., Petroleum Processing, McGraw Hill p. 123.
2. Martin & Wills, Advances in Petroleum Refining and Technology edited by Kenneth Kobe, Interscience Vol. II, p. 359.
3. (a) Brooks, B.T., Kurtz, S.S., Boord, C.E. & Schmerling. The Chemistry of Petroleum Hydrocarbons Reinhold Vol. I, p. 40.
   (b) Ibid.Vol. II, p. 12.
   (b) & 47 Nelson, W.L., Petroleum Refinery Engineering, McGraw Hill p. 676, 652, 684.
5. Genisse & Reuter, I.E.C., 24, 1932, p. 219.
6. Aluddin, Md., Rao, B.N.B. & Banerjee T.S., Chemical Age of India 1970.
7. Charles L. Thomas & Edward J. McNelis, 7th WPC, IB, p. 161-169.

443

8. Sittig, M. Pet. Ref. 31, 9, 1952, p. 263.
9. (a) Voltz, S.E., Mace, D.M., Jacob, S.M. & Weekman, W.W., IEC, Proc. Des. Dev. 11(2), 1972, p. 261.
   (b) Estwood, S.E., Sailor, R.A. & Schwartz, A.B. 7th WPC, Vol. 4, p. 90.
10. Costantinides, G & Arich, G., Fundamental Aspects of Petroleum Geochemistry (B. Nagy & U. Colombo,) Elsevier 1967, p. 109.
11. John B. Rush, HCP Sept. 1981, p. 113.
12. & 51 Frank G. Ciapetta, 7th WPC, Vol. IB, p. 141-143.
13. Bailey, W.A. & Morse, N.L., Oil & Gas J., 1966, 64 (No. 11) p. 110-114.
14. Plank, L.J., Rosonski, E.J. & Hawthrone, W.P. IEC Prod. Res. Vev. 3, 1964, p. 165.
15. Smith, W.M., 7th WPC, Vol. 4, p. 83.
16. Conn, A.L., Meehan, W.F. & Shankland, R.V., CEP, 40 1950, p. 176-186.
17. Murphy, J.R., Oil & Gas J. 68 (47), 1970, p. 72.
18. Whittington, E.L., Murphy, J.R. & Lutz, I.H. Oil & Gas J. 70(44), 1972, p. 49.
19. Bunn Jr. D.P, Gruenke, G.F. & Jones, H.B. CEP, 65 (6), 1966, p. 86.
20. (a) & (b) HCP, Sept. 1972, p. 136, 137.
21. Petroleum, March 1953, p. 75-80.
22. (a) Pohlenz, J.B., Oil & Gas J. 61 (3), 1963, p. 124.
    (b)   -do-74(48), 1976, p. 63.
23. Montgomery, J.A. Oil & Gas J. 70 (50), 1972, p. 81.
24. Paul B. Venuto & Habib Jr. E.T., Fluid Cracking with Zeolits Catalysts Marcel Dekker, p. 65.
25. Ewell, V.R.W., Gadmer, G. & Turk, W.J., HCP, Sept. 1981, p. 103.
26. Pierce, W.L., Souther, R.S., Kaufman, T.G. & Ryan, D.F., HCP, May 1972, p. 92.
27. Strother, C.W., Vermillion, W.L. & Conner, A. J., HCP, May 1972, p. 89.
28. Henz Heinmann, Harold Shalit & Briggs, W.S., IEC, April 1953, p. 800.
29. Haensel, V. & Addison, E.C. 7th WPC, Vol. 4, p. 117.
30. Conn, A.L., CEP. Dec. 73, p. 11.
31. (a) Coggins & Petersen, API, Panel Discussion HCP, April 1972, p. 131.
    (b) Jacob, R.R. Ibid. p. 151.
32. Seebold, J.E., Bertetti, J.W., Snuggs, J.F. & Bock, J.A. Pet. Ref. May 1952, p. 114.
33. Payne, J.W., Evans, L.P., Bergstrom, E.V. & Bowles, V.O. Pet. Ref. May 1952, p. 117.
34. 57 & 76 James G. Speight. The Chemistry and Technology of Petroleum Marcel Dekker, p. 359, 344, 202.
35. Law, C., International Petroleum Abstracts No. 1, Vol. 9, 1981, p. 21.
36. Edmister, HCP, July 1973, p. 128.
37. Sittig, M. Handbook of Catalyst Manufacture Noyes Data Cor, p. 205, 141.
38. Samuel, D. Burd Jr. & John Maziuk, HCP, May 1972, p. 97.
39. Bernard J. Cha. Ronald Hein, Hugo Van Landeghem and AndreVidal. HCP, May 1873, p. 98.
40. Ascrizzi, J.M. HCP, April 1967, p. 140.
41. Steiner, H. Introduction to Petroleum Chemicals Pergamon Press p. 56.
42. Chakravarty, S. & Arun Kumar. Proc. of All India Seminar on Carbon Tech., Korba, (BALCO) March 1978, p. 11.
43. Rao, B.K.B. & Sen, P. - Ibid-p. 33.
44. & 73 James H. Gary & Glen E. HandWerk, Petroleum Refining Marcel Dekker, p. 58.
45. Nelson, W.L. Oil & Gas J. 54, 3, 1956, p. 129.

444

46. Stekhum, A.I., Mustafina, S.A., Korchagina, T.U. & Korchagina, R.N. - Ibid- p. 143.
48. & 59 Kett, T.K., Gerard C. Lahn & N. Schutte, William. Process Technology and Flow sheets, McGraw Hill, 1975, p. 204.
49. Indian Petroleum Handbook, 1973, p. 213.
50. Mantell, C.L., Carbon and Graphite Handbook, Intr. Science 1968, p. 149.
52. Carter, C.P. HCP, May 1981, p. 96.
53. Scott, J.W. & Patterson, N.J. 7th WPC, Vol. 4, p. 105.
54. Chemistry of Petroleum Extraction II; MSS Information Corporation ₀55, Madison Avenue, 1976, p. 17, 12.
55. Light, S.D., Bertran, R.V. & Ward J.W. HCP, May 1981, p. 93.
56. Boevink, J.E., Foster, C.E. & Kumar, S.R. HCP, Sept. 1981, p. 123.
58. Beavon, D.K. 7th WPC, Vol. 6, p. 28.
60. Clarance, D. Chang & Anthony J. Silverstri, IEC, Process Des. Dev. 1976, Vol. 15, No. 1, p. 161.
61. Stephen, M. Oleck & Howard S. Sherry, IEC, Process Des. Dev. 1976, Vol. 16, No. 4, p. 52.
62. Hung, C.W. & Wei, J., IEC, Process Dec. Dev. April 1980, 19(2) p. 250-263.
63. Mohammed, A.H.A.K. & Aboulgheit, A.K., HCP, Sept. 1981, p. 145.
64. Arie de Bruijn, Itaru Naka & Sonnemans, J.M. IEC, Process Des. Dev. 1981, No. 1, 20, p. 40.
65. Sivasubramanian, R. & Crynes, B.L, IEC, Prod. Res. Dev. 1980, No. 3, 19, p. 456-459.
66. Kellett. T. F., Sartor, A.F. & Trevino, C.A., HCP, May 1980, p. 139.
67. Bull, S. & Marnin, A., 10th WPC, 1979, Vol. 4, p. 221.
68. Yanik, S.J., Frayer, J.A. & Hauling, G.P., HCP, May, 1977, p. 97.
69. Billion, A., Franck, J.P. & Peries, J.P., International Petroleum Abstracts, 1981, No. 2, Vol. 9, p. 21.
70. Rao, B.K.B., Banerjee, T.S. & Sen, P., 2nd LAWPSP Symposium IIT, Bombay.
71. Buiter, P., Vain, P., Spijkar, J. & Vanzoonen, D., 7th WPC vol. 4, p. 127.
72. Hofman & Scheritshein, J.American Chem. Soc. 1962, 84, p. 954.
74. Doshi, B. & Albright, L.E. IEC, Proc. Des. 1976, Vol. 15, No. 1, p. 53.
75. & 78 Kenneth A. Kobe & John A. McKetta., Advances in Petroleum Chemistry and Refining Vol. 1, p. 343, 297-301, Vol. 7, p. 297.
76. Dunning, H.N., IEC, March 1953, p. 552.
77. Langlois, G.E. & Walkay, J.E., Pet. Ref. August 1952, p. 79.
79. Henry Martyn Noel Petroleum Refining Mannual, Reinhold Pub. N. York, p. 125. 80-83 James, G. Speight
The Desulphurization of Heavy Oils and Residua.
Marcel Dekker, New York, p. 96.
81. Bohdan W. Wojciehwoski & Avelimo Corma
Catalytic Cracking Marcel Dekker, New York, 1986, p. 55.
82. HCP, Sept. 88, p. 69.
84. D.S.J. Jones, Elements of Petroleum Processing; John Wiley Pub., New York, 1995.
85. Hydrocarbon Processing: June 2000, 32.
86-87. T.S.R. Prasadarao and G. Muralidhar: Recent Advances in Basic and Applied Aspects of Industrial Catalysis: Elsevier; 1998; 5, 211.

Exercise

1. Relative velocity of a cracking reaction of a heavy oil at 490°C was found to be 18.8. To operate at 70 kg/cm^2 pressure the coil volume was increased by 10% per barrel. The temperature correction at this pressure was 0.38. Calculate soaking factor. Find the additional volume of the coil for cracking 100 lbs. of oil per hour.

   (Ans. S.F. 0.071, Vol. 24 liters)

2. Explain briefly the effect of steam in a naphtha cracker.
   (b) Full range naphtha (45-180°C) is cracked at a rate of 5 tons/ hr. Assuming different steam to hydrocarbon ratios estimate the optimum ratio of steam to hydrocarbon for producing maximum olefins.

3. How do the different homologous series influence cracking pattern. If the feed stock meant for thermal cracker is wrongly delivered to a hydrocracker, what sort of repurcussions may be anticipated.

4. A delayed coker is designed to process 1000 kg/hr of a feed stock consisting of atmospheric bottoms and lube extracts. The average CCR was found to be 31.3%. Calculate the following.
   (a) The amount of coke and gas produced per day.
   (b) The CCR of products is 10, find CDE.

5. Calculate heat of combustion of a mole of methane from the available data.

Heat of formation at 0°C	kJ/ kg mole
Methane	−72,000
Water (1)	−272,500
Carbon dioxide (g)	−376,220

   (Ans. 212 K cals/gm mole)

6. Find the heat of decomposition per kg. gasoline produced in vis breaker which handles 7000 bbs/day of a charge of properties given below:

Feed stock	C_{20}-C_{30}
ρ at 30°C	0.825
Heat of combustion	36 M^{2}J/kg

	Products	Yield %	Heat of Combustion M^2J/kg	Mol. wt.
1.	Gas	2.2	38.5	32
2.	Gasoline	12	44.5	80
3.	Middle fraction	28	41.3	190
4.	Fuel oil	29	38.2	295
5.	Residue	25	30.5	480
6.	Rest, losses.			

(Ans. 3.14 M^2J/Kg)

7. From the data (Prob. 6) find the activation energy per Kg. gasoline produced
   Time of cracking is 20 minutes at 400°C. (C=30)

(Ans. 52,303 K cals)

8. The following reaction was studied in a hydrocracker. The equilibrium constant at 25°C and one atmospheric pressure was found to be 0.03; find the equilibrium const. at 400°C, pressure remaining same.

$$C_2H_5C_6H_5(gas) \xrightarrow{H_2} C_2H_6 + C_6H_6(g)$$
$$\Delta H = -2,600 \text{ K cal/Kg mole.}$$

(Ans. 2.49 × 10^{-3})

6

Asphalt Technology

6.1 SOURCE OF ASPHALT (BITUMEN)

Asphalt is obtained as the ultimate bottom product of a vacuum distillation column. The residuums may still contain some oil, but further distillation serves no use at all. These residuums are rich in asphalts and form the starting materials for air blown bitumen. Asphalts are also obtained in large scale, from deasphalting units. Reduced crudes, waxy distillates, lube oils and long residuums are usually deasphalted for getting quality lube oil base stocks. Naphthenic crudes leave more asphaltic bodies in high boiling fractions, hence deasphalting operations are necessary for producing bright stocks and quality lube oils.

Cat. cracker feed stocks are also necessarily subjected to deasphalting operations.

6.1.1 Chemical Structure of Asphalt

Asphalts do not have general representative formula and are very complex in nature exhibiting inconsistent properties. Asphalts are colloidal in nature and contain asphaltenes in oil which are stabilised by resins to maintain the colloidal form. A sample of asphalt contains:

Oil	—	35 to 50%
Resins	—	5 to 20%
Asphaltenes	—	20 to 30%
Acids etc.	—	upto 10%

Asphalts obtained from vacuum distillation units shall have less pour point (usually between 30 to 60°C) where as asphalts from deasphalting operations show a high pour point (45 to 90°C). Asphaltic materials are usually responsible for some of the physical properties of crude oils. High sulfurous crudes give more asphaltic materials.

Asphaltic materials are black in colour and display agglutinating properties and hydrophobic tendencies. These are the properties which are industrially explored for road making and water proof works.

Structure of asphaltic materials has been the subject for numerous investigations. These investigations have revealed the significant details of the structure.

Most of the asphaltenes contain carbon to an extent of 82 ± 3% and hydrogen to 8.1 ± 0.7%. It can be seen that the carbon hydrogen ratio is greater than 9 but surprisingly the ratio for various asphalts remains same. Nitrogen, oxygen and sulfur are the usual hetero elements present in asphalts in varying amounts. With increasing aromaticity and hetero elements, the molecular weight of the material also increases. Oxygen content vary from 0.3 to 4.9 and sulfur content reach upto 10%, while variation in nitrogen content is not appreciable-goes upto 3% only.

The structure of asphalts has been open to question for some time.[1] The present NMR studies show that they are condensed aromatic structures of sixteen carbon atoms or more. The idea of such big molecule may be difficult to conceive. X-ray analysis has furnished the valuable information regarding the dimensions of interlamellar distance, layer diameter and height of unit cell.

Indeed, the solubility of asphalt in different solvents varies. This advantage has been capitalised in precipitating different grades of asphalts by the use of solvents.

Sulfuric acid[2] is quite destructive to asphaltenes. Some salts like aluminium chloride can successfully remove the last traces of asphaltenes. The facile interactions of metal chlorides with asphaltenes may prove as a possible means of petroleum deasphalting in future.

The molecular weight of asphalts vary considerably from 900 to 2000 or even more, but no structural investigations have emphasised the nature and location of the inorganic materials in these asphalts.

6.1.2 Action of Heat on Asphalt

Thermal cracking of asphalts result in more or less the same products as contemplated in heavy oil cracking. Thermal decomposition of asphaltenes is a good method for identifying and estimating the hetero atoms. A spectrum of compounds like saturates, unsaturates, and oxidation products are obtained during thermal decomposition of asphalts. Some of the decomposition products of asphalt are:

H_2, CO, CO_2, H_2S, SO_2, H_2O.

C_2H_4, C_2H_2; RNH_2, CH_4, C_2H_6, C_3H_6

C_4H_{10}, C_5H_{12}, C_6H_{14}

Dealkylation of aromatic rings at low temperatures; and formation of benzene and naphthalene structures at high temperatures are quite contrary but pronounced.

However, the oxidation of asphaltenes by common oxidising agents like acetic acid, peroxides etc. is exceedingly slow. Phenolic and carboxyl groups are formed during oxidation. There is some up-take of oxygen by the molecule which effectively results in decreasing the solubility of asphalt in heptane. During oxidation dehydrogenation also takes place, with the result, unsaturates and aromatics tend to increase. Blowing air through asphalts brings about orientation in molecular structure, usually decreasing the molecular weight but producing necessary functional groups to give more agglutinating properties. Moschopedis[3] et al. have shown that during oxidation of asphaltenes oxygen to carbon ratio increases with increase of time, further the colour and consistency are progressively increased.

6.1.3 Types of Asphalts

Asphalt is categorised into three distinct species, depending upon the source. The primary source is the residuums of vacuum distillation unit, second source is from deasphalting operations. Third is a mixed source comprising all the above two and solvent extracts. The necessary qualities like ductility, penetration index and API gravity are not available with these raw stocks; hence are to be imbeded. Some qualities are enhanced by air blowing of asphalt, the product of which is commercially known as air blown bitumen.

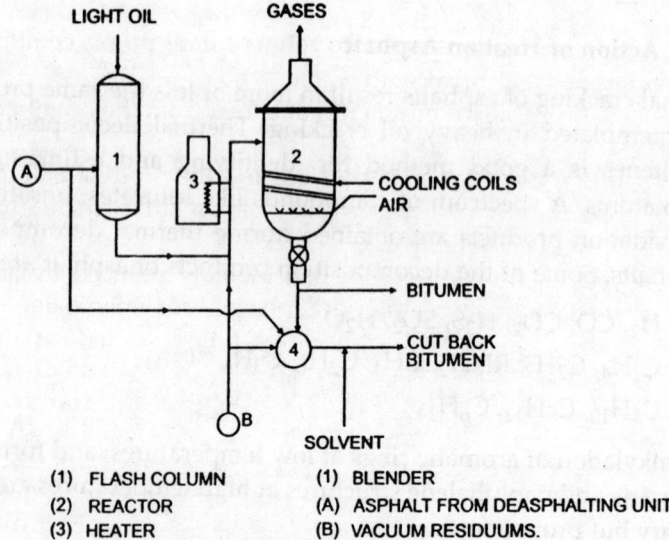

(1) FLASH COLUMN   **(1) BLENDER**
(2) REACTOR   **(A) ASPHALT FROM DEASPHALTING UNIT**
(3) HEATER   **(B) VACUUM RESIDUUMS.**

Figure 6.1   Bitumen Blowing.

6.2 AIR BLOWING OF BITUMEN

Air blowing is done mostly in batch units although continuous air blowing of bitumen is also in existence. In this operation, asphalts from various sources are mixed and heated to a temperature of 200 to 210°C and sent into the reactor. Figure 6.1 shows the outlines of the operation. The reactor may be horizontal or vertical type, usually made of mild steel, of a capacity to hold 1000 tons of charge stock. These reactors are fitted with air distributors at the bottom and also cooling and heating coils. During oxidation the temperature may increase rapidly, hence as a precautionary measure provision for cooling is made available.

The reactor should be maintained at a temperature of 200 to 320°C, and air rate of 0.5 to 1.5 m^3 at a pressure of 1.1 to 1.2 kgs/cm^2 per hour, per ton of charge stock is necessary.

Blowing time lasts for 10 to 14 hours depending upon the required consistency of bitumen. Heat of reaction[4] is dependent upon the consistency of bitumen as shown below:

Initial softening point °C	Final softening point °C	Heat of reaction kJ/kg.
41.4		16.9
21.7	120	170.75

Gases are allowed to escape into refinery flare up or/combustion system after being stripped with water.

The product, air-blown bitumen is obtained from the bottom of the reactor.

With sulfur also oxidation is possible and the product is found to have the comparable qualities as with air blown bitumen; but the cost of such process is exorbitant.

The time of air blowing can be considerably reduced by adding certain chemicals like ferric chloride[5], phosphorous pentoxide etc. A concentration of 0.3% ferric chloride is found to reduce the air blowing period by half.

Cut back bitumens are not really manufactured ones, but blended ones. These are obtained by blending various asphalts with a suitable solvent; in fact these are liquid bitumens. Their action prevails after the solvent is dried off. These are mostly cheap grade bitumens. Simple blending of asphalts also may serve common needs.

Penetration index, the most important property of bitumen judges the quality of bitumen. Arbitrary estimation of penetration index is possible[6] if the softening point (Ball and Ring method) is known; by the following correlation:

$$\frac{20-P.I.}{10+P.I.} \times \frac{1}{50} = \frac{log\ 800 - log\ P.I.\ at\ T°C}{Softening\ point\ at\ T°C}$$

This equation is based upon the assumption[7], that penetration index of all asphalts at softening point is approximately 800.

6.3 UPGRADATION OF HEAVY CRUDES

Heavy crudes are different from light crudes in many ways. They possess high gravity, high viscosity, high asphalt and metal contents.

The distillation characteristics usually show a very small amount of recovery of light and middle distillates. These properties are definitely different from the crudes of normal category; hence they cannot be processed by the conventional refineries. To feed these crudes into normal refineries a sort of pretreatment is required which goes by the name upgradation.

Thermal cracking, hydrocracking and deasphalting in combinations are suitable for the upgradation of heavy crudes. Visbreaking and separation of lights and then deasphalting the remaining portion, then feeding into light hydrogenation and cracking reactors may be worthy. If the asphalt rich crude can be

TABLE 6.1 Results in Different Upgradation Processes.

Property	Heavy crude oil	Visbreaking	Visbreaking after deasphalting	Deasphalting after visbreaking	Visbroken-Hydrocracking	Hydrocracking after deasphalting
AFI	17.5	21.5	30.6	27.1	33	33
Viscosity (CPs)	740	50.	18.0	30.0	10	10
Pour point (°C)	8.5	-1.5	-5.5	-6.0	-12	-15
Distillation %						
IBP-110°C	2	6.7	13.0	7.5	7.3	7.0
110-200°C	3.2	17.0	15.0	17.0	13.0	13.0
200-250°C	4.4	15.9	16.5	17.5	29.6	30.6
250-300°C	6.8	18.5	16.0	19.5	18.6	21.6
300-350°C	4.6	6.1	8.0	5.0	5.3	8.6
+ 350°C	79.0	31.5	33.5	26.2	26.2	19.14

deasphalted and fed into the hydrocracker, thermal cracking may be eliminated but processing may become uneconomical. It is better to reduce the load on hydrocracker. Hence visbreaking, distillation, deasphalting of heavy fractions and then feeding to the hydrocracker may result in an artificially excellent crude.

The processing of heavy crudes in future is a must not only for our country but for others too, in view of the huge deposits available as heavy crudes and shale oils. Table 6.1 shows some results on upgradation of heavy crude available from north Gujarat zones.

REFERENCES

1. James G. Speight. The Chemistry and Technology of Petroleum Marcel Dekker, p. 202.
2. Kalchivesky, V.A. & Kenneth A. Kobe. Petroleum Refining with Chemical, Elsevier Pub., 1956, p. 24.
3. Moschopedis, S.E. & Speight, *J.G. Fuel* 1978, p. 57, 239.
4. Douglas, S., Smith, B. & Berbert E. Schweyer. *IEC* Vol. 2, No. 3 July 1963, p. 209.
5. Pal Zakar, Asphalt, Chemical Publ. Co. Inc. N. York, 1971 p. 85.
6. Edwin J. Berth, Asphalt, Science and Technology, Gordon and Breach Science Pub., N. York, p. 272.
7. Herbert Abraham, Asphalt & Allied Substances, D. Van Nostrand Comp., Vol. IV, p. 130.

Bio Fuels

7.1.1 BIO FUELS (DIESEL)

Biodiesel is a domestic, renewable fuel for diesel engines. Made from soybean oil, natural oils, and yellow greases. Bio diesel is mono-alkyl esters of long chain fatty acids derived from vegetable oils or animal fats, designated B100. It is an advanced biofuel and good to be called than as biodiesel, If it is biodiesel it must meet the specifications of ASTM D 6751 like petroleum diesel.. Biodiesel can be used in any blend with petroleum , meeting the requirements of ASTM D 6751.

India, because of shortage of land and edible oils/ non edible-oils, biofuel development hangs mainly around the cultivation and processing of Jatropha plant seeds which are very rich in oil (40%). Jatropha oil has been used in India for several decades as lighting oil-later as biodiesel for the diesel fuel requirements of remote rural and forest communities; jatropha (Karanj, (Pongamia Pinatta Mahua, Neem), are other oil bearing seeds , these grow all over India,.[Karanj Satish Lele (www.svlele.com (1)]oil can be used directly after in diesel generators and engines like any oil. India's total biodiesel requirement is projected to grow to 3.6 Million Metric Tons in 2011-12, in par with the demands of the domestic automobile industry.Analysis from Frost & Sullivan, Strategic Analysis of the Indian Biofuels Industry, Analyst Hari Krishnan , reveals that the market is an emerging one and has a long way to go before it catches up with global competitors. To set up a biodiesel plant, one should have own source of land for cultivation of oil from Jatropha plantation. To set up 1,000 liters per day plant, a plantation in 200 hectares is needed which is very difficult in India..Hence the explorers have to collect seeds from a number of farmers. As the government biodiesel purchase price of Rs 26.5 (48 cents) per liter is still below the estimated biodiesel finished production cost (Rs 35 to Rs 40 per liter

In India, there are five large plants, set up with a capacity of 300,000 liters per day, 4 medium size plants, with a capacity of 30,000 liters per day, and a number of small plants, with a capacity of 1,000 to 3,000 liters per day. Practically all plants are running at very low capacities, or closed due to lack of oil. 4 more large plants are coming up to face the same fate. Diesel demand in the country is growing at an annual rate of 8%. At this rate India will need a 9-MMT capacity refinery every year, India will become diesel-deficit by 2016. According to Satish Lele the specifications of jatropha or any oil to be converted into Bio Diesel must have these specifications

FFA (preferably) : <: 2.0% w/, Water content : <: 1000 ppm
Phosphorus : <: 20 ppm w/w, Sulphur : <: 50 ppm Iodine Value (mg I2/100g) : <: 120, Saponification Number : >: 190 (mg KOH/g) pecific Gravity : 0.840 - 0.920,

Fatty Acid Profile

Myristic Acid : 0.38 %,	Palmitic Acid : 16.0 % max.
Palmitoleic Acid : 1 - 3.5 %,	Stearic Acid : 6 - 7.0 %
Oleic Acid : 42 - 43.5 %;	Linoleic Acid : 33 - 34.4 %
Linolenic Acid : >0.80 %;	Arachidic Acid : 0.20 %
Gadoleic Acid : 0.12 %	

Bio Fuels differ from the conventional mineral fuels that they have narrow boiling point range and compositions /The need of Bio Fuels to contain GHGs is top priority in many countries These are derived from the available bio resources

7.1.2   History and significance and Relative Importance

Bio Fuels have passed Three generations. Each generation is better than the previous one.

First-generation biofuels are made largely from edible sugars and starches.

Second-generation biofuels are made from nonedible plant materials.

Third-generation biofuels are made from algae and other microbes.

The first generation bio fuels (bio diesel from vegetable oils, ethanol from sugar, corn starch) are costly. Not only that raising food costs, forcing limited land availability for food crops etc are main concerns. In USA main food crop is corn ,with the diversion for alcohol the food cost increased by more than 100% and the affect else where was reflective.

The second generation biofuels based on gasification of bio mass waste. Cellulosic biomass can be separated from lignin and subjected to enzymztic conversion.. Alternatively bio mass can be gasified to CO and hydrogen which can react to form wide variety of hydrocarbons through FT process.(Bio mass to liquids) Coal conversion to liquids is another route (GTL), but require several treatments for quality fuels. Catalytic pyrolysis is a good approach. Laborious and but out dated Coal gasification and subsequent chemicals production through FT(2) reactions is even today a reality in Sasol (SA Africa) plant. Even India was also having coal based fertilizers but abandoned in favour of petroleum due to the cost factor. But with barrel at 100$, coal gasification will be ideal.. Third generation resource is Algae.

7.2.1a Algal biofuels

According to Michael Briggs(3a) at the UNH Biofuels Group , hopes to replace all vehicular fuel with biofuels by using algae that have a natural oil content greater than 50%, Briggs opined algae can be grown in ponds at wastewater treatment plants.: The University of New Hampshire Biodiesel Group also provided the following information on their Algae ponds:

"Micro algaes present the best option for producing biodiesel in quantities sufficient to completely replace petroleum. While traditional crops have yields of around 50-150 gallons of biodiesel per acre per year, algaes can yield 5.000-20,000 gallons per acre per year. Algae grow best off of waste streams. Agricultural, animal, or human. Some other studies have looked into designing raceway algae ponds to be fed by agricultural or animal waste. We are now pursuing funding to investigate redesigning wastewater treatment plants to use raceway algae ponds as the primary treatment phase. With the dual goal of treating the waste and growing algae for biodiesel extraction. We also plan to investigate the possibility of using the algae mush (what is left after extracting the oil) as a fertilizer."

This oil-rich algae can then be extracted from the system and processed into biofuels, with the dried remainder further reprocessed to create ethanol. The production of algae to harvest oil for biofuels has not yet been undertaken on a commercial scale, but feasibility studies have been conducted to arrive at the above yield estimate. In addition to its projected high yield, alga culture, unlike crop-based biofuels - does not entail a decrease in food production, since it requires neither farmland nor fresh water. New method of production is from the freshwater alga Chlorella vulgaris, grown using flue gas from a gas-fired power station as the carbon source. Cultivation using a two-stage method was considered, whereby the cells were initially grown to a high concentration of biomass under nitrogen-sufficient conditions, before the supply of nitrogen was discontinued, whereupon the cells accumulated triacylglycerides. Cultivation in typical raceways and air-lift tubular bioreactors was investigated, as well as different methods of downstream processing. Results from this analysis showed that, if the future target for the productivity of lipids from microalgae, such as C. vulgaris, of 40 tons hactare^{-1} year^{-1} could be achieved.

"Algae can be cultivated and harvested in support of a wide array of biofuel products. In addition, algae biofuels systems hold promise to enable rapid production of high quality, high throughput biofuels systems in support of carbon emissions reductions targets, and in support of clean fuel production, (3b).

At the International Aerospace Exhibition in Berlin in June 2010, the aircraft manufacturer EADS presented a pioneering development in biofuels , when a twin-engine plane powered by diesel made from microalgae took to the skies for the first time. The experiment was a success and surprisingly consumed less fuel than if conventional aviation fuel. In 2009 more than 16 billion liters of biodiesel and over 76 billion liters of bio ethanol were produced, and the output is increasing at double-digit rates every year. The U.S. biodiesel industry was producing more than 1 billion gallons of fuel in 2011, The total volume of nearly 1.1 billion gallons is by far a record for the industry New York City residents use the oil heat as the cleanest oil, since Oct 2012,, every gallon of oil heat in the city contained at least two percent biodiesel. The blend is known as Bioheat® fuel, a greener heating oil that is gaining popularity in many states.. Microalgae, which typically consist of only one cell, can grow in any dirty and salty water and therefore don't need high-quality arable land or water. In addition, they produce up to ten times as much oil as other plants on the same amount of space. The focus here is on lipids, which are formed by means of photosynthesis and have a molecular structure similar to that of petroleum products.This is relatively simple to turn such lipids into fuels. In 2010, worldwide biofuel production reached 105 billion liters (28 billion gallons US), up 17% from 2009,and biofuels provided 2.7% of the world's fuels for road transport, a contribution largely made up of ethanol and biodiesel.

The total oil content in algae can be up to 70% of their dry weight.

Micro-algae are capable of producing more than 30 times the amount of oil (per year per unit area of land) when compared to oil seed crops.

Micro-algae are much more efficient converters of solar energy than any known plant, because they grow in suspension where they have unlimited access to water and more efficient access to CO_2 and dissolved nutrients. Micro-algae are the fastest growing photosynthesizing organisms. They can complete an entire growing cycle every few days.

Upto 120 tons of oil/hectare/year can be produced from algae.

Carbon-di-oxide from atmosphere is to mop it up by fast growing Algae on barren and desert lands, which can fix up to 30 tons of Carbon-di-oxide per acre per year. (www.svlele.com)

7.2.1b Green diesel

Green diesel is produced through hydrocracking biological oil feedstocks, such as vegetable oils and animal fats . The cracking of big molecules into shorter hydrocarbon chains used in diesel engines. It may also be called renewable diesel, hydrotreated vegetable oil Green diesel has the same chemical properties as petroleum-based diesel Green diesel is being developed in the Louisiana and Singapore by many companies like, Neste Oil, Valero, Dynamic Fuels, and Honeywell UOP

Some of the economics of comparison

Crude	60$
Ethanol	54$ (from mollases)
Diesel (jatropha)	156$
Bio pyrolysis	120$
Bio crude (BTL)	72$

Analysis:
ASTM D6751-07 is the standard specification for bio-diesel fuel blend stock with middle distillate fuels in USA. Blend is 0-40% and is determined by using FT-IR spectroscopy. Methanol traces cannot incluence Vapour pressure but can influence flash point (10% can reduce FP from 120 to 60°C

New Leaf a company producing bio diesel from used vegetable oils is doing wonders. New Leaf is planning an expansion that would allow it to more than double output to 2.5 million gallons a year in 2011. New Leaf Biofuel is doing excellent service to contain emissions.

7.2.2 Processing of bio diesel

Fats and oils are triacylglycerols separated by the slightly arbitrary distinction of solid or liquid state at room temperature. In triacylglycerols of vegetabl· origin, fatty acids esterfied onto position 2 significantly differ from those esterified onto positions 1 and 3 - which exhibit little overall difference in substitution pattern, whereas in products of animal origin random substitution seems to predominate. Animal and plant fats and oils are composed of triglycerides, which are esters containing three free fatty acids and the tri hydric alcohol, glycerol. In the trans esterification process, the alcohol is first de-protonated with a base to make it a stronger nucleophile. Commonly, ethanol or methanol are used for esterification. As can be seen, the slow reaction is between triglyceride and the alcohol. Heat, as well as catalysts (acid and/or base) are used to speed up the reaction. Common catalysts for trans esterification consist of sodium hydroxide, potassium hydroxide, and sodium methoxide.

Almost all biodiesel is produced from virgin vegetable oils using the base-catalyzed technique as it is the most economical process for treating virgin vegetable oils, requiring only low temperatures and pressures and producing over 98% conversion yield (provided the starting oil is low in moisture and free fatty acids). However, biodiesel produced from other sources or by other methods may require acid catalysis, the process is however much slower. Since the base-catalyzed transesterification process is the predominant method for commercial production, this is described below.

Triglycerides [1] are reacted with an alcohol such as ethanol [2] to give ethyl esters of fatty acids [3] and glycerol [4]: The reaction sequence is

$$
\begin{array}{c}
\left[\begin{array}{l} O_2CR^1 \\ O_2CR^2 \\ O_2CR^3 \end{array}\right. \quad + \quad 3 \quad \diagup\diagdown_{OH}
\end{array}
\longrightarrow
$$

1   2   3   4

R^1, R^2, R^3 are radicals

Potassium Hydroxide is the catalyst; about 3-5% in methanol . The catalyst is added to 60 -70 liters vegetable oil in a tank reactor. The oil is kept at 45 to 65 0 C. The mixture is stirred to esterify. The schematic production is shown in Figure 2.

An alternative, catalyst-free method comprises the one-stage trans-esterfication with super critical methanol (4) The reaction of transesterification of triglycerides is carried out under supercritical conditions, i.e. at temperatures higher than the critical temperature of methanol. Raw materials for the reaction are methanol and triglycerides with any amount of free fatty acids thus enabling the production of biodiesel from cheap feedstocks such as beef tallow or high acidity yellow grease. The reaction proceeds without the aid of alkaline or acid catalysts, thus eliminating the need for neutralization steps downstream the reactor. In order to minimize the heat consumption and pumping power which are usually very high in the one-reactor configuration of all reported supercritical processes, two medium-pressure successive reactors with intermediate glycerol removal are used and a heat recovery scheme composed of heat exchangers and adiabatic flash drums is proposed.. Glycerol is cyclically retained in adsorption beds, desorbed in a swing step and recycled to the first reactor. Design parameters are obtained both from experimental data and estimation. Advantages are no process water effluents are produced . The whole system is essentially "dry" and only small water is produced and glycerol purification is simplified by the absence of catalyst and low water content. Usual high temperatures (Tc=235 °C) and high pressures 35-40mᵱa, the oil and methanol are in a single phase, and reaction occurs spontaneously and rapidly.[4] The free fatty acids are converted to methyl esters instead of soap, so a wide variety of feed stocks can be processed . Also the catalyst removal can be eliminated.(5)

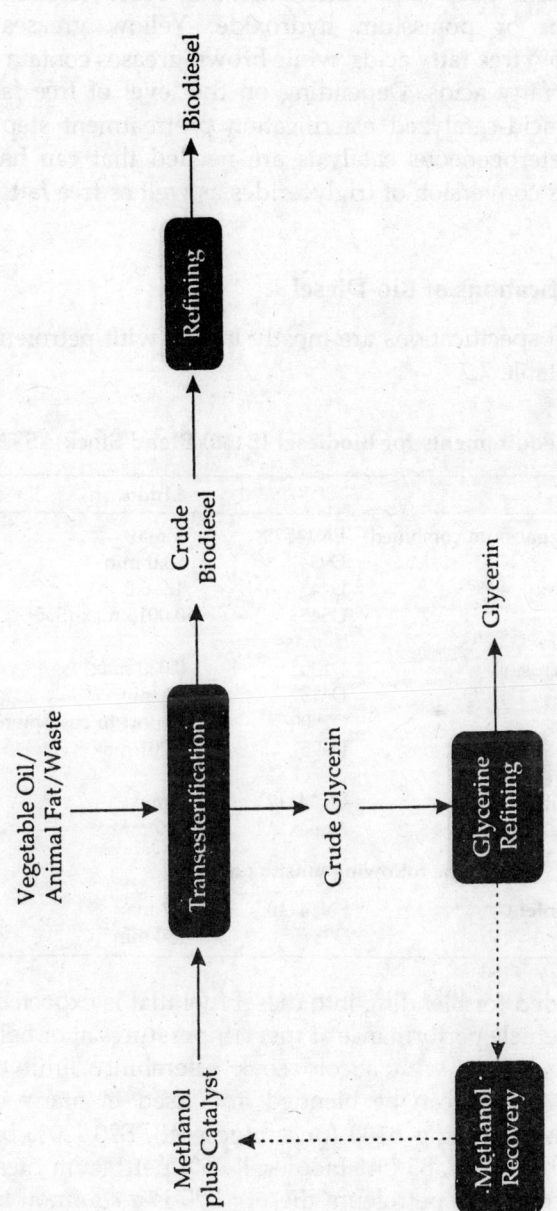

Figure 7.1   Schematic production of biodiesel

Complications arise when the triglycerides contain free fatty acids which will form "soap" and water when they react stoichiometrically with sodium or potassium hydroxide. Yellow greases contain between 5-15% free fatty acids, while brown greases contain in excess of 15% free fatty acids. Depending on the level of free fatty acids present, an acid-catalyzed esterification pretreatment step may be required. Heterogeneous catalysts are needed that can handle the simultaneous conversion of triglycerides as well as free fatty acids to FAME.

7.2.3 Specifications of Bio-Diesel

Bio-Diesel oil specifications are mostly in line with petroleum diesel as shown in table 7.2

Table 7.2 Requirements for Biodiesel (B100) Blend Stock ASTM D6751

Property		Limits	Units
Calcium and magnesium combined	EN14538	5 max	Ppm
Flash point	D93	93.0 min	°C
Kinematic viscosity, 40°C	D445	1.9-6.0	Mm^2/s
Sulfur	D5453	0.0015 max (S15)	
0.05 max (S500)	% mass		
Copper strip corrosion	D130	0.020 max	-
Cetane number	D613	47 min	-
Cloud point	D2500	Report to customer	°C
Phosphorus content	D4951	0.001 max	% mass
Oxidation stability	EN14112	3 min	Hours
Cold Soak filterability	Annex A1	360 maxc	seconds
Alcohol control - One of the following must be met:			
(1) Methanol content	EN14110	0.2 max	vol %
(2) Flash point	D93	130 min	°C

B100 intended for blending into diesel fuel that is expected to give satisfactory vehicle performance at fuel temperatures at or below 10°F (-12°C) shall comply with a cold soak filterability limit of 200 s maximum. Biodiesel can be blended and used in many different concentrations, including B100 (pure biodiesel), B20 (20% biodiesel, 80% petroleum diesel), B5 (5% biodiesel, 95% petroleum diesel), and B2 (2% biodiesel, 98% petroleum diesel). B20 is a common biodiesel blend in the United States.

Table 7.2 Requirements for Biodiesel (B100) Blend Stock ASTM D6751

Property	Europe	Limits	Units	Diesel
CCR 100%	0.05 max (Germany)	0.05 max(USA)		
Carbon residue (10%dist.residue)	0.3 max / 0.3 max	carbon residue shall be run on the 100% sample.		0.3 max (diesel)
Sulphated ash	0.02 max	0.02 max		
Oxide ash	0.03 max	0.02 max		0.1 max (diesel)
Water	300 max / 500 max	500 max		
Ester content	96.5 min			
Free glycerol	0.02 max / 0.02 max	0.02 max		
Total glycerol	0.25 max / 0.25 max	0.24 max		
Iodine value	115 max / 120 max			
Cetane number	49 min / 51 min	47 min		51 min (diesel)
Acid value	0.5 max / 0.5 max	0.5 max	mgKOH/g	
Density 15°C	0.875-0.90 / 0.86-0.90		g/cm³	0.82-0.845(diesel)
Viscosity 40°C	3.5-5.0 / 3.5-5.0	1.9-6.0	mm²/s	2.0-4.5 (diesel)
Distillation		90%,360°C	% @ °C	85%,350°C - 95%,360°C (diesel)
Density 15°C	0.875-0.90 / 0.86-0.90		g/cm³	0.82-0.845(diesel)
Viscosity 40°C	3.5-5.0 / 3.5-5.0	1.9-6.0	mm²/s	2.0-4.5 (diesel)
Distillation		90%,360°C	% @ °C	85%,350°C - 95%,360°C (diesel)

464

Advatages

Produced from non-petroluem, renewable resources
Can be used in most diesel engines, especially newer ones
Less air pollutants (other than nitrogen oxides)
Less greenhouse gas emissions (e.g., B20 reduces CO2 by 15%)
Biodegradable, Non-toxic (?), Safer to handle (?)
Disadvatages
Use of blends above B5 not yet approved by many auto makers
Lower fuel economy and power (10% lower for B100, 2% for B20)
Currently more expensive
B100 mostly not suitable for use in low temperatures

7.3.1 Bio ethanol

Ethanol since last two decades has been used in gasoline. Ethanol fuel has a "gasoline gallon equivalency" (GGE) value of 1.5 US gallons (5.7 L), which means 1.5 gallons of ethanol produce the energy of one gallon of gasoline. E10 is a low-level blend composed of 10% ethanol and 90% gasoline. Environmental Protection Agency (EPA) has legalised for use in any gasoline-powered vehicle E15 is a low-level blend composed of 15% ethanol and 85% gasoline. E85 is a high-level gasoline blend containing 51% to 83% ethanol, depending on geography or season, and qualifies as an alternative fuel under EPA act. E85 can be used in flexible fuel vehicles (FFVs), run on either E85 or gasoline. E85 cannot be legally used in conventional gasoline-powered vehicles. Ethanol can influence vapour pressure of gasoline hence the blend must be checked for Vapour Pressure. It is tested using ASTMD4806-06c like regular gasoline for stability, antioxidants, metal deactivators and biocides (not different from petroleum fuels) are added. Additives which can do water shedding , foam reduction are special requirement.

The use of E10 was spurred by the Clean Air Act Amendments of 1990 (and subsequent laws), which mandated the sale of oxygenated fuels in areas with unhealthy levels of carbon monoxide.

7.4. 1 Biofuel Mission of India

The country's energy demand is expected to grow at an annual rate of 4.8 per cent over . Projected requirement of biofuel for blending under different scenario are given in table 7.2&3.

Table 7.3 Shows requirements for 6% biodiesel (B6) to 20% biodiesel (B20) as listed in ASTM D7467.

Requirements for Biodiesel B6-B20 ASTM D7467			
Property	Test Method	Limits	Units
Acid number	D664	0.3 max	mg KOH/g
Viscosity at 40°C	D445	1.9-4.1a	mm2/s
Flash point	D93	52b	°C
Cloud point	D2500	Report to customer	°C
Distillation temperature, 90% evaporated	D86	343	°C
Ramsbottom carbon residue on 10% bottoms	D524	0.35 max	% mass
Sulfur	D5453	0.0015 max (S15) 0.05 max (S500)	% mass
Cetane number	D613	40 min[c]	–
Ash content	D482	0.01	% mass
Water and sediment	D2709	0.050 max	% mass
Copper corrosion 3 h at 50°C	D130	No. 3	–
Biodiesel content	DXXXXd	6-20	% (V/V)
Oxidation stability	EN14112	6 min	Hours
Lubricity at 60°C	D6079	520 maxd	Micron

One of the following must be met:

(1) Cetane index	D976-80	40 min	–
(2) Aromaticity	D1319-88	35 max	–

Table 1: Projected demand for petrol and diesel and biofuel requirements

Ethanol blending requirement (in metric ton) : Biodiesel blending requirement (in metric ton) Demand

Year	petrol Demand MMMton	Ethanol blending requirement in ton			Diesel demand MMMton	Biodiesel blend tons		
		5%	10%	20%		5%	10%	15%
2006-07	10.07	0.50	1.01	2.01	52.32	2.62	5.23	10.46
2011-12	12.85	0.64	1.29	2.57	66.91	3.35	6.69	13.38
2016-17	16.40	0.82	1.64	3.28	83.58	4.18	8.36	16.72

Source: Planning commission Govt. of India, 2003

ASTM D975 Diesel Fuel Test

D975 - Specification for Diesel Fuel
D93 - Flash Point

D2709 - Water and Sediment
D86 - Distillation
D445 - Kinematic Viscosity at 40°C
D482 - Ash Content
D3120 - Sulfur Content (D4294 if S500 or > fuel)
D130 - Copper Strip Corrosion
D2500 - Cloud Point
D976 - Cetane Index
D4052 - API Gravity @ 15°C
D524 - Ramsbottom Carbon Residue on 10%
D6079 - Lubricity by HFRR D524 - Sulfated Ash
D93 - Flash Point
D2709 - Water and Sediment
D1160 - Vacuum Distillation
D445 - Kinematic Viscosity at 40°C
D482 - Ash Content
D3120 - Sulfur Content (D4294 if S500 or > fuel)
D130 - Copper Strip Corrosion
D2500 - Cloud Point
D613 - Cetane Number
D4951/1158 - Additvies By ICP, Metals
D524 - Carbon Residue
D6584 - Glycerin in BioDiesel By GC
D664 - Acid Number

India's vehicular pollution is estimated to have increased eight times over the last two decades. This source alone is estimated to contribute about 70 per cent to the total air pollution. With 243.3 million tons of carbon released from the consumption and combustion of fossil fuels in 1999, India ranked fifth in the world behind the U.S., China, Russia and Japan. India's contribution to world carbon emissions would increase in the coming years due to the rapid urbanisation, shift from non-commercial to commercial fuels, increased vehicular usage A jatropha seed contains 31 to 37 per cent extractable oil. A jatropha plantation over 100,000 hectares is expected to yield 250,000-300,000 tons of crude jatropha oil per annum. It is estimated that an initial 100,000-hectare jatropha farm will yield revenues of $100 million per annum. Reliance is also in talks with Maharashtra, Gujarat, Andhra Pradesh and Rajasthan Governments, to get access to land for contract farming. the Jatropha cultivation for a long time.

The contracted ethanol supply for current 2011/12) fiscal would be just sufficient to meet 2 percent blending. Production of biodiesel in India is insignificant and will not soon be commercially deployed as a economically viable biofuel. The government's plan to blend diesel fuel with 20 percent content of biodiesel by fiscal year (April-March) 2011/12 is improbable mostly due to unavailability of high yielding, drought tolerant jatropha seeds to produce biodiesel by the end of 12th FiveYear Plan (2017). India produces over 4 billion liters of rectified spirit (alcohol) per year in addition to 1.5 billion liters of fuel ethanol. , about 140 distilleries have the capacity to distill around 2 billion liters [7]of conventional ethanol per year and could meet the demand for 5percent blending with gasoline; This all goes well if the can production is good.

Some oils and analysis

	Saturated	Unsaturated		Flashpoint
		Mono	Poly	
Almond	8%	66%	26%	221 °C
Avocado oil	12%	74%	14%	271 °C
Butter	66%	30%	4%	150 °C
Ghee, clarified butter	65%	32%	3%	190-25 °C
Canola oil	6%	62%	32%	204 °C
Coconut oil, (virgin)	92%	6%	2%	177 °C
Rice bran oil	20%	47%	33%	
Corn oil	13%	25%	62%	236 °C
Cottonseed oil	24%	26%	50%	216 °C
Grape seed oil	12%	17%	71%	204 °C
Lard	41%	47%	2%	138-201 °C
Margarine, hard	80%	14%	6%	150 °C
Mustard oil	13%	60%	21%	254 °C)
Margarine, soft	20%	47%	33%	150-160 °C (302-320 °F)
Diacylglycerol (DAG) oil	3.05%	37.95%	59%	215 °C
Olive oil (virgin)	14%	73%	11%	215 °C
Ghee	65%	32%	3%	
Palm oil	52%	38%	10%	230 °C
Peanut oil / groundnut oil	18%	49%	33%	231 °C
Pumpkin seed oil	8%	36%	57%	121 °C
Safflower oil	10%	13%	77%	265 °C
Sesame oil	14%	43%	43%	177 °C

Canola oil is low in saturated fat and contains both omega-6 and omega-3 fatty acids in a ratio of 2:1.[7] If consumed, it also reduces

Low-density lipoprotein and overall cholesterol levels, and as a significant source of the essential omega-3 fatty acid is associated with reduced all-cause and cardiovascular mortality Rapeseed is known for containing high amounts of acid, so much so that it can hurt or be fatal to humans. Thus, after years, it has been genetically modified into a less acidic, more palpable oil. "organic" canola oil, the plant was genetically modified.. Some more information on saturation of oils

| Oil | Saturation | Unsaturation | | Unsaturation 6 | 3 | 9 ω | Flash |
		Mono	poly			acids	point
Canola (rapeseed)	7.365	63.276	28.142	-	-	-	(204 °C)
Coconut	91.00	6.000	3.000	-	2	6	(177 °C)
Corn	12.948	27.576	54.677	1	58	28	(232 °C)
Cottonseed	25.900	17.800	51.900	1	54	19	(216 °C)
Flaxseed/ Linseed (European)	6–9	10–22	68–89	56–71	12–18	10–22	(107°C)
Olive	14.00	72.00	14.00	-	-	-	(193 °C)]
Palm	49.300	37.000	9.300	-	10	40	(235 °C)]
Peanut	16.900	46.200	32.000	-	32	48	(225 °C)]
Soybean	15.650	22.783	57.740	7	54	24	(238 °C)]
Sunflower (<60% linoleic)	10.100	45.400	40.100	0.200	39.800	45.300	(227 °C)]

The European Union Renewable Energy Directive (RED), mandates that 20% of the EU's energy consumption consists of renewable sources by 2020. As part of the EU RED, all member states are required to derive 10% of energy in the transportation sector from renewable energy sources by 2020;t,will be derived from biofuel may not all. . With the EU projected to become the largest importer of biofuels by 2020 - with anticipated annual imports of 15.9 billion litres compared to 10.8 billion litres by the United States (OECD/FAO 2010, Bowyer 2010)

The sustainability criteria as laid by EU can be summarised as follows[8]

1. Greenhouse gas (GHG) emission savings from biofuel/ bioliquids consumption should be at least 35%, increasing to 50% by 2017. Installations that commence production after 1 January 2018 are required to reduce emissions by 60%.

2. Biofuels/bioliquids cannot be produced from raw materials obtained from land with high biodiversity value. This

includes land that in or after January 2008 had the following status: (a) primary forest, (b) designated as natural protected area, and (c)highly biodiverse area.

India imports ethanol only to meet shortfalls during years of low sugar production. Demand is mostly for consumption across the potable liquor and chemical industries, [9] .Total ethanol supply for ethanol blending program during current fiscal year (2011/12) is anticipated to be just suficient to meet the 2 percent blending target while production of biodiesel from jatropha in India is commercially at present insignificant Though there are no quantitative restrictions on import of biofuels; high impor duties sometimes make economically unviable. The GOI does not provide any financial assistance for exports of biofuels (biodiesel or ethanol)..However, current trade regulations allow duty free imports of feedstocks for re-export by certified export oriented units.

Jatropha plantation is a subject for state governments.Public sector petroleum companies and private sector firms have entered into memoranda of understanding with state governments to establish and promote jatropha plantation on government wastelands or to contract with small and medium farmers... Slow progress in jatropha planting has resulted in lower availability of jatropha seeds to be used as feedstock for biodiesel production and hence most of the biodiesel units are not operational most of the year.There are about 20 large capacity biodiesel plants (10,000 to 200,000 metric tons per year) in India that produce biodiesel from alternative feed stocks such as edible oil waste (infact there is no residue left over as the oil is consumed to the end because of cost), animal fat and inedible oils (which are again field dependent).Presently, commercial production and marketing of jatropha based biodiesel in India is small, with estimates varying from 140 to 300 million liters per year.. Important thing is that the government biodiesel purchase price of Rs 26.5 (48 cents) per liter is far below the estimated biodiesel production cost (Rs 35 to Rs 40 per liter , hence there is no charm in the trade. It should be 3-4 Rs less than regular diesel. Though the amount of biomass available for industry is more than 500MMT from forestry, agricultural wastes and industrial wastes(food processing ,textiles, breweries ,etc) not active initiations have come up Around 1525 MW of power is proposed to be harnessed from agricultural residues and plantations by end of 2012.

470

REFERENCES

1. Biodiesel and Jatropha Cultivation; Satish LELE(Riddhi International
2. B.K.Bhaskararao. A Text On Petrochemicals, Khann Pub. 5th Eed 3a. Michael Briggs, NHUniversity, Energy Independence, *American energy independence.com*, 3b .**September 24-27, 2012:** 2012 Algae Biomass Summit - Denver, Colorado, USA
3. Analysis from Frost & *Sullivan, Strategic Analysis of the Indian Biofuels Industry,*
4. C.R. Vera*, S.A. D'Ippolito, C.L. Pieck, J.M. Parera Instituto de Investigaciones en Catálisis y Petroquímica, INCAPE (Facultad de Ingeniería Química, Universidad Nacional del Litoral, CONICET)
5. 2nd Mercosur Congress on Chemical Engineering, 4th Mercosur Congress on Process Systems Engineering. Rio de Janeiro. Retrieved 2007-12-20. 2ndMercosur Congress on Chemical Engineering , 4th Mercosur Congress on Process Systems Engineering6 Chemical Reactions of Oil, Fat and Fat Based Products (3)
6. Department of Chemical Engineering, Instituto. Superior Técnico, Lisbon (Portugal), October 1997 http://www.ist.utl.pt/
7. "Fats, Oils, Fatty Acids, Triglycerides". Scientific Psychic (R). All values for ω-3, ω-6, ω-9 fats (not hydrogenated) are from Scientific Psychic (R) unless otherwise cited.
8. Department of Science And technology GOI 2012 ": Annual report of MNRE 2011-12 GOI
9. GAIN Report IN2081, dated June 20th 2012. p15

Further Reading

1. "Nutrient database, Release 24". United States Department of Agriculture. All values in this column are from the USDA Nutrient database unless otherwise cited.
2. Katragadda, H. R.; Fullana, A. S.; Sidhu, S.; Carbonell-Barrachina, Á. A. (2010). "Emissions of volatile aldehydes from heated cooking oils". *Food Chemistry* 120: 59. doi:10.1016/j.foodchem.2009.09.070. edit
3. Wolke, Robert L. (May 16, 2007). "Where There's Smoke, There's a Fryer". *The Washington Post*. Retrieved March 5, 2011.
4. Department of Science And technology GOI 2012 , Annual report of MNRE 2011-12 GOI

8

Green House Gases

8.1.1a Presence of GHGs

The greenhouse effect is a warming effect caused by certain gases that retain heat from sunlight .Without such gases, the average surface temperature of the Earth would be below freezing "The natural greenhouse effect is not only real; it is a blessing. As a result of this effect, the Earth is about 33°C warmer than it would be without it. Without it, the average temperature of the Earth's surface would be below 0°C, and life, as we know it, would not exist."(1) The presence of green house gases in atmosphere has reached alarming situation during the last one and a half century.Although we cannot avoid these gases on the Earth, we can manipulate. Some are natural greenhouse gases such as water vapor, (causes about 36–70% of the greenhouse effect). CO_2 is the next important (causes 9–26%); followed by methane (CH_4), (causes 4-9%) and ozone, (causes3–7%). Due to many reasons and activities methane, concentration in atmosphere cannot be reduced . The yearly emissions are shown in Table.1

Table 1 Gases left into atmosphere by different fossil carbon

Pollutant	Hard coal	Brown coal	Fuel oil	Other	Oil
CO_2 (g/GJ)	94600	101000	77400	74100	56100
SO_2 (g/GJ)	765	1361	1350	228	0.68
NO_x (g/GJ)	292	183	195	129	93.3
CO (g/GJ)	89.1	89.1	15.7	15.7	14.5

Non methane organic (g/GJ)	4.92	7.78	3.70	3.24	1.58
Particulates. g/GJ	1203	3254	16	1.91	0.1
Flue gas volume total (m3/GJ)	360	444	279	276	272:

EPA Home (Global Emissions Source: IPCC (2007); based on global emissions

Total C02 Emissions are contributed by different attributes:

Forests and soil	100×10^6 tons
Human activity	30 giga tons
Geological	3000 -3,200 giga tons
Oceans	1400- 2,000 giga tons

Brown coal emits 3 times as much CO2 as natural gas, black coal emits twice as much CO2 per unit of electricenergy. Coal produces 1.7 to 2.2 times CO2 by wt/wt Gas produces 2.7- 3.0 oil about 3-4 timeswt/wt.

The most ubiquitous and interesting one is carbon dioxide (which is being vented almost 3 billion tons per year) which is well with in our reach to capture to some extent. Many other gases also are GHGS. Infact any triatomic or more than that is a GHG Oxides of nitrogen, (N_2O, NO_2, N_2O_3) and sulfur, are increasing in concentration owing to human activity such as agriculture, power houses and automobiles. The atmospheric concentrations of CO2 and methane have increased by 31% and 149% respectively above pre-industrial levels since 1750;. (Source: Keeling, C.D. and T.P. Whorf. 2005. Atmospheric CO2 records from sites in the SIO air sampling network. In Trends: A Compendium of Data on Global Change. Carbon Dioxide Information Analysis Center, Oak Ridge National Laboratory, U.S. Department of Energy, Oak Ridge, Tenn., U.S.A.)

(Figures 1 & 2) These levels are considerably higher than at any time during the last 650,000 years, the period for which reliable data has been extracted from ice cores and the concentration of CO2 is likely to increase if it is not harnessed (present CO2 values high were last attained 20 million years ago?) Fossil fuel burning was responsible, ie. three-quarters of the increase in CO2 from human activity over the past 20 years. Other part is due to elimination of matured forests and other biological activities which cannot be eliminated. Fossil fuels-coal, oil and natural gas-release carbon dioxide during production and consumption. Fossil fuels are contributing to a rising quality of life in many parts of the world, particularly in developing countries. Based on current projections of population and economic growth, the world's demand for energy will increase substantially over the next 25 years. The majority of that energy will be provided by fossil fuels, even as lower - carbon alternatives continue to emerge. UNEP said that emission levels, driven by the burning of fossil fuels, need to drop by 14 per cent by 2020 for the world to reach a pathway that could keep the global temperature rise below 2 C, compared with pre-industrial levels,.

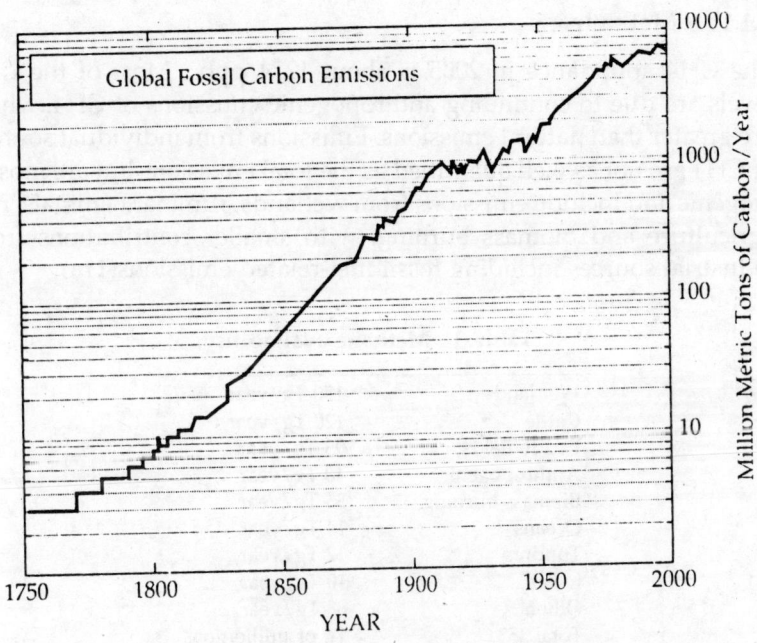

Figure 8.1   Fossil carbon emission.

"Of the current 100 billion tons of carbon [gigatons of carbon only] emitted annually as CO_2 (37.5 billion tons of anthropogenic CO_2) into the atmosphere by human activities, only around 40% remain in the atmosphere, while the rest is absorbed by the oceans and the land biota to about equal proportions "(2) GH gases presence in atmosphere attributed to human activity or from nature and their contribution to approximate green house effect is presented below:

Table 3   Gases contribution to trapping radiation

Gas- and its presence (vol) In atmosphere		% greenhouse effect - that would be absent if All gas were removed from Earth's atmosphere	% in atmosphere due to human activity
water vapor-	1-3%	36%	0%
clouds		14%-	
carbon dioxide-	0.04%	12%	26%
ozone <ppm	3%	not known	
methane-	2ppm	?	60%

8.1.1b   Methane

The CH4 abundance in 2005 is about 1774 ppb . Most of the CH4 levels are due to continuing anthropogenic emissions of CH4, which are greater than natural emissions. Emissions from individual sources of CH4 are not as well quantified as the total emissions but are mostly biogenic and include emissions from wetlands, ruminant animals, rice agriculture and biomass burning, with smaller contributions from industrial sources including fossil fuel-related emissions(11a)

Table 3   Methane availability

Wet lands	150 Tg/year
Cattle	120 Tg/year
Paddy fields	95 Tg/year
Anthropogenic	40 Tg/year
Biomass burning	35 Tg/year
Oceans	14 Tg/year
Tundras	12 Tg/year
Lakes	10 Tg/year
Others	88 Tg/year
Total 553	Tg or milliontons

8.1.1c   Water Vapour

The main natural greenhouse gas is water vapour. Water vapour is always present throughout the lower atmosphere, even if sometimes at a very low level. Water is constantly transferred between the oceans, atmosphere and land in the global hydrological cycle, or the water cycle. When condensed as liquid or ice droplets, water is the main constituent of clouds.

Although human activities affect the water cycle, they do not appear to have directly changed the concentration of water vapour globally. As will become clear below, water vapour is therefore not measured as part of anthropogenic-human-generated-greenhouse gas emissions. It is also worth noting that although water as a gas, traps heat in the lower atmosphere, when it is in the form of suspended droplets (essentially clouds), it can also act to cool the surface of the earth.

8.2.1 Global-warming potential (GWP)(5)

It compares the amount of heat trapped by a certain weight of a gas to the amount of heat could be trapped by the same quantity of carbon dioxide. Most commonly GWP is calculated over a specific time, commonly 20, 100 years and then . GWP is expressed as a factor of carbon dioxide (CO_2 is taken as 1). For example, the 100 year GWP of methane is 21, means that if 21 times carbon dioxide of methane mass were introduced into the atmosphere both will trap same amount of heat over 100 years.

* GWP	Lifetime (years)	20 years	100 years
Methane	12	72	25
Nitrous oxide	114	289	298
HFC-23 (hydrofluorocarbon)	270	12,000	14,800
HFC-134a (hydrofluorocarbon)	14	3,830	1,430
Sulfur hexafluoride	3200	16,300	

* GWP values and lifetimes from 2007 IPCC AR4 p212

Radiation Trapping by various constituents really add to this heat trapping in a peculiar way.Partly because the infrared bands of the various components overlap, the contribution of the individual [radiation] absorbers have no additive property..(donot exercise full quantum of absorption) do not linearly. See in Table 4 how the percentage of [radiation] trapping that would remain after a particular absorber was removed from the atmosphere. It is known that the clouds only contribute 14 per cent (seeTable3) to the trapping heat when, with all other species present, but would trap 50 per cent if the other absorbers were removed. Similarly Carbon dioxide adds 12 per cent to the trapping of the present atmosphere:((, it is a less important trapping agent than water vapor or clouds) On the other hand, on its own CO_2 would trap three times as much as it actually does in the Earth's atmosphere.

Table 4. Contribution of absorbers to atmospheric thermal trapping

Species removed	Percentage trapped radiation remaining
None	100
O_3	97
CO_2	88
clouds	86
H_2O	64
H_2O, CO_2, O_3	50
H_2O, O_3, clouds	36
All	0

(Data of V. Ramanathan and J.A. Coakley,Rev. Geophys. & Space Phys., 1978, 16, 465).

The N_2O concentration in 2005 was 319 ppb. The increase in N_2O is due primarily to human activities, particularly agriculture and associated land use change

8.2.2 Extreme Weathers

The greenhouse effect is a warming effect caused by certain gases that retain heat from sunlight .Without such gases, the average surface temperature of the Earth would be below freezing, and as explained by the Encyclopedia of Environmental Science, "life, as we know it, would not exist." Increased ocean temperatures cause average sea levels to rise because water expands as it becomes warmer. Per a 2006 paper in the journal Nature, this thermal expansion is calculated to have the largest current influence on average sea level changes. The second largest influence is calculated to be the melting of glaciers and mountain icecaps. Per a 1958 New York Times article, the melting of sea ice does not alter sea levels by a single millimeter because ice has more volume than water Worldwide, sea level is not evenly distributed. For instance, the sea level in the Indian Ocean is about 330 feet below the worldwide average, while the sea level in Ireland is about 200 feet above average. Such variations are caused by gravity, winds, and currents; and the practical effects of these phenomena are dynamic. For example, between 1992 and 2010, sea level rose by about 6 inches in the tropical Western Pacific while falling by about the same amount in San Francisco In his 1993 book, Earth in the Balance, Al Gore wrote: About 10 million people in Bangladesh will lose their homes and means of sustenance because of the rising sea level, due to global warming, in the next few decades. 60 percent of the present population of Florida may have to be relocated. On the contrary between 1993 and 2011, the population of Bangladesh increased from 119 million to 159 million people (34%)and between 1990 and 2006, the coastal population of Florida increased from 10.1 million to 13.8 million people (37%).

In 2011, Ph.D. biologist Richard Hilderman wrote: Fossil fueled power stations are major emitters of CO2. This is responsible for global warming observed over the last 100 years. Further to state coalfired power plants generate 1000kg/MWh of C02 while NG fired plants produce less than 400 kgCO2 /MWh. Modern power plants work on combined cycle power plants hence average CO2 generated may be taken as 550-600kg/MWh,. Electric compressors generate less emissions is a myth (3) while the primary source of motor coal fired power plants plus one MW motor drive produces 770 kg/hr of CO2 together 1.5 kgNOx and 1.5 kg of Sox. While NG powered turbine compressors give only gives only 575 kg/hr CO2 and 0.26kg NOx for the same operation.Hence NG operated turbine compressors can reduce pollutants.

According to David J. C. MacKay,(4) the burning of fossil fuels sends seven gigatons (3.27 percent) of carbon dioxide into the atmosphere each year., while the biosphere and oceans account for 440 (55.28 percent) and 330 (41.46 %) gigatons, respectively... The 2007 global fossil-fuel carbon emission estimate, 8365 million metric tons of carbon, represents an all-time high and a 1.7% increase from 2006 , as shown below:

BY source	By Gas
Energysupply 26%	CO2 fossil fuels 57%
Transport 13%	Methane 14%
Residential & commercial 8%	Nox 8%
Agriculture 14%	Water vapor & FO/Clouds 1%
Forestry 17%	Deforestation Decay 17%
industry 19%	other sources 3%

In comparison, a 1,000 MW nuclear plant will generate about 30 short tons of high-level radioactive solid packed waste per year. It is estimated that during 1982, US coal burning released 155 times as much uncontrolled radioactivity into the atmosphere as the Three Mile Island incident. The collective radioactivity resulting from all coal burning worldwide between 1937 and 2004 is estimated to be 2,700,000 curies or 0.101 Ebq During normal operation, the effective dose equivalent from coal plants is 100 times more that from nuclear : just the Chernobyl nuclear disaster released, I-131 alone, an estimated 1,76 EBq . of radioactivity, a value one order of magnitude above this

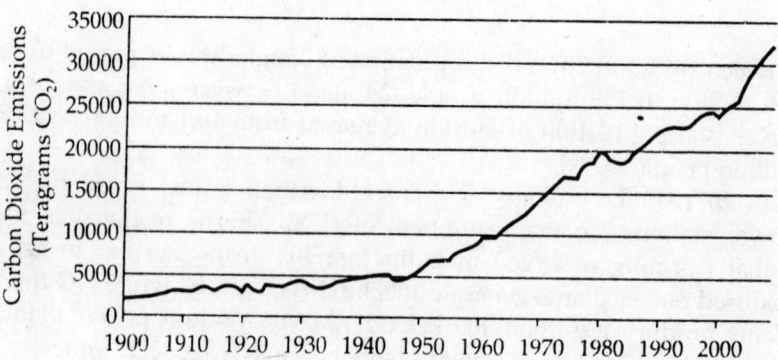

Figure 8.2 **Global Carbon Dioxide (CO2) emissions from fossil-fuels 1990-2008**

(Source of data: Boden, T.A., G. Marland, and R.J. Andres (2010). Global, Regional, and National Fossil-Fuel CO2 Emissions.) Carbon Dioxide Information Analysis Center, National Laboratory, U.S. Department of Energy, , Tenn., U.S.A. Oak Ridge

value for total emissions from all coal burned within a century. It shall also be understood that the I-131, the major radioactive substance has a half life of just 8 days. Hence, it is not going to cause as much as damage as the Uranium and Thorium which are released from Coal fired power plants as they higher half lives. Also, I-131 can be easily removed by eating normal Iodine tablets. In fact, the contribution of decreasing cosmic ray activity to climate change is almost 40 per cent, argues Dr. U.R.Rao in a paper in Current Science,The IPCC model, on the other hand, says that the contribution of carbon emissions is over 90 per cent. Although no single country comes close to the 2.8 billion tons of CO2 produced annually by the U.S. power sector, other countries collectively account for three-quarters of the power-related CO2 burden. China comes second after the U.S. with 2.7 billion tons; followed by Russia -- 661 million tons; India -- 583 million tons; Japan -- 400 million tons; Germany -- 356 million tons; Australia -- 226 million tons; South Africa -- 222 million tons; the United Kingdom -- 212 million tons; and South Korea -- 185 million tons the biggest power producer (,NTPC LTD India lets off 182,000,000 tons CO2)

'Unidimensional focus on carbon emissions put additional pressure on countries like India'. Because other greenhouse gases besides CO2 such as methane (CH4), nitrous oxide (N2O), and the chlorofluorocarbons (CFCs), are also increasing, Thus should be de꜀ided by an 'effective CO2 doubling' . This has been defined as the combined radiative forcing of all greenhouse gases having the same forcing as doubled CO2 (usually defined as ~600 ppm). Level of CO2 is important when estimating potential impacts on crops, because crop growth and water use have been shown to benefit from increased levels of CO2 (Cure and Acock, 1986). The effects of climate change, such as increasing temperatures, shrinking ice caps and expanding deserts, are known, but climate change it is also having an evolutionary effect on animals. Modern studies have shown animals are adapting to changing higher global temperatures, and have been evolving to deal with the averse conditions.

Recent scientific evidence shows that major and widespread climate changes have occurred with startling speed. For example, roughly half the north Atlantic warming since the last ice age was achieved in only a decade, and it was accompanied by significant climatic changes across most of the globe. Similar events, including local warmings as large as 16°C, occurred repeatedly during the slide into and climb out of the last ice age. Human civilizations arose after those extreme, global ice-age climate jumps. Severe droughts and other regional climate events during the current warm period have shown similar tendencies of abrupt onset and great persistence, often with adverse effects on societies. Abrupt climate changes were especially common when the climate system was being forced to change most.

Science Daily (Nov. 14, 2007) - Now for the first time, the CO2 emissions of 50,000 power plants worldwide, the globe's most concentrated source of greenhouse gases, have been compiled into a massive new data base, called CARMA--Carbon Monitoring for ActionRankings of the 4,000 electric power companies in the world show which are the biggest carbon polluters, globally, nationally, and at sub-national levels.

8.3.1 Kyto Protocol

The base of Kyoto Convention was originate from The Intergovernmental Panel on Climate Change (IPCC) is a scientific body established in 1988 by the United Nations and World Meteorological Organization. It is the "leading international body for the assessment of climate change," and its "work serves as the key basis for climate policy decisions made by governments throughout the world,. United Nations Framework Convention on Climate Change (UNFCCC) and its Kyoto Protocol are the establishment of a global response to climate change, stimulation of an array of national policies, and the creation of an international carbon market and new institutional mechanisms that may provide the foundation for future mitigation efforts. And UN Framework Convention on Climate Change (UNFCCC), which was signed by nearly all nations at the 1992 mega-meeting popularly known as the Earth Summit. Accordingly Developing countries (Non-Annex I Parties) report in more general terms on their actions both to address climate change and to adapt to its impacts - but less regularly by the Least Developed Countries.

The Kyoto Protocol is an international agreement linked to the United Nations Framework Convention on Climate Change, which commits its Parties by setting internationally binding emission reduction targets. The protocol, initially adopted in Kyoto, Japan, in 1997, is aimed at fighting global warming Recognizing that developed countries are principally responsible for the current high levels of GHG emissions in the atmosphere as a result of more than 150 years of industrial activity, the Protocol places a heavier burden on developed nations under the principle of "common but differentiated responsibilities." 28 articles the convention The Kyoto Protocol was adopted in Kyoto, Japan, on 11 December 1997 and entered into force on 16 February 2005. The detailed rules for the implementation of the Protocol were adopted at COP 7 in Marrakesh, Morocco, in 2001, and are referred to as the "Marrakesh Accords." Its first commitment period started in 2008 and ended in 2012.

Earlier "Montreal Protocol" was initiated after the ozone whole was discovered. Montreal Protocol, 16 September 1987 was connected with the Substances that Deplete the Ozone Layer The draft was made for stopping the use of ozone deplets and adopted in Montreal on and as subsequently adjusted and amended.

They must also submit an annual inventory of their greenhouse gas emissions, including data for their base year (1990)and all the years since. That treaty was finalized in Kyoto, Japan, in 1997, after years of negotiations, and it went into force in 2005. Nearly all nations have now ratified the treaty, with the notable exception of the United States. Developing countries, including China and India, weren't mandated to reduce emissions, given that they'd contributed a relatively small share of the current century-plus build-up of CO_2.

The ultimate objective of the Convention is to stabilize greenhouse gas concentrations "at a level that would prevent dangerous anthropogenic (human induced) interference with the climate system." It states that "such a level should be achieved within a time-frame sufficient to allow ecosystems to adapt naturally to climate change, to ensure that food production is not threatened, and to enable economic development to proceed in a sustainable manner. Keeps tabs on the problem and what's being done about it.

The 240 megatonne reduction target for the first commitment period (2008-2012), which is 6 per cent below 1990 levels, is extremely ambitious for an energy-producing country like Canada. In fact, it would put Canada in the anomalous position of being the only major energy-producing country in the world that is required to reduce its GHG emissions below 1990 levels.

Accordingly industrialized countries (Annex I) have to report regularly on their climate change policies and measures, including issues governed by the Kyoto Protocol (for countries which have ratified it). Under Kyoto, industrialised nations pledged to cut their yearly emissions of carbon, as measured in six greenhouse gases, by varying amounts, averaging 5.2%, by 2012 as compared to 1990. That equates to a 29% cut in the values that would have otherwise occurred. However, the protocol didn't become international law until more than halfway through the 1990-2012 period. By that point, global emissions had risen substantially.

In principle The Kyoto protocol does not cover the world's largest two emitters, the United States (American lawmakers have been able to set their own targets for reducing carbon emissions that are not always in line with the goals maintained by the U.N. in the name of Clean Air Act) and China, and therefore cannot work, was the view of Canadian Government , hence though signed never impemented; later with drew (2011) from the convention "strong political will to urgently combat climate change in accordance with the principle of common but differentiated responsibilities and respective capabilities" is the theme of UN in regulating GHG "That to prevent dangerous anthropogenic interference with the climate system, recognizes "the scientific view that the increase in global temperature should be below 2 degrees Celsius", in a context of sustainable development, to combat climate change.

Copenhagen Accord briefly: Endorses the continuation of the Kyoto Protocol. initial deadline set 31 January 2010 to date, countries representing over 80% of global emissions have engaged with the the Accord to submit emissions reductions targets, as shown below. All are for the year 2020. States that "enhanced action and international cooperation on adaptation is urgently required to... reduce vulnerability and build.. resilience in developing countries, especially in those that are particularly vulnerable, especially least developed countries (LDCs), small island developing states (SIDS) and Africa" and agrees that "developed countries shall provide adequate, predictable and sustainable financial resources, technology and capacity-building to support the implementation of adaptation action in developing .

Except very few countries Countries agreed to limit emissions: Compared to 1990:

EU: 20% - 30%

1. Japan: 25%
2. Russia: 15% - 25%
3. Ukraine: 20%

Compared to 2000:

4. Australia: 5% - 25%
5. Compared to 2005:
6. Canada: 17%
7. US: 17%

1. Brazil: 36.1% - 38.9%
2. Indonesia: 26%
3. Mexico: 30%
4. South Africa: 34%
5. South Korea: 30%
6. Carbon intensity compared to 2005:
7. China: 40% - 45%
8. India: 20% - 25%

China also promised to increase the share of non-fossil fuels in primary energy consumption to around 15% by 2020, and increase forest coverage by 40 million hectares and forest stock volume by 1.3 billion cubic meters by 2020 from the 2005 levels. The Protocol was adopted by Parties to the UNFCCC in 1997, and entered into force in 2005. The first commitment period applies to emissions between 2008-2012, and the second commitment period applies to emissions between 2013-2020. The protocol was amended in 2012 to accommodate the second commitment period, but this amendment has (as of January 2013) not entered into legal force. Developing countries do not have binding targets under the Kyoto Protocol, but are still committed under the treaty to reduce their emissions.

8.3.2b Results vary

According to Henry Jacoby(MIT Panel Discussion Feb.2010) to his analysis, assuming that the pledges submitted in response to the Accord (as of February 2010) are fulfilled, global emissions would peak around 2020.. In March 2010, Nicholas Stern in his assessment, to have a reasonable chance of meeting the 2 °C target, the preferred emissions level in 2020 would be around 44 gigatons. The voluntary pledges made in the Accord (at that date) would, according to his projection, be above this, nearer to 50 gigatons. that without the accord, emissions might have been above 50 gigatons in 2020. A study published in the journal Environmental Research Letters found that the Accord's voluntary commitments would probably result in a dangerous increase in the global average temperature of 4.2° C over the next century..

The International Energy Agency (IEA) publication, World Energy Outlook 2010, it is assumed that these pledges at Copenhagen, are acted on cautiously, reflecting their non-binding nature., GHG emission trends follow a path that is consistent with a stabilization of GHGs at 650 parts per million (ppm) CO_2-equivalent in the atmosphere. In the long-term, a 650 ppm concentration could lead to global warming of 3.5 °C above the pre-industrial global average temperature level.

A preliminary assessment published in November 2010 by the United Nations Environment Programme (UNEP) suggests a possible "emissions gap" between the voluntary pledges made in the Accord and the emissions cuts necessary to have a "likely" (greater than 66% probability) chance of meeting the 2 °C objective.. The UNEP assessment takes the 2 °C objective as being measured against the pre-industrial global mean temperature level. To having a likely chance of meeting the 2 °C objective, assessed studies generally indicated the need for global emissions to peak before 2020, with substantial declines in emissions thereafter.

8.3.2c Bright future

In Doha, Qatar, on 8 December 2012, the "Doha Amendment to the Kyoto Protocol" was adopted. The amendment includes: UN Framework Convention on Climate Change (UNFCCC), which was signed by nearly all nations at the 1992 mega-meeting popularly known as the Earth Summit. Uner which

- New commitments for Annex I Parties to the Kyoto Protocol who agreed to take on commitments in a second commitment period from 1 January 2013 to 31 December 2020;

- A revised list of greenhouse gases (GHG) to be reported on by Parties in the second commitment period; and

The Ad Hoc Working Group Platform Durban for Enhanced Action, ADP is to complete its work as early as possible, but no later than 2015, in order to adopt this protocol, legal instrument or agreed outcome with legal force at the twenty-first session of the Conference of the Parties and for it to come into effect and be implemented The 240 megatonne reduction target for the first commitment period (2008-2012), which is 6 per cent below 1990 levels, is extremely ambitious for an energy-producing country like Canada. In fact, it would put Canada in the anomalous position of being the only major energy-producing country in the world that is required to reduce its GHG emissions below 1990 levels from 2020

8.4.1 Petroleum Disasters:

On April 20, 2010 an explosion on the Deepwater Horizon rig in the Gulf of Mexico killed 11 men and led to the largest ever accidental offshore oil spill. One year on ,the oil industry and BP From April 20th until July 15th, 2010, 200 million gallons of black crude oil spilled, affecting hundreds of miles of coastline and killing thousands of animals. The Gulf oil spill was the worst of its kind in U.S. history

January 1969, a natural gas blowout on an oil rig miles off the coast of Santa Barbara, California, spilled 80,000 gallons of oil into the Pacific Ocean and onto surrounding beaches., In March 1989, the Exxon Valdez oil tanker struck a reef and spilled 10.4 million gallons of oil into Prince William Sound, Alaska, affecting 1,300 miles of shoreline. Natural flora and fauna being destroyed by such spills. Public became confrontational on seeing the images of spilled oil bubbling to the ocean's surface and covering birds and other wildlife and argue that offshore drilling is dangerous, that it inflicts tremendous environmental harm, and that its costs are not worth its benefits. . U.S. obtains about 25 percent of the nation's natural gas production and about 24 percent of its oil production have become thus linked to environmental degradation. Since 1975, offshore drilling in the Exclusive Economic Zone (within 200 miles of U.S. coasts) has a safety record of 99.999 percent, meaning that only 0.0001 percent of the oil produced has been spilled.(6) between 1993 and

2007 there were 651 oil spills, releasing 47,800 barrels of oil. Specifically, for the years 1964-99, the authors calculated the following oil-spill rates: (a) 0.32 spills per billion barrels of oil handled for OCS platform spills greater than or equal to 1,000 barrels; (b) 0.12 spills per billion barrels of oil handled for OCS platform spills greater than or equal to 10,000 barrels; and (c) 1.33 spills per billion barrels of oil handled for OCS pipeline spills greater than or equal to 1,000 barrels., (7) According to Cheryl McMahon Anderson and Robert P. Labelle,the authors, said that eleven platform spills and sixteen pipeline spills greater than or equal to 1,000 barrels occurred in the OCS between 1964 through 1999, while total production was estimated to be 12 billion barrels of crude oil and condensate during the same period. Worldwide, from 1974 through 1999, there were 278 crude-oil spills greater than or equal to1,000 barrels from self-propelled crude-oil carriers, while an estimated 239.67 billion barrels of crude oil moved worldwide during the same period. Forty-six crude-oil tanker spills greater than or equal to 1,000 barrels occurred in U.S. coastal and offshore waters (including U.S. territorial waters) from 1974 to 1999, while tankers moved an estimated 44.5 billion barrels of oil in U.S. waters during the same period.Given 7.5 billion barrels of oil produced during that period, one barrel of oil has been spilled in the OCS per 156,900 barrels produced.(8) .Hence contribution by human activity is neglisible.

Average annual contribution to oil in the ocean (1990-1999) from major sources of petroleum in kilotonnes(9)

Natural seeps	600×10^3 tons
Spills blow outs, during production	3 8...,
oiltankers,pipelines leakage,voluntory	160...
Wastes and runoffs from marine/road vehicles	480

The Bureau estimates of oil and gas resources (2006) in undiscovered fields on the OCS (outer continental slope) a total 86 billion barrels of oil and 420 trillion cubic feet of gas. These volumes represent about 60 percent of the oil and 40 percent of the natural gas resources estimated to be contained in remaining undiscovered fields in the United States. Revising previous estimates first published in 1994, the authors analyzed data through 1999 and concluded that oil-spill rates for OCS platforms, tankers, and barges continued to decline. The number of oil spills from platforms, tankers, and pipelines is small, relative to the amount of oil extracted and transported. Even so, oil spills remain an unpleasant reality of offshore oil drilling. Certainly, any amount of oil spilled into the ocean is undesirable, but offshore oil operations contribute neglisible amount of the oil that enters ocean waters each year. See the fig 7.3 ocean floors naturally seep more oil into the ocean than do oil-drilling accidents and oil-tanker spills combined. This is not easily spotted as seepage generally does not rise to the surface or reach the coastlines According to the National Academies' National Research Council, natural processes are responsible for over 60 percent of the petroleum that enters North American ocean waters and over 45 percent of the petroleum that enters ocean waters worldwide.[.9-1] Thus, in percentage terms, North America's oil-drilling activities spill less oil into the ocean than the global average, suggesting that the American companies drilling is comparatively safe for the environment. Also ironically, drilling can actually reduce natural seepage, as it relieves the pressure that drives oil and gas up from ocean floors and into ocean waters. It is illuatrated that natural seepage in the northern Santa Barbara Channel was significantly reduced due to oil production. As per the researchers documented that natural seepage declined 50 percent around Platform Holly over a twenty-two-year period, concluding that, as oil was out of the reservoir, the pressuredecreases resulting in less seepage ... An estimated 2,000 to 3,000 gallons of crude oil are released naturally from the ocean bottom every day just a few miles offshore from a beach California Coast near Santa Barbara.(10) A study by the University of California Energy Institute reports that, during the 1990s, "natural seeps annually emitted an estimated 600,000 tons of oil into the ocean, approximately half the annual total (1,300,000 tons) entering the ocean. By comparison, spills from marine vessels accounted for 100,000 tons, terrestrial run-off 140,000 tons, and pipelines just 12,000 tons. In North America, seeps emit an estimate of 160,000 tons per year" (Ira Leifer, Jim Boles, and Bruce Luyendyk, "Measurements of Oil and Gas Emissions from a Marine Seep," University of California Energy Institute, January 2007.

8.4.2 Personal safety

Safety on off shore platforms is very much important. Many things can happen Landing of helicopters, walking on platforms all are riskey. Non slippery floors made out of ceramic or synthetic materials are required on off shore plat forms. These decks are always layered with moisture and few oil drops make the floors more slippery. Anti-oil Mats are suitable for humid and oily floors as they offer excellent anti slip effect.. Also corrosion resistant mats are preferred YSBEL spill control pallets are designed to hold drums of oil and fuels. These pallets have higher size and larger capacity. Easy escape ways in case of accident. Proper alarms, fire fighting equipment, (Water pumps, Foams, Chemicals like carbon tetra chloride to contain fires).masks for containing gaseous/liquid chemicals. Proper first aid facilities are important requirements.

Some more Saftey precautions are listed below:

Explosion Proof Alarm, Signal, Control and Communications Equipment
Marine and Industrial Signalling
Fall Protection, Head Protection and Body Protection Equipment
Fire and Explosion Proof Doors for Offshore Platforms and Ships
Life-Saving Equipment for Marine Protection; Life boats and davits
Personal Fall Protection Equipment
Survival Suits, Re-Breather Systems
Technical Safety Equipment for Hazardous Working Environments
Worker Safety Solutions and Personal Protective Equipment
Gas Safety Equipment
Information about the gases and chemicals used/stored
Portable Multi-Gas Detector
Nitrogen Gas Purging
Gas masks

8.4.3a Cleaning Spills

The principle behind the cleaning of oil spills is like picking up pearls from pebbles. It is based up on two characters of oil that follows:

- **Specific Gravity:** Most crudes have a lower specific gravity than water. Thus oil floats on water to the surface. These oils are known as LNAPL's, Light Non-Aqueous Phase Liquid. Organic or inorganic liquids or compounds that sink in water have a higher specific gravity and are known as DNAPL's, Dense Non-Aqueous Phase Liquid.
- **Surface Tension and Affinity:** Normally, oil bonds more tightly to itself i.e.organic phase likes organic phase and other materials than to water. This affinity, and differences in surface tension between oil and water, cause oils to adhere to a skimming medium.

8.4.3b   Oil Spill Equipments

Oil Booms
Skimmers
Different accessories
Portable sump/ spill pallets
Polyethylene Sump Pallets

8.4.3c   Oil Skimming

Usually two methods are in operation a. suction like a vacuum cleaner
b. physical adherence to medium. Based on these two principles
designes are made to operate on light oil to heavy oils and adhering
gums and greases and in varying quantities. This is no surprise if
these skimmers are used even to purify waste waters to get rid off
floating oils ?(as low as 0.5%). These skimmers are in two varieties
Vauum type, absorbent(attraction) type

8.4.3d   Portable Vacuum Unit

Vacuum Systems for Industrial and Emergency Response. These
systems recover all types of liquids in a broad range of viscosities;
solids and debris are easily recovered by these rugged, easy to use
vacuum systems. Vacuum systems are used for many industrial
applications. They utilize an engine to power a blower/vacuum
pump, creating a large movement of air and vacuum that draws
liquids and solids through hoses. The collected material is deposited
in a tank or other holding/storage device. The size of engine and
blower determine the power of the vacuum system and its
capacity.Oil can be removed from the surface of water with a vacuum
truck - however this method is somewhat inefficient. A vacuum truck
will typically draw in much or more water than oil when used in the
traditional way. However, oil skimmer can be used to selectively
remove the oil from the water and greatly improve the efficiency of
the process. Oil can be recovered from the surface of water with 97%
efficiency (3% water). When the recovered liquid has filled the
vacuum storage tank, a hydraulic transfer pump can be used to empty
liquids from the tank.

The ELASTEC MiniVac (Elastec/American Marine) manufactures
portable vacuum systems for industrial and oil spill cleanup..These
systems can recover all types of liquids with a broad range of
viscosities, sludge and dense solids.. It is ideal for working in remote
areas as well as industrial locations. The MiniVac can recover a wide

range of liquids, oils and sludges with solids up to 2 inch/50mm diameter can be used in pipelines as well as industrial locations. The diesel driven vacuum pump quickly generates suction - oils and liquids are recovered into standard oil drums (not included) or into a hopper. Material recovered by MiniVac is deposited into common open head drums, by drum filling vacuum head with an automatic shut off and adjustable vacuum relief valve the drums are filled up. A sight glass shows the level of liquid collected. Any oil can be removed by sucking as the oil floats. Most of these skimmers are based on the principle, only the method of sucking varies. Second type is by applying suitable absorber which preferentially absorbs oil from water or either. Even from sumps of water contaminated with oil droplets can be used.

8.4.3e Brush Skimmer

Brush skimmers are a relatively new concept, the rows of bristles (stiff short hairs) gather more oil in the right conditions , as oil is trapped between the bristles. These devices can be configured for light or heavy oil and are available as small units for inland spills or large units for offshore applications. The skimmer head incorporates two side by side brush wheels covered with thousands of lyophilic bristles which rotate with the flow. The large surface area created by the bristles makes the unit very effective and as the wheels rotate the water flows off and the oil is scraped off into a collection hoper before being pumped to a recovery tank. Powered by a hydraulic power pack, the unit comes out in various sizes, effectively creating large surface area.

8.4.3f Disk Oil Skimmers

These oil skimmers rotate a disk shaped medium through the liquid. Oil is wiped off and discharged into a collection container in a manner similar to belt oil skimmers. It is important to consider reach, the portion of the disk that actually gets immersed, when looking at a disk oil skimmer. Less disk in the fluid means less oil removed. Obviously, fluctuating fluids can be a real problem for disk oil skimmers.

8.4.3g Drum/Barrel Styles

These are similar to the disk type, but use a rotating drum shaped medium. Compared to disk types, they are usually more rugged and have higher removal capacity. Depending on the designs. Husky Oil

490

Booms are designed without compromise,high tensile strength strong chains they can bde towed and spread quickly in emergency response situations.

8.4.3h Absorbent Boom

Absorbent Booms work on a different principle. Oil absorbent booms will float on water even if saturated. These booms are often referred to as sorbent booms, marine booms, marina booms, sea booms, ocean booms or oil spill booms can be used on calm water when you have debris and oil to contain for use on inland waterways, marinas and harbors.

REFERENCES

1. Encyclopedia of Environmental Science. Edited by David E. Alexander and others. Kluwer, 1999. Topic: "Greenhouse Effect." By Richard A. Houghton. Page 303:
2. Wolfgang Knorr. Geophysical Research Letters, November 7, 2009. (http://ruby.fgcu.edu/courses/twimberley/envirophilo/knorrarticle.pdf
3. Rainer Kurz , Solar Turbines Inc.World , Pipe Lines,April2013,110
4. Atmospheric Chemistry. By Ann M. Holloway and Richard P. Wayne. Royal Society of Chemistry, 2010. Page 17
5. B.K. Bhaskara Rao, A Text On Petrochemicals , Khanna Publishers, 5th ed,
6. EIA, "Offshore Petroleum and Natural Gas," http://www.eia.doe.gov/kids/energyfacts/sources/non-renewable/offshore.html#oilgas. See also Minerals Management Service, "What About an Oil Spill?," http://www.gomr.mms.gov/homepg/offshore/egom/spill.html. Research published in 2000
7. EIA, "Offshore Petroleum and Natural Gas," http://www.eia.doe.gov/kids/energyfacts/sources/non-renewable/offshore.html#oilgas. See Cheryl McMahon Anderson and Robert P. Labelle, "Update of Comparative Occurrence Rates for Offshore Oil Spills," U.S. Minerals Management Service, Spill Science & Technology Bulletin 6, nos. 5-6 (2000): 303-21, http://www.gomr.mms.gov/homepg/offshore/egom/spill.html.
8. 'Snake Oil': Debunking Three 'Truths' about Offshore Drilling," editorial, Washington Post, August 12, 2008, http://www.washingtonpost.com/wp-dyn/content/article/2008/08/11/AR2008081102145.html. For more on drilling safety, see Minerals Management Service, "Safety and Oil Spill Research,"
9. From Oil In The Sea, Ocean Studies Board and Marine Board of the National Academy of Sciences (2003).
10. See "Oil in the Sea III: Inputs, Fates and Effects," National Academies' National Research Council, http://www.nap.edu/catalog.php?record_id=10388.
11a. [IPCC's Third Annual Report] (Dlugokencky et al., 2005).) "Global Warming Facts." By James D. Agresti and Schuyler Dugle. Just Facts, August 15, 2011. Revised 7/20/12. http://www.justfacts.com/globalwarming.asp 33a) Paper: "Is the airborne fraction of anthropogenic CO emissions increasing?" By Wolfgang Knorr. Geophysical Research Letters, November 7, 2009. http://ruby.fgcu.edu/courses/twimberley/envirophilo/knorrarticle.pdf

Further Reading

1. Global Change and Oceanic Primary Productivity: Effects of Ocean-Atmosphere-Biological Feedbacks - A. J. Miller etal., 2003
2. Chester, Bronwyn (20 April 2000).. Source: IPCC (2007) based on global emissions from 2004. Details about the sources included in these estimates can be found in the Contribution of Working Group I to the Fourth Assessment Report of the Intergovernmental Panel on Climate Change Beijing China 2013.
3. See EIA, "Offshore Petroleum and Natural Gas," http://www.eia.doe.gov/kids/energyfacts/sources/non-renewable/offshore.html#oilgas. See also Minerals Management Service, "What About an Oil Spill?," http://www.gomr.mms.gov/homepg/offshore/egom/spill.html.
4. National academic press 2013 National Academy of Sciences. Natural Climate Variability on Decade-to-Century Time Scales
   Encyclopedia of Earth :Air Pollution Emissions Published: August 21, 2008, 7:26 pm
   Updated: April 25, 2013, 11:59 pm
   Lead Author: Sjaak Slanina
5. Chandrasekar (2010). Basics of Atmospheric Science. PHI Learning. ISBN 81-203-4022-1
6. Robert V. Rohli and Anthony J. Vega (2011). Climatology, 2nd Edition. Jones and Bartlett Learning. ISBN 0-7637-9101-6.
7. Chemical Engineering Design, R.K.Sinnott,Vol 6.Elsevier

Appendix 1

Speciality Products

A. INDUSTRIAL GREASE

The consumption of grease bears a direct relation, with industrialisation, however in our country it is far from the truth as the importance of application of grease ignored. Hence causes disruption in service of the equipment or machinery. Commercial advantage of the lubrication must be thoroughly explained to users, Though different types of lubricants are employed for the purpose, it is not followed as per specification. Grease application is still not received adequate importance. In fact grease is nothing but soap dispersed in oil. Calcium grease is the most versatile grease and is largely consumed in the industry.

Manufacture of Calcium Grease

Grease is manufactured by two processes (a) Continuous (b) Batch. Batch process description is given below:

Batch process can be followed by two operational adoptions namely either open kettle or closed kettle. The process begins with the saponification of oil and hydroxide. Usually non edible oils are employed for this purpose and cheap hydroxide like calcium hydroxide. However the quality of lime is important. It is first hydrated with 2 to 3 times of water and allowed to pass through screens of 60 to 100 mesh size grizlies. The lime is now reacted with equivalent vegetable oil (fat) and the temperature is slowly raised to 100 + 20°C. The addition of fat continues till the neutrality is viscible.

Saponification lasts for four hours. During saponification water separates out which is evaporated. Now addition of pale oil equal to the weight of the fat is done. The whole mixture is vigorously agitated keeping the temperature around 100°C. Anti oxidants colours can be added at this stage. The whole mixture is now drained for kneading operation.

During kneading which lasts about an hour some more pale oil can be added. Any additives, fillers can be added, after the temperature is brought down to 56°C and the mixture transferred into small drums. Calcium grease is suitable for most of the industrial non specialised type of application. Silicone grease is useful for vacuum and moderate temperature ambients. Molybdenum and Tungsten greases are also available for pressure applications and high temperature operations. It is essential to see the greases are non corrosive i.e. essential neutral, Greases are tested for load bearing capacity water washability, penetration, melting point and oil separation etc. If the grease is water soluble it can not be used in water or steam atmospheres. Similar considerations are given priority for selection of grease application.

B. LIQUID PARAFFIN

Depending upon the boiling point range the paraffins can be classified as light or heavy. Most usually light paraffins are used for medicinal purposes such catharacting, hair oils, body oils, ointments, sprays etc. Heavy oils are more suitable for industrial applications like paints, solvents, ointments, chemical preparations etc. These are also known as white oils.

Production of Paraffins

The boiling point range for light liquid paraffins is 200 to 250°C, free from wax. The fraction is first isolated and tailored if required depending upon the density and viscosity. The fraction must be free from fluorescence otherwise it has to be eliminated. The stock is to be completely free from aromatics. If the aromatics are more than 5% it is better to remove them by physical extraction like phenol extraction or furfurol extraction. If the material is having pugnacious materials like high sulphur and nitrogen components it is even better to go for hydrotreating operation. The cost factor only has to be judged. After removing the solvent from the feed it is treated with concentrated (above 98.4%) strength sulphuric acid till the coloration disappears.

Finally it may be given oleum treatment may be upto 2%. Sulfuric acid requirement may vary from 100 to 200%. The treated stock is now extracted with propanol-alcohol mixture. This extract forms good detergent and can be employed for house-hold phenyl making or for cheap detergency actions.

The extracted feed is now thoroughly washed with water and dried by passing air or adding molecular sieves or silica gel. After drying the feed is allowed to pass through a column of bauxite, fuller earth or silica gel of 60 mesh. The finished product will be colorless transparent, odourless and tasteless and does not react with con. sulfuric acid; usually a flash point of 100°C. It should be nongreasy oil with a viscosity not more than 30 cst. at 100°F. Pour point should be as low as possible, perhaps 5°C is sufficient. However it should satisfy U.V. reading not more than 0.1 at 240-280 ηm. for a 2% solution in iso octane. The stock being free from aromatics and resins its refractive index will be in the paraffin range, i.e. less than 1.44. For medicinal applications U.V. reading of the sample has to be passed.

C. PETROLEUM JELLYS

Jellys are fine dispersions of wax in oils. These are like greases though there is no saponification. Quality oil i.e. paraffinic oil is preferred as the viscosity variation with temperature is less. Microcrystalline wax is dispersed and to maintain the viscosity and viscosity index it is mixed with polymers like poly butene, poly ethylene liquids. Usually about 5 to 6% polymer is needed. In addition to the polymer an anti oxidant is added. These jellys are thixotropic and non melting. Phase changes do not take place appreciably over temperature variations. These are meant for cable filling operations and act as highly water repelents.

Usually the density of these gels is between 0.9 to 0.83. Cone penetration at 23°C for 3 days is 270 to 310 mm at -10°C, 250 to 270 mm. Oil separation is less not more than 6%. Flash point of these gels is above 250°C. Vaselens are also a sort of jellys but pure paraffinic dispersions sometimes with moisturisers, colorants and perfumes. These can also form as the bases for ointments.

Appendix 2

Some Information About Present Status in Refining

Petroleum Refining capacity is likely to increase by 2001 AD as shown below:

Existing capacity	51.8 M²TPA
Expansion work undertaken or finished	5.2 M²TPA
IOC (Six Refineries)	24.4 M²TPA
HPCL (Two Refineries)	10.0 M²TPA
BPCI	6.0 M²TPA
Cochin	7.5 M²TPA
Madras	6.5 M²TPA
Narimanam	0.5 M²TPA
Bongaigon	2.35 M²TPA

New refineries to be added

Mangalore	3.00 M²TPA
MRL	0.5 M²TPA
IOC, Karnal	6.0 M²TPA
IBP, Assam	3.0 M²TPA
Koyali Expansion	3.0 M²TPA
East-West Refinery	6.0 M²TPA
Central Region Refinery	6.0 M²TPA
West Line	6.0 M²TPA
Private sector Reliance	29.0 M²TPA
IOC - Kuwait, Paradip	6 MMT*

Nippon-Denro Ispat Ltd., Paradip	6 MMT*
Bhatinda Punjab	6 MMT
Panipat	12 MMT
Bina M.P BPL	6 MMT
Total	119.55 M²TPA

* These two refineries are to produce mostly the middle distillates. Initially these will be producing the products as shown below:

	IOC	Nippon	
LPG	262	281	MT
MS	600	806	MT
Naphtha	22	442	
ATF + Kerosene	4056	2000	
HSD	2287	2786	
Fuel coke	405		
Sulfur	93	93	
Fuel oil	1446		
Bitumen	1902		

Kuwait in collaboration with BPC going to start a refinery at Paradip of 6 MMton capacity, while with Oman another refinery will be set up by ONGC in MP.

BRPL coker = 0.5 M² per year. Assam Refinery Numaligar will have a delayed coker.

Karnal Visbreaker is soaker type.

Needle coke plant has been planned for Barauni refinery from phenol extracts of lube plant.

To decrease lead in gasoline as low as possible i.e. 0.15 gm/lt. Gujarat, Mathura, Barauni, Digboi refinery will be equipped with Catalytic reformers.

8th Plan ending demands for petroleum products were anticipated to grow upto 79.4 MMT.

LPG extraction plants at Hazira, Ankleswar, Uran, Duliajan, Bijapur (GAIL plant of 4 lakh ton capacity).

Bakoria in Gujarat of 0.75 lakh ton capacity against 2.7 MMTA of consumption.

Khandla-Bhatinda HBJ line diversion to augment power projects and petrochemical industries based on gas.

Natural gas outlets for vehicles are already working at Bombay, Surat, Bharuch. Natural gas supply in pipe line in Surat is already in service arranged by Mafatlal Group.

ONGC the biggest producer of the crude in India is reeling under hysterisis in Bombay High. The production is likely to touch 29 MMTA. Neelam field has a potential of 63 MMT of oil and 9 billion cu. meters of gas. Present production from this is around a million ton and likely to touch 5 millions when fully harnessed.

Heavy oil deposits in Bikaner Nagur sediments classified as palaeozoic deposits is 779 M^2T (Recoverable).

Gas deposits 718 BM^3 reserves.

Neelam-2 production 2000 Bbl/day

Iran - Qatar line to supply gas 5.5 m^2/day

The cost of this project is \$4.5 billion likely in operation by 2005. The present demand for gas is 240 million cu.m. per day. The indegenous production is only upto 60 million cu.m. per day.

Energy Prospects in Ninth Plan in India

Energy is a critical input for economic development and the development experience all over the world is associated with a massive increase in energy requirement. The fifty years since independence have seen an expansion in the total energy use in the country with a shift from non-commercial to commercial sources. The use of commercial energy has increased ten fold over this period. Nevertheless, per capita energy use in India remains very low and growth in future requires a large increase in commercial energy. This calls for optimising the capacity to expand, domestic production of commercial energy and the ability to do so will be a crucial constraint upon future growth. Even with the best efforts in this area India will remain energy deficient and import of energy in the form of crude oil and petroleum products and also coal will continue. Efforts at managing energy demand through rational energy pricing will be especially in the years ahead.

Non-Commercial Primary Energy Resources

India had long back (in fifties) realized the importance of tapping new and renewable energy which popularized solar cookers, heaters and solar energy based utilities. However, the start was good but follow up was not good till the first oil shock was felt. Now the Government had taken up the issue as a major one and created even separate ministry in the name of Non-Conventional Energy Sources. The estimates show about 100,000 MW renewable energy potential is available but only a mere 1378 MW energy contribution i.e. less than 1.6% of total power generation capacity of the country is made available. In this direction biomass and bagas projects (28) with a capacity of 141 MW and another 27 projects with 180 MW capacity are under the process of commissioning. Similarly wind power potential of about 6000 MW capacity has been identified however about 1000 MW installation has been completed till 1998.

The industrial sector consumes the largest share of energy in Developed (37%) and Developing countries (27%). Industrialization has led to increased economic activity that forced a larger demand of transport services. Transportation is mainly credited to the development of oil industry. It is almost accounting 60% oil consumption in Developed Nations and about 50% in Developing Nations. Perhaps direct energy translation is in the form of gasoline, diesel, and natural gas while production of electricity and then utilizing for transportation is secondary form.

OIL

Oil accounts for about 30% of India's total energy consumption. The majority of India's roughly 4.7 billion barrels in oil reserves are located in the Bombay High, Upper Assam, Cambay, Krishna-Godavari and Cauvery basins. The offshore Bombay High field is by far India's largest producing field, with production of 210,000 barrels per day (bbl/d) in 1999. India's average crude oil production level for the first four months of 2001 was estimated at 652,000 bbl/d. India imported over 1.2 million bbl/d in 2000.

Future oil consumption in India is expected to grow rapidly, to 3.4 million bbl/d by 2010 from 1.9 million bbl/d in 1999. India is attempting to limit its dependence on oil imports somewhat by expanding domestic exploration and production. To this end, the Indian government is pursuing the New Exploration Licensing Policy (NELP), announced in 1997, which permits foreign involvement in exploration, an activity long restricted Indian state-owned firms. While the initial response to the 1999 tender was disappointing, with no bids received from the major multinational oil companies (causing an extension of the deadline for submission of bids), India proceeded with the award of 25 oil exploration blocks in early January 2000. The largest winner in the bidding round was India's domestic Reliance Industries, in partnership with independent Niko Resources of Canada, which received 12 blocks. British independent Cairn Energy, Russia's Gazprom, the U.S. firm Mosbacher Energy, and Geopetrol of France were all awarded single blocks in partnership with Indian firms. India's state-owned Oil and Natural Gas Corporation (ONGC) was awarded eight blocks, three of which it will hold in partnership with other public-sector Indian firms. A second round of bidding, with a total of 25 blocks offered, concluded in March 2001. Sixteen of the blocks have been awarded to ONGC, and four blocks to Hardy Oil of the United Kingdom, in partnership with India's Reliance Petroleum. The others were either awarded to smaller independent firms or failed to receive bids. As with the first round, no bids were received from major international oil companies. A third round of bidding is planned in the near future.

Low drilling recovery rates are a major part of the oil supply problem for India. Recovery rates average only around 30% in currently producing Indian fields, well below the world average. It is hoped that allowing foreign investment will bring in technology that is not available to Indian state firms, thereby increasing overall recovery rates. ONGC currently is undertaking a project to increase recovery rates in the Bombay High offshore field, which will involve the drilling of 140 new wells.

NATURAL GAS

Indian consumption of natural gas has risen faster than any other fuel in recent years. From only 0.6 trillion cubic feet (Tcf) per year in 1995, natural gas use was nearly 0.8 Tcf in 1999 and is projected to reach 1.3 Tcf in 2005 and 1.8 Tcf in 2010. Increased use of natural gas in power generation is to account for most of the increase, as the Indian government is encouraging the construction of gas-fired electric power plants in coastal areas where they can be easily supplied with liquefied natural gas (LNG) by sea.

NINTH FIVE YEAR PLAN (1997-2002)

Review of the Eighth Plan

The major objectives for Petroleum and Natural Gas sector during the Eighth Plan included maximization of indigenous production of crude oil and natural gas, a reasonable level of accretion of new hydrocarbon reserves through intensification of exploration activities, augmentation of domestic refining capacity with emphasis on cost-effective debottlenecking or expansion of existing refining capacity, encouragement to private sector participation etc.

Demand for Petroleum Products

The actual consumption of petroleum products was 79.16 million tonnes as against the Eighth Plan projected demand of 81.19 million tonnes in 1996-97. The compound average annual growth rate during the Eighth Plan period was 6.8% as against the projection of 6.9% envisaged at the time of the formulation of the Eighth Plan. The cumulative consumption of petroleum products during the Eighth Plan was however lower at 341.69 million tonnes than the projected demand of 346.78 million tonnes due to lower than anticipated increase during the first two years of the Plan. These consumption figures include product sales through parallel marketing system (PMS). The share of imports of petroleum products by private sector was less than 3%. The consumption of petroleum products during the Eighth Plan is given in table 1.36.

India is poor in hydrocarbon deposits; in contrast the consumption of petroleum is increasing at 1.2 to 1.3 times GDP growth rate. The indigenous production of crude dwindles or is stagnating around a mysterious figure of 34 MMTPA and doubts arise if this figure can be increased at all. Imports or more than 200% of native crude production i.e. total imports in petroleum consumption will be around 75% and the trend increases in coming years. Till few years ago the products of mass consumption were listed as kerosene and diesel, now the growth in MS consumption cannot be sidelined. The average annual consumption growth rate in this decade is as follow:

MS	0.225 MMTPA
Kerosene	0.22 MMTPA
Diesel	0.22 MMTPA
Total light distillates	0.65 MMTPA
Total middle distillates	2.2 MMTPA

With gasoline and diesel consumption increasing it certainly foredains increased imports of oil where the population is phenomenon.

Share of LPG, SKO, MS and Diesel in POL Consumption

Percentage Share in Consumption (excluding RBF)

	1970-71	1980-81	1990-91	1996-97
LPG	1.0	1.3	4.4	5.4
SKO	18.3	13.7	15.3	12.8
MS	8.1	4.9	6.4	6.3
Diesel	27.5	37.1	41.1	45.9
Total	54.9	57.0	67.2	70.4

The share of the four oil products shown in the above Table AP.1 has increased from 54.9% in 1970-71 to 70.4% in 1996-97. If these trends continue in future as well it will give rise to severe balance of payments problems since all these products, except motor spirit, in addition to crude oil are imported at the margin.

Long-term Energy Scenario-some Key Issues

Based on end-use analysis and past income elasticities, it is estimated that the requirement for primary energy is likely to increase from 374 MTOE at the end of Eighth Plan to around 475-500 MTOE by the end of the Ninth Plan i.e. an annual growth rate of 4.9%. Of this, the share of commercial energy is expected to increase from 66 percent at the end of Eighth Plan to 75 percent at the end of Ninth Plan while that of non-commercial energy will decline from 34 percent to 25 percent. The annual growth rate of commercial energy is therefore 6.8%.

The energy import dependence is also expected to increase from about 25 percent at the end of Eighth Plan to 28 percent at the end of Ninth Plan. This raises the issue of energy security. In view of stagnant domestic oil production, higher oil imports appear inevitable. Uncertainties regarding prices and availability make the developing countries like India more vulnerable than the developed countries. Both refining and marketing operations have been opened to the private sector. A new exploration licensing policy for making exploration and production competitive has been announced. The growing requirement of coal and inability of the domestic production to meet this requirement may necessitate increased imports of coal even for power generation. Allowing the entry of private sector in coal production, without the captive consumption restriction, will go a long way in increasing the availability of coal.

The need for energy conservation and other demand management measures can not be over-emphasised. Some efforts are already being made in this direction. However, the results achieved have not been to the desired level. There is a need for R&D and technology development in this field also.

The demand for oil is increasing faster than the addition to hydrocarbon reserves in the country. The efforts made during the last few by the oil industry have not yielded the expected results. There is a need for enhancing the pace of exploration and development in the hydrocarbons sector in order to add to the recoverable resources. Concerted efforts are required in this area through the adoption of better techniques of oil exploration. Similarly, adoption of latest technology would also improve the recovery from existing wells.

With the growth of industry transportation also increases. The use of petroleum fuels bring emission, pollution problems. Accordingly the new formulations have to be produced. Some of the metros are using LNG, electrical vehicles for transportation. While unleaded gasolines with reduced benzene quantities are being used. Oxygenated gasolines have not reached for city consumptions. Some of the formulations to meet the Clean Air Act demands are shown in Table AP. 2 & 3.

Most of the country's primary energy is derived from coal as shown in Figure AP1, NG, LPG consumption in the country is increasing not only for house hold sector but in transport sector too.

TABLE AP 2 : New Specifications for Diesel Fuel.

Diesel Properties	Sweden Urban 1 Jan. 1992	Finland Reform. July 1993	United States Oct. 1993	European CEN-19 Oct. 1996
Sulphur, wt-ppm max	10	50	500	500
Cetane Index, min	–	47[1]	40[2]	46
Cetane Number, min	50	47[1]	–	49
Aromatics, % vol max	5	20	35[2]	–
Polycyclic Aromatic Hydrocarbons, % vol max	0.02	–	–	–
Density, kg/m³ range	800 to 820	820-860	876 max	820 to 860
API Gravity, range	41.1 to 45.5	33.0 to 41.1	30.0 min	33.0 to 41.1
Viscosity, cSt @ 40°C (104°F)	1.2 to 4.0	2.0 to 4.5	1.9 to 4.1	2.0 to 4.5
Distilled, °C (°F)				
Initial Boiling Point, min	180(356)	–	–	–
85% vol, max	–	350 (662)	–	350 (662)
90% vol, max	–	–	338 (640)	–
95% vol, max	285 (545)	370 (698)	–	370 (698)

1. Either the cetane index or cetane number must be 47 minimum.
2. Either the cetane index or aromatics limit must be met.

TABLE AP 3 : Indian Specification of Motor Gasoline (IS 2796 : 1995).

Characteristics	Requirements		
	87 Octane Unleaded	87 Octane Leaded	93 Octane Leaded
Colour, Visual	Colourless	Orange	Red
Copper Strip Corrosion for 3 hours at 50°C	Not more than No. 1		—
Density at 15°C, kg/m³	Not limited but reported		—
Distillation :			
Initial Boiling Point, °C	Not limited but reported		
Recovery up to 70°C, % vol min	10-45	10-45	10-45
Recovery up to 100°C, % vol min	40-70	40-70	40-70
Recovery up to 180°C, % vol min	90	90	90
Final Boiling Point, °C max	215	215	215
Residue, % vol max	2	2	2
Octane Requirements :			
Research Octane No., min	87	87	93
or			
Antiknock Index, (RON + MON)/2, min	82	82	88
Potential Gum, g/m³ Max	50	50	50
Existent Gum, g/m³ Max	40	40	40
Sulphur Content, total, % wt. max	0.20	0.20	0.20
Lead Content (as Pb), g/l max	0.013	0.56	0.80
Reid Vapour Pressure at 38°C, kPa	35-70	35-70	35-70
Water Tolerance of Gasoline-Alcohol Blends, Temperature for Phase Separation, °C max			
Summer	10	10	10
Winter	0	0	0
Vapour Lock Index (VLI) (VLI = 10RVP + 7E70), max			
Summer	750	750	750
Other months	950	950	950

Note: Potential gum test is to be carried out at the refinery end only, and the limit for the test is meant for products prior to addition of multi functional additives, if used. Use of MFA does not exempt the manufacturer from meeting this requirement prior to addition of MFA.

The share of the different of oil products in consumption is present in Table AP. 1.

Against this background the Figures AP 2 and 3 show the World production of oil by region and the Developing Asia is likely to

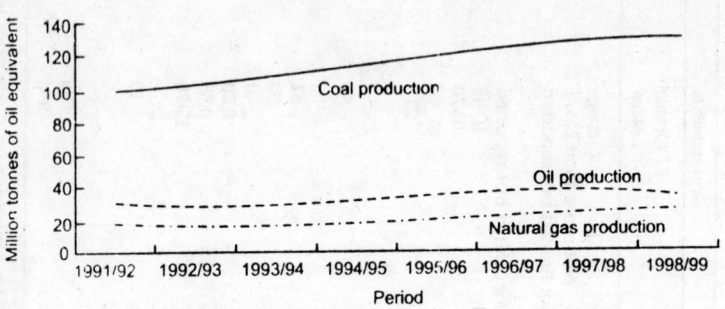

Figure AP 1 Trends in Production of Coal, Oil and Natural Gas in India.

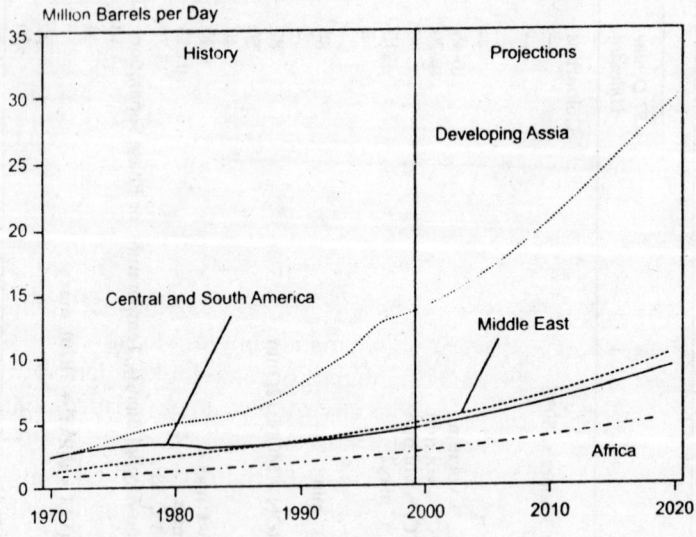

Sources History : Energy Information Administration (EIA), Office of Energy Markets and End Use, International Statistics Database and *International Energy Annual 1999*, DOE/EIA-0219(99) (Washington, DC, January 2001). Projections: EIA, World Energy Projection System (2001).

Figure AP 2 Oil Consumption in the Developing World by Region, 1970-2020

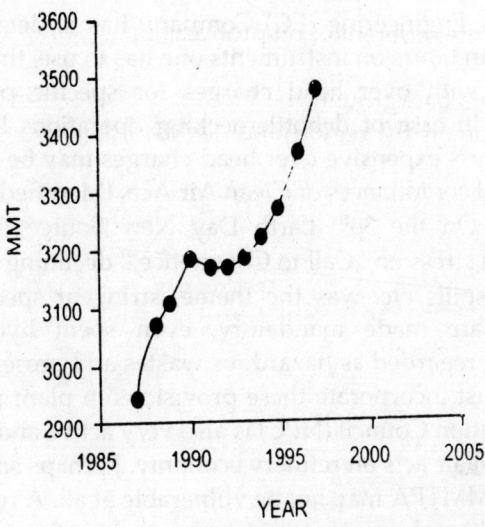

Fig. AP 3 : World Oil Production.

consume more oil than others. Whatever may be consumption and conservation the fossil fuels have limited existence specially oil and gas are going to be exhausted by the end of 21st century. Only mankind should be wise enough to prolong.

From Hydrocarbon Technology: May 1996 p. 24, 26.

Management of the Refinery

Earlier the management of the refinery was purely financial comprising of a. Purchase b. Distribution & Sales c. Operations d. R & D e. Maintenance. Resource management, Measurement analysis, process modeling have become part of the establishment. But present day refinery has to follow the envirnmental compliances. Safety and reliability are again important aspects of operation and more so in the process of flammable installations. Safety Instrumented Systems (SIS) are increasingly used in petroleum industry to complement process controls SIS using Plant Level Control (PLC) aid in reduction of risks. Smart analog transmitters specificaliy designed for SIS application now offer increased risk reduction factors at the sensor level. Advanced process Control (APC) on line data reconciliation systems (DRS) implemented to optimize ADU operation. Safety valves and their actuating systems are essential part in the process of preventing calamities. Linear programming was credited as an improtant tool of process operation, it aids in financial forecasts to the management system. Every Engineering (EC) Company has to detail the specific number of man hours on instruments one has to use, thus hourly cost of operation with over head charges for specific person is well documented. In case of debottle-necking operations less expensive worker with less expensive over head charges may be opted. Earlier environmental compliances or Clean Air Acts (Modified in 1997) were not in force. On the 30th Earth Day New Source Review (NSR) standards laid stress on "Call to Conscience," declining air emissions, green gases, spills etc. was the theme; stringent specifications for future fuels are made mandatory, even spent hydroprocessing catalysts were regarded as hazardous wastes and present installation of refinery must incorporate these provisions in plant projects itself. National Pollution Council (NPC) is also very active and studying the impact of clean air acts on refinery economy. Perhaps small refineries of less than 5 MMTPA may not be vulnerable at all. A rough estimate shows an additional investment of 4.4 cents per gallon of gasoline or 5.8 cents per gallon of diesel to reduce the sulfur to 30 ppm is required. Information Technology (IT) has brought down the distance between points and the business out lets are connceted closely. Industrial IT according to @ McAllister "should provide real - time automation information connecting across the enterprise for informed production decisions", Real-time information connectivity across the enterprise, consistent enterprise infrastructure for operations, data configurations, maintenance etc. are all significant gains. Industrial IT enabled applications for e-commerse for plant control, optimization, waste managmement and every plant has adopted to some or major extent and some times automated. Every improtant unit or instrument or control is meticulously scanned and documented in e-files.

Business to business (B2B), simplified e-business transactions between companies. This kind of application leads to the success of B2B transactions such as a. to reduce cost of making corporate purchases, b. standardize maintenance repair, overhaul, travel and other items c. bring buyers and sellers together d. comparison of products and purchases in the various market places. Impact of e-commerce on information, control technology focuses on the methods of energy consumed, changing demands and price sturcture and any change desired for efficient running of the plant. Earlier management fell short of environmental compliances, now future pollution programs concerned with sound, light-radiation etc. cananot be ignored. Storage and distrubution shoud be done adroity and waste material management, ground and surface water pollution, disposals of toxics, hazardous materials be given high priority. Even thougth industry is trying to control sulfur pollution in air, water and surface, it has not been fully successful. Waste water disposals were given high priority in the last three decades. Today with the help of IT from the similar industries operating world wide, every refinery and petrochemical plant has to give high priority for safery to the plant and surroundings because of its inflammable nature.

Ap 2 AUTO FUEL POLICY

Schedule for introducing improved quality fuels as per Auto Fuel Policy.

(i) Euro-lll Petrol & Diesel has been introduced from 1-4-2005 in all 11 identifies cities (Delhi/National Capital Region Mumbai, Kolkata, Chennai, Bangalore, Hyderabad, Ahmedabad, Pune, Surat, Kanpur & Agra) in line with Auto Fuel Policy.

BS-II Petrol throughout the country has been introduced w.e.f 1-4-2005 in line with Auto Fuel Policy

BS-II Diesel in all states except Rajasthan, West U.P, Uttranchal, M.P., Punjab, H.P and Jammu & Kashmir has been introduced from 1-4-2005 as per Auto Fuel Policy.

As per the revised programme BS-II diesel has been introduced in Rajasthan from 1-6-2005 and in West U.P and Uttaranchal from 1-7-2005.

Introduction of BSII Diesel is proposed in a phased manner as per the revised programme as under: a, M.P. from 1-9-2005 b) Punjab, H.P. and J & K from 1-10-2005

Appendix 3

API	Density	lbs per gal
10	1	8.328
11	0.933	8.270
12	0.9861	8.212
15	0.9659	8.044
17	0.9529	7.935
20	0.934	7.778
23	0.9159	7.627
25	0.9042	7.529
28	0.8871	7.387
30	0.8762	7.296
35	0.8498	7.076
40	0.8251	6.8700
45	0.8017	6.675
50	0.7796	6.490
60	0.7389	6.151
70	0.7022	5.845
80	0.6690	5.565
90	0.6388	5.320
100	0.611	5.086

Physical Properties of Some Pure Components

Compound Formula	Mol wt.	B.P.°C	M.P.°C.	Critical psia	Const. ok	Density g/cc 15°C
Methane	16.04	141.52	-182.47	667.8	190.55	0.415 (-164°C)
Ethane	30.07	-88.58	-182.8	707.8	305.43	0.485
Ethylene	28.05	-103.7	-169.2	50.8a	9.9c	0.57(-120°C)
Propane	44.1	-42.06	-187.68	616.3	369.82	0.507
						(0.585/-45°C)
Propylene	42.07	-47.7	-185.3	45.9a	91.9c	0.609 (-47)
Cyclo Propane	42.07	-32.8	-127.4	54.6a	124.9c	
n-Butane	58.12	-0.49	-138.46	550.7	425.16	0.585

Cyclo butane	56.1	12.5	-90.7	48.46a	190.5c	0.703
i-butene	56.1	-6.3	-185.4	39.7a	146.2c	
i-butene	58.12	-11.72			408.2	0.5631
i-butene	56.1	-6.9	-140.35	39.5a	144.7c	
i-Butadiene	54.09	-4.4	-108.9	43.0a	152.0c	
i-Pentane	54.09					0.6248
n-Pentane	54.09					0.632
Cyclo pentane	70.14	49.6	-94.0	44.8a	238.6c	
Cyclo hexane	84.16	80.74	6.56	49.7a	280.3c	0.779 (20°C)

Appendix 4

Basic Energy Units

1 joule (J)	= 0.2388 cal	
1 calorie (cal)	= 4.1868 J	
(1 British thermal unit [Btu]	= 1.055 kJ =	0.252 kcal)

WEC Standard Energy Units

1 tonne of oil equivalent (toe)	= 42 GJ (net calorific value)
	= 10 034 Mcal
1 tonne of coal equivalent (tce)	= 29.3 GJ (net calorific value)
	= 7 000 Mcal

Note: The tone of oil equivalent currently employed by the International Energy Agency and the United Nations Statistics Division is defined as 10^7 kilocalories, net calorific value (equivalent to 41.868 GJ).

Volumetric Equivalents

1 barrel	=	42 US gallons	=	approx. 159 litres
1 cubic metre	=	35.315 cubic feet	=	6.2898 barrels

Electricity

1 kWh of electricity output = 3.6 MJ = approx. 860 kcal

Representative Average Conversion Factors

1 tonne of crude oil	=	approx. 7.3 barrels
1 tonne of natural gas liquids	=	45 GJ
1000 S.C.M of NG	=	36 GJ
1 tonne of Uranium	=	10,000 – 16,000 toe.

Appendix 5

The New Theory of Formation Petroleum

INTRODUCTION

The origin and formation of petroleum was full of controversies. Inorganic Theory of formation of petroleum initiated by Berthelot and *Mendeleev* in the nineteenth century, was displaced by better Organic Theory announced by Engler. Even Cosmic Hypothesis of Sokolov (Ref. P. 2) was also given a hearing. Like that many schools of thoughts prevailed one displacing the other one. However due to strong evidence of plant residues and porphyrins in relatively young Gambian crudes prompted to launch API Project 43 under the Director ship of Zobell. The essence of which established that the formation of petroleum was genuinely Biogenic and the origin being accumulated debris of lower plant organisms like planktons in sea oozes. It soundly explained all the qualities of petroleum, hence was accepted universally. That was the history of those times.

Today there are many space probes and remote sensing mechanisms to explore new possibilities. In this direction a new theory ABIOGENIC was announced by scientists mostly from USA and Russia. The inconclusive explanations have already given cracks to hither to believed Biogenic Theory. The orthodox biogenic theories of petroleum concluded that petroleum accumulations within the crust were remnants of buried plant and animal life. The debris were converted into petroleum hydrocarbons via diagenesis and metamorphism. Organic matter consisting of dead plants and animals is continuously deposited and buried in sea oozes. Accumulating sediments compress the material over aeons. At a depth of several hundred meters, heat and pressure convert it to kerogens. Kerogens are the mother substances for the fossil fuels. Depending upon the kerogen content the rock can become an oil bearing rock or shale oil rock or just humic coal. Time and temperature are responsible for conversion of kerogens into hydrocarbon series in a process called catagenesis. Thus the plant remains, such as terpenoids, terpenes, and porphyrins, which are large, chelating molecules in the same family as heme and chlorophyll show their presence as biomarkers;

Abiogenic Theory

The hypothesis of **abiogenic petroleum origin** holds that petroleum was formed by primordial non-biological processes deep in the earth's crust and mantle. It looks like the revival of old inorganic theory initiated by Berthelot and Mendeleev Though there is growing consensus on abiogenic origin petroleum due to some evidence most modern geologists do not support this for the vast majority of petroleum deposits within the Earth. The deep biogenic petroleum theory proposes, mostly after the work of Thomas Gold that the "deep hot biosphere" may be the source of some petroleum. According to Gold "hydrocarbons are not biology reworked by geology (as the hard line would hold) but geology reworked by biology.

The hypothesis is founded primarily upon;

a. The large amounts of constituents of petroleum within the solar system and extra terrestrial systems
b. Mechanisms of abiotically chemically synthesizing hydrocarbons within the crust
c. Interpretations of the chemical composition of natural petroleum
d. The presence of oil within non-sedimentary rocks upon the Earth
e. Modern thermodynamic equilibrium models and experiments show that methane compressed to 30 or 40 kbar yields hydrocarbons having properties similar to petroleum
f. No investigator has ever produced anything resembling petroleum in the laboratory by the application of heat and pressure to plant debris.

Biogenic theory pronounces that petroleum accumulations within the crust are remnants of buried plant and animal life. This is converted into petroleum hydrocarbons via diagenesis and the incipient stages of metamorphism. Organic matter consisting of dead plants and animals is deposited and buried. Accumulating sediments compress the material over geologic time scales. At a depth of several hundred meters, heat and pressure convert it to kerogens. Time and temperature modify kerogens into the hydrocarbon series or petroleum in a process called, catagenesis (geocatalytic activity).

Mechanism of Abiogenic Theory

Abiogenic theory considers that within the mantle much carbon exists as hydrocarbon molecules, chiefly methane, also as carbon dioxide and carbonates. The full suite of hydrocarbons found in petroleum is generated at depth by abiogenic processes and therefore shallow petroleum deposits come from the upward displacement of those hydrocarbons. When this material passes up through temperatures at which thermophilic microbes can survive some of it is consumed and modified. The material continues to migrate upwards through the crust until they are trapped by impermeable strata, where it forms reservoirs. Some experts of abiogenic origin theory sees microbial life strictly as a contaminant. The inorganic reactions taken into consideration to produce hydrocarbons are bascically listed by Fischer-Tropsch synthesis

Spinel polymerization mechanism

The crucial application of Fe-Spinel group metals like Magnetite, *chromite* and *ilmenite* found in certain rocks to act as catalysts to either for cracking methane etc or for F.T. reactions projecting simply as hydrothermal events. Such as:

$$Methane + Magnetite \rightarrow Ethane + Hematite \qquad ... 1$$

Over active iron, ruthenium, cobalt, nickel oxide catalysts/at 250-370°C/ at near normal pressure with activators FT reactions take place

$$nCO + mH2 \rightarrow R=CH_2/RCH_3/RCH_2OH/RCHO/RCOOH$$

With difficulty above reactions result in *n*-alkane hydrocarbons, alcohols, aldehydes, ketones, aromatics, and not much cylisation results.

The upgrading of methane to higher n-alkane hydrocarbons is via dehydrogenation of methane in the presence of catalyst transition metals (e.g Fe, Ni) can be termed spinel hydrolysis.

Evidence from petroleum geochemistry

The geochemistry of petroleum deposits has been thoroughly studied by oil companies and academia for many years in order to explain the origin of petroleum and develop simulated scientific models. However some findings of this research show that both the theories are justified. Thus include biomarker chemicals, the optical activity of oils, chirality and the trace metal abundance of oils etc on one hand, methane and carbon presence in Earth's mantle for Abiogenic.

Isotopic evidence

Large amounts of methane in crustal fluid and gas gives rise to depletion of observed carbon isotope with depth in the crust favours for abiogenic oil. However, diamonds, which are definitively of mantle origin, are not as depleted as methane. Helium isotope geochemistry is a clear indicator of mantle source within gases.

Odd-number carbon abundance

It is seen that odd carbon series are the result of debris of organism but it is shown linear carbohydrate molecules in living systems exhibit the same preference for odd carbon numbers. All mixtures of linear hydrocarbon chains synthesized or otherwise, exhibit the same tendency. It arises from the geometry of the covalent bond in linear molecules.

Reduced carbon

Sir Robert Robinson studied the chemical makeup of natural petroleum oils, and concluded that they were mostly far too hydrogen-rich to be a likely product of the decay of plant debris as hydrogen content of plants donot exceed 5%. He believes that biomarkers are intruders into petroleum

Example abiogenic deposits

Supergiant fields such as the Athabasca Tar Sands (Canada), Orinoco Heavy Oil Belt (Venezuela) and the Ghawar Field (Saudi Arabia) are good examples that have been interpreted as having been formed by abiogenic oil The White Tiger oil field in Vietnam has been proposed as an example of abiogenic oil because it is found 1000 meters within crystalline basement rock. However, others argue that it contains biogenic oil which leaked into the basement horst from conventional source rocks at depth within the Cuu Long basin.

Geological framework and contradictions

In essence the abiogenic theory takes the advantage of lab made F.T reactions to explain the theory. It is very difficult to carry out F.T. Reactions. Further the suitable catalyst, conditions and proper ratio of CO and Hydrogen to yield various alcohols, aldehydes, acids, hydrocarbons and waxes. Why in hot bed of earth CO should go to hydrogen and both going in search of suitable catalyst to give F.T. Series.? How can the formed series degenerate in to carbon and hydrogen at such high temperatures? These are some obvious questions. If this reaction can be so easy why the scientists are unable to produce ethane by taking methane from atmosphere where Carbondioxide and methane are potential green gases. However as mentioned earlier (Ref. p. 7.) many reactions are possible each may contribute different type of oils in the Earth. Why petroleum should migrate upward only? What causes petroleum to differentiate into different BASES? Further the luminescence in petroleum has to be explained as contamination of biomarkers, the % of which has not been mentioned. What happened to the organic matter in sedimentary rocks and Earth's crust? The hydrocarbons specially heavier one are heat sensitive, they decompose very easily the ultimate product of cracking is Carbon not hydrocarbon series.

The geological observations proposed for the abiogenic theory are presented below, followed by investigation of several key deposits on a case by case basis to evaluate their genesis. In oil and gas geology, research on carbon isotopes of organic matter and bitumen not only generally corroborated the traditional views on the organic concept of the genesis of oil and gas, but also provided significant criteria for separating stages of hydrocarbon generation of various composition in agreement with the stages of sedimentary rocks transformation in diagenesis and catagenesis.

Bibliography

1. Mendeleev, D., 1877. L'origine du petrole. Revue Scientifique, 2e Ser., VIII, p. 409-416.
2. Gold, Thomas (1999). *The deep, hot biosphere*. Copernicus Books.
3. Kenney, J.F., Karpov I.K., et al. 2002. The Prohibition of Hydrocarbon Genesis at Low Pressures. *Unpublished research*.

Index